Dahlem Workshop Reports
Life Sciences Research Report 31
Microbial Adhesion and Aggregation

The goal of this Dahlem Workshop is:
to explore the mechanisms and consequences
of microbial adhesion and aggregation

Life Sciences Research Reports
Editor: Silke Bernhard

Held and published on behalf of the
Stifterverband für die Deutsche Wissenschaft

Sponsored by:
C. H. Boehringer Sohn Chemische Fabrik Ingelheim
Deutsche Forschungsgemeinschaft
Senat der Stadt Berlin
Stifterverband für die Deutsche Wissenschaft

Microbial Adhesion and Aggregation

K. C. Marshall, Editor

Report of the Dahlem Workshop on
Microbial Adhesion and Aggregation
Berlin 1984, January 15 – 20

Rapporteurs:
J. A. Breznak · G. B. Calleja · G. A. McFeters
P. R. Rutter

Program Advisory Committee:
K. C. Marshall, Chairperson · W. G. Characklis
Z. Filip · M. Fletcher · P. Hirsch · G. W. Jones
R. Mitchell · B. A. Pethica · A. H. Rose

Springer-Verlag
Berlin Heidelberg New York Tokyo 1984

Copy Editors: K. Geue, J. Lupp
Text Preparation: M. Böttcher, J. Lambertz, M. Lax, D. Lewis
Photographs: E. P. Thonke

With 4 photographs, 86 figures and 35 tables

ISBN-13:978-3-642-70139-9 e-ISBN-13:978-3-642-70137-5
DOI: 10.1007/978-3-642-70137-5

CIP-Kurztitelaufnahme der Deutschen Bibliothek
Microbial adhesion and aggregation :
Report of the Dahlem Workshop on Microbial Adhesion and Aggregation,
Berlin, January 15–20, 1984 / K. C. Marshall, ed. Rapporteurs: J. A. Breznak ...
[Held and publ. on behalf of the Stifterverb. für d. Dt. Wiss.
Sponsored by: C. Boehringer Sohn, Chem. Fabrik Ingelheim ...].
– Berlin ; Heidelberg ; New York ; Tokyo : Springer, 1984.

(Life sciences research report ; 31)
(Dahlem Workshop reports)

NE: Marshall, Kevin C. [Hrsg.]; Breznak, John A.

2131/3020-5 4 3 2 1 0

Table of Contents

The Dahlem Konferenzen

Founders
Recognizing the need for more effective communication between scientists, especially in the natural sciences, the Stifterverband für die Deutsche Wissenschaft*, in cooperation with the Deutsche Forschungsgemeinschaft**, founded Dahlem Konferenzen in 1974. The project is financed by the founders and the Senate of the City of Berlin.

Name
Dahlem Konferenzen was named after the district of Berlin called "Dahlem", which has a long-standing tradition and reputation in the arts and sciences.

Aim
The task of Dahlem Konferenzen is to promote international, interdisciplinary exchange of scientific information and ideas, to stimulate international cooperation in research, and to develop and test new models conducive to more effective communication between scientists.

Dahlem Workshop Model
Dahlem Konferenzen organizes four workshops per year, each with a limited number of participants. Since no type of scientific meeting proved effective enough, Dahlem Konferenzen had to create its own concept. This concept has been tested and varied over the years, and has evolved into its present form which is known as the *Dahlem Workshop Model*. This model provides the framework for the utmost possible interdisciplinary communication and cooperation between scientists in a given time period.

**The Donors Association for the Promotion of Sciences and Humanities*
***German Science Foundation*

The main work of the Dahlem Workshops is done in four interdisciplinary discussion groups. Lectures are not given. Instead, selected participants write background papers providing a review of the field rather than a report on individual work. These are circulated to all participants before the meeting to provide a basis for discussion. During the workshop, the members of the four groups prepare reports reflecting their discussions and providing suggestions for future research needs.

Topics
The topics are chosen from the fields of the Life Sciences and the Physical, Chemical, and Earth Sciences. They are of contemporary international interest, interdisciplinary in nature, and problem-oriented. Once a year, topic suggestions are submitted to a scientific board for approval.

Participants
For each workshop participants are selected exclusively by special Program Advisory Committees. Selection is based on international scientific reputation alone, although a balance between European and American scientists is attempted. Exception is made for younger German scientists.

Publication
The results of the workshops are the Dahlem Workshop Reports, reviewed by selected participants and carefully edited by the editor of each volume. The reports are multidisciplinary surveys by the most internationally distinguished scientists and are based on discussions of new data, experiments, advanced new concepts, techniques, and models. Each report also reviews areas of priority interest and indicates directions for future research on a given topic.

The Dahlem Workshop Reports are published in two series:
1) Life Sciences Research Reports (LS), and
2) Physical, Chemical, and Earth Sciences Research Reports (PC).

Director
Silke Bernhard, M.D.

Address
Dahlem Konferenzen
Wallotstrasse 19
1000 Berlin (West) 33

Microbial Adhesion and Aggregation, ed. K.C. Marshall, pp. 1-3. Dahlem Konferenzen 1984. Berlin, Heidelberg, New York, Tokyo: Springer-Verlag.

Introduction

K.C. Marshall
School of Microbiology, University of New South Wales
Kensington, N.S.W., Australia

Microbial adhesion and aggregation can mean many things to many people. Why is this so? These processes of adhesion and aggregation are manifest in many forms and are studied by researchers from a wide variety of disciplines. Adhesion and/or aggregation of microorganisms are involved in certain diseases of humans and animals, in dental plaque formation, in industrial processes, in fouling of man-made surfaces, in syntrophic and other community interactions between microorganisms, and in the activity and survival of microorganisms in natural habitats. Different approaches to the study of these phenomena have developed depending on whether the work has been carried out by microbiologists, biotechnologists, physical chemists, or engineers.

The goal of this workshop was to explore the mechanisms and consequences of microbial adhesion and aggregation. Both of these processes involve the interaction between microorganisms and some type of surface. These surfaces may be inert materials, the exterior of other, usually larger, organisms, or other microorganisms of the same or different species (aggregation). Several symposia and a number of books have been devoted to the subject of adhesion and aggregation in recent years, but they have not provided an adequate forum for the exchange of ideas and concepts from the different interested disciplines. This workshop involved the bringing together of microbial physiologists, ecologists and geneticists, industrial microbiologists, physical and polymer chemists, and bioengineers in an attempt to assess the current status of the field and, hopefully, to develop new approaches and concepts as a result of the interdisciplinary discussions.

The first really serious study on microbial adhesion was that of ZoBell (2), who recognized the possibility of macromolecular conditioning films modifying surfaces prior to microbial adhesion and suggested that adhesion was a biphasic phenomenon, with an initial reversible phase being followed by a firm, irreversible binding of microorganisms to surfaces. Progress in all aspects of adhesion and aggregation processes was very slow, until a reawakening of interest in the early 1970s. At that time, research workers in various fields became conscious of the almost universal association of microorganisms with surfaces or with each other (1). Research on microbial adhesion and aggregation has now reached almost explosive proportions and, consequently, it seemed timely to stand back and reassess the directions in which this research might move in the future. Because of the major contributions made to research on adhesion and aggregation by microbiologists, chemists, and engineers, it was deemed essential that all of these fields be represented at the workshop in order that the subject be considered from a very broad viewpoint.

The workshop was planned to consider four major aspects of the main topic, namely: a) the mechanisms whereby microorganisms adhere to surfaces, b) the development of biofilms and the consequences of such films, c) the activities of microorganisms on surfaces, and d) the process and consequences of microbial aggregation. Each aspect was considered by a group consisting of representatives of different disciplines, and each group aimed to examine old concepts and attempted to evolve new concepts based on the context of its discussions. Of course, one of the ultimate aims in studying microbial adhesion and aggregation is to evolve the means for manipulating these processes to our own advantage. This may be to control the fouling of ship hull or heat exchanger surfaces, to control dental plaque or periodontal disease, to control cholera and other intestinal diseases, to enhance aggregation in activated sludge systems, to enhance biofilm development in fixed film fermenters, or to modify microbial associations in nature to establish more effective consortia.

Finally, some brief statements regarding certain confusing terminology that has developed in this field. Adhesion is a process that can, in essence, be explained on a physicochemical basis. Consequently, the word adhesion should be used in place of the term adherence (which is best used in the sense of, say, adherence to a principle). There is a need to clarify the usage of the terms substrate and substratum. A substrate (pl.: substrates) is a material utilized by microorganisms, generally as a source of energy. A substratum (pl.: substrata) is a solid surface to which a microorganism may attach.

REFERENCES

(1) Marshall, K.C. 1976. Interfaces in Microbial Ecology. Cambridge, MA: Harvard University Press.

(2) ZoBell, C.E. 1943. The effect of solid surfaces upon bacterial activity. J. Bacteriol. 46: 39-56.

Standing, left to right:
Paul Rutter, Rolf Freter, Mike Silverman, Ian Robb, Hinrich Mrozek, Frank Dazzo, David Gingell

Seated, left to right:
Garth Jones, Stanislawa Tylewska, Staffan Kjelleberg, Ellen Rades-Rohkohl, Kevin Marshall

Microbial Adhesion and Aggregation, ed. K.C. Marshall, pp. 5-19. Dahlem Konferenzen 1984. Berlin, Heidelberg, New York, Tokyo: Springer-Verlag.

Mechanisms of Adhesion
Group Report

P.R. Rutter, Rapporteur
F.B. Dazzo
R. Freter
D. Gingell
G.W. Jones
S. Kjelleberg
K.C. Marshall
H. Mrozek
E. Rades-Rohkohl
I.D. Robb
M. Silverman
S. Tylewska

INTRODUCTION

The subject, Mechanisms of Microbial Attachment, has proved to be a fertile source of argument between both microbiologists and physical chemists. In order to find common ground for discussion, the group endeavored to come to a consensus regarding a number of definitions pertinent to the subject. Some of these are described in the text and others will be found at the end of this report.

The group restricted its discussions to the interactions between microbial cells and substrata in close proximity. The general topic of deposition was not covered in any detail. It was noted at the outset that the adhesion of microbial cells to animate and inanimate surfaces and the mechanism by which they adhere can be influenced directly by changes in environmental conditions. Many examples of this will be found in Rutter and Vincent, Kjelleberg, Jones, Dazzo, and Silverman et al. (all this volume).

WHAT IS ADHESION?

It was agreed that microbial adhesion can be defined unambiguously only in terms of the energy involved in the formation of the adhesive junction. Thus, the strength with which a microbial cell may be said to have adhered to a substratum can be quantified as the work required to remove the cell to its original isolated state. It is important that removal of the cell from the substratum does not leave fragments of the cell wall or cell surface polymers remaining on the substratum. If this is so, then both adhesive and cohesive failure will have occurred, giving smaller energies of adhesion than the true value. All other definitions can lead to confusion, although they might be useful to researchers working with particular, well characterized systems. In addition to the definition of adhesion given above, it was felt that a more functional description of adhesion was also required. This prompted the use of the term "association" which is defined at the end of this report.

MEASUREMENT METHODS

Having agreed on what was meant by adhesion and association, a number of methods of measuring these two phenomena were discussed. The experimental details of a number of assays designed to quantify microbial association in particular systems are described in the papers by Rutter and Vincent, Kjelleberg, Jones, and Dazzo (all this volume).

The Direct Measurement of Adhesion by Methods Involving the Separation of Cell and Substratum

Physicochemical techniques are now available for measuring the force of adhesion between molecularly smooth surfaces. It may be possible to measure the adhesion of bacteria grown as monolayer cultures on well-defined surfaces by a similar technique (4). Other techniques involving micromanipulation or well-defined shear conditions might also be devised.

Deposition Studies under Controlled Hydrodynamic Conditions

These methods quantify the association of microbial cells with substrata. It is possible to devise experimental systems where a suspension of microbial cells undergoes laminar flow over a surface. Under these circumstances it is often possible to calculate the number of collisions between the cells and substratum from the bulk properties of the system, e.g., cell concentration, viscosity, temperature, flow velocity, etc. Comparison between the observed cell deposition rate and calculated collision rate enables the deduction of a collision efficiency (deposition rate/collision rate) which may be used to quantify the association under study. The advantages of this type of technique are that many of the often poorly defined quantities in experiments designed to compare the interactions of microbial cells with different surfaces, e.g., cell concentration, settling under gravity, etc., are encompassed by the collision rate calculations (2).

Techniques Involving Contact Angles

Good correlations have been obtained between certain association phenomena and interaction energies calculated from deduced surface energies of the apposing surfaces. These techniques do not measure adhesion but calculate the energy involved from contact angle measurements and compare this with the observed interaction of cells with different surfaces, including those of phagocytes (1). The striking fact about these studies is the good correlation observed. However, it was felt that there were a number of problems arising from the measurement of microbial cell surface contact angles, particularly with respect to the effect of the diagnostic liquids on the cell surface components and the effect of removing the cells from their original environment, especially if drying is involved.

Mammalian Cell Techniques

It was pointed out that techniques involving the deformation of mammalian cell aggregates under gravity and the response to electrolytes of red blood cells held in secondary minimum contact with a substratum have been used to deduce adhesive forces. It may be possible to adapt these techniques to the study of certain types of microbial adhesion, although doubts were expressed because of the apparent rigidity of bacterial cell walls.

Measurement of the Association of Microbial Populations with Surfaces

It was recognized that the proportion of bacteria within a population which remain attached to a surface in a particular experimental environment is a useful and convenient measure of association (see, for example, Dazzo, Jones, and Kjelleberg, all this volume).

SPECIFIC AND NONSPECIFIC ADHESION

Discussion of the terms "specific" and "nonspecific" occupied a considerable amount of the group's time but eventually led to definitions which the group considered to be helpful.

Before arriving at a definition, it was necessary to come to some agreement as to what constituted a microbial cell, host cell, and inanimate substratum surface. Animate cell surfaces are invariably coated with complex macromolecules extending in various ways into the surrounding environment. It was also the experience and opinion of the group that inanimate surfaces, although they might be clean and well-defined initially, will rapidly adsorb some form of surface layer (often referred to as a conditioning film) which may or may not be complete, as soon as they are immersed in an environment containing microbial cells. Apparently, even thoroughly washed cells will release components to the surrounding environment. The approaching surfaces in any microbial adhesion event therefore will almost certainly consist of a concentration gradient of complex molecules with a certain thickness. However, bare patches (i.e., the original surface) may also exist.

During a collision between a microbial cell and a substratum, the cell surface macromolecules may interact with the substratum surface or, more likely, with macromolecules which are adsorbed to or form part of the substratum. A number of classes of interaction may be involved (ionic, dipolar, hydrogen bonds, hydrophobic interactions). The number and strength of these interactions vary considerably from system to system so that the level of adhesion of a given microbe to a variety of surfaces (or a range of microbes to a given surface) will also vary considerably. This in itself does not imply specificity. An additional requirement for specificity is some form of stereochemical constraint which brings into contact more than one (normally several) pairs of neighboring interacting groups on the microorganisms and substratum. An example would be sugar residue-protein interactions on two complementary polymers, i.e., a lock-and-key mechanism is required.

Monomers and oligomers which sufficiently mimic the combining sites of either interacting surface polymer can occupy the combining sites and competitively block specific attachment, reducing it to the background level of adhesion characteristic of that system.

This definition implies that a range of specificities are possible, depending upon the number of interacting pairs which can be brought into apposition when cell and substratum collide. Similarly blocking molecules will present a range of activities, depending upon how closely they imitate the full sequence and stereochemistry of the interacting molecules (i.e., the adhesin and receptor molecules). In certain special cases it may be possible to have receptor-like molecules on both cell and substratum surfaces and free adhesin-like molecules which bind the two surfaces together (see Dazzo, this volume).

While emphasizing the distinction between specific and nonspecific mechanisms of adhesion, the group was aware of the fact that a precise determination of the adhesive mechanism is not necessarily the most important parameter describing the functional aspects of a given ecosystem. For example, the rate at which the association occurs (as determined by concentration, collision efficiency, motility, etc.) and the degree of reversibility are among the ultimate factors determining how many bacteria will actually associate with a given substratum. There are, indeed, additional characteristics which must be determined if one is to understand a given adhesive interaction. There may be high or low levels of attachment with either mechanism of adhesion. Nevertheless, detailed elucidation of the various adhesive mechanisms is finally necessary because these are the ultimate determinants of adhesion in complex biological systems.

THE IDENTIFICATION OF SPECIFIC ADHESION INTERACTIONS

To be useful, any definition should be capable of application, consequently considerable discussion centered round the question of how to determine whether or not an adhesive interaction was specific.

It was agreed that specificity at the cellular level is of little help in distinguishing specific from nonspecific adhesion. For example, the selective (i.e., nonrandom) accumulation of bacteria at a particular site could be a consequence of nutrient advantage, toxin tolerance, lack of competitive accumulation by other organisms, or specific adhesion.

Assays may also be performed which show that although bacteria will preferentially select one surface rather than another, they will also associate with surfaces which are totally alien to them. For example, although certain oral organisms will select human rather than rat epithelial cells, they will also stick to glass.

Specificity at a molecular level, however, is a more useful concept since it may be tested for by using blocking molecules. Care should be taken to ensure that the blocking molecules are not so large as to alter properties of the cell surface other than the specific interaction sites.

Modern tools of molecular genetics have already been applied to studies of the structure and regulation of the adhesins by which some bacteria adhere specifically to surfaces. Structural genes encoding some adhesins have been cloned and at least partially sequenced. Amino acid sequences and, to some extent, secondary structures of the adhesins have been deduced from the genetic (DNA) sequences. Sequences of particular amino acids in the putative binding regions have been recognized by such methods. In some cases, the number of genes encoding a particular adhesin have been identified and information gained about how the genes are regulated. The approach and methods are extremely powerful and are often more incisive than conventional methods for studying how cells adhere and what inhibits adhesion.

An important aspect of specific adhesion is its often apparent reversibility. This will again depend upon the number, stereochemistry, and bonding energy of the interactions involved. If the interactions involve a limited number of "interacting pairs," then it may well be possible to "unzip" the pairs by using a high concentration of competitive blocking molecules. However, the more interactiong groups that are in a lock-and-key interaction, the more difficult this will be. For example, cellulose fibers are held together by many hydrogen bonds acting cooperatively, which prevents their solution in water. Hydrophobically substituted celluloses are more soluble because they have less hydrogen bonding capability.

IDENTIFICATION OF NONSPECIFIC ADHESION

The group felt strongly that nonspecific adhesion was not an unimportant phenomenon, even in the adhesion of microorganisms to living surfaces. In many respects nonspecific adhesion is more difficult to identify than specific adhesion, since the cells may utilize different mechanisms to adhere to different surfaces. This has led to a considerable literature describing examples of the adhesion of cells to various substrata under various conditions.

Experiments have been performed which show that the interfacial energy of a series of substrata can be related to the number of cells associated with them (1). In some cases the Langmuir equation has been used to fit experimental data, and affinity constants have been deduced for the association of bacteria for "treated" and "nontreated" surfaces. It is also clear that the presence or absence of fimbriae and exopolymer layers has a considerable effect on the association of an organism with a particular surface (Kjelleberg, this volume).

A difficulty arises in attempting to decide whether the interactions involved in these associations are predominantly hydrophobic, dependent on surface charge or polymer adsorption, or a combination of these factors. The situation is further compounded by the possibility of surface macromolecules specifically involved in one type of adhesion interacting nonspecifically with another surface.

As discussion continued it became clear that before mechanisms responsible for a particular adhesive phenomenon could be identified, the system needed to be defined in a very detailed way. For example, it could be misleading to ascribe the interaction of a hydrophobic cell with a hydrophobic surface as simply being due to hydrophobic interactions. The average properties of the two contacting surfaces may not be directly responsible for their adhesion. In the case of the substratum, it may be the material adsorbed from the bulk environment (conditioning film) which determines the subsequent interaction with a microbial cell. If the substratum adsorbs cell surface components which undergo little denaturation, the number of cells which subsequently attach might be very small. The influence of surface charge is also difficult to identify. Although the two surfaces may exhibit an overall negative charge, small areas of positive charge could still be present. Selectively removing or blocking charged groups on cell and substratum surfaces or the use of large changes in ionic strength could also alter the conformation of surface macromolecules and hence lead to ambiguous results.

Perhaps the least controversial way of investigating nonspecific adhesion would be to use substrata previously coated with defined molecules to produce small changes in, for example, surface charge or wettability. However, it would be necessary to continuously monitor the properties of the substratum surface during the course of the adhesion experiment to be certain of identifying the controlling adhesive mechanism. This, of course, may be very difficult to achieve in practice.

A word of caution was raised concerning the use of the Langmuir adsorption isotherm to quantify adsorption energy. Particles adsorbing reversibly at a fluid interface can be described by Langmuir's isotherm, and an energy of adsorption can be calculated from the initial slope of the isotherm. This energy is only meaningful if the particles are truly reversibly bound. Langmuir-shaped isotherms can be obtained from heterogeneous distributions of particles, all of which may individually adsorb essentially irreversibly (e.g., to dilution) by a preferential adsorption mechanism. For these cases, use of the slope of the isotherm to obtain an energy is quite wrong and misleading. Adsorption energies for particles calculated from Langmuir isotherms can only be meaningful if adsorption has previously been shown to be reversible in the strict thermodynamic sense of the word (Rutter and Vincent, this volume).

THE TIME SEQUENCE OF ADHESION

During the discussions of adhesion and association it became clear that time was an important factor in these events. In some cases, for example, bacteria may have preformed extracellular polymer layers and will immediately adhere to a substratum surface. In other cases, organisms may adhere reversibly at first and synthesize polymeric material to achieve irreversible adhesion. There is also evidence to show that polymer synthesis by bacteria can lead to detachment (Kjelleberg, this volume). In one latter case, the hydrophilic polymer can be washed off to restore the cell's original hydrophobic surface and adhesive properties (3). Polymer synthesis, however, is not the only mechanism which can bring about an increase in adhesion with time. The adhesion of certain rubbers, for example, will increase with time as a result of viscous flow and, hence, increased contact area (Robb, this volume). Following this idea, it was pointed out that the appearance of polymer material at the initial site of contact between a microbial cell and substratum may be due to the movement or migration of polymer molecules already on the cell surface. Such a mechanism would be broadly analogous to the well-known ligand-induced patching or capping which occurs in many animal cells.

In addition to sequential changes occurring in the microbial cell, the substratum can play a time-dependent role in adhesion, particularly in the case of living cells. For example, degradation of the bacterial cell surface polymers may be caused by enzymes secreted by the host cell.

HOW DO MICROORGANISMS SENSE CONTACT WITH A SURFACE?

It seems that many microorganisms change their behavior after association with a surface. This may result in the formation of smaller cells than those remaining in the bulk suspension during the process of starvation survival (Kjelleberg, this volume). These changes may also be influenced by the type of surface with which the cell is associated. Other examples are the slime layers produced by some gliding bacteria when brought into contact with a surface and the production of lateral flagella by Vibrio parahaemolyticus (Silverman, this volume).

The discussion centered round the issue of how the microbial cell could detect the proximity of a surface in order to effect these changes. It was suggested that in some cases it might be due to differences between the surface and bulk pH, redox potential, or water structure and the effects these might have on the potential across the cytoplasmic membrane of microorganisms. It was also possible that association with a surface and subsequent changes were simply part of a sequence which begins before adhesion with cells suspended in the bulk. Cells may change in response to a stimulus which causes them to alter their adhesive properties.

Although there are examples where the causal link between association with a substratum and a subsequent change in microbial activity is doubtful, the group felt that in many cases microbial cells sense whether they were stuck to a surface or not. This was considered to be somewhat similar to the ability of microbial cells to detect nutrient gradients, light, magnetic fields, etc. It is not precisely known, however, how this detection of the substratum might be achieved. One suggestion was that the stresses set up in the cell membrane caused by the forces involved in adhesion, whilst not great enough to visibly deform the cell, might result in changes in membrane permeability. There has been, however, as yet no experimental evidence for this suggestion.

THE IMPORTANCE OF GENETIC REGULATION IN MICROBIAL ADHESION

Since the benefit which a microorganism derives from the synthesis of a particular adhesin is often limited to specific circumstances, many mechanisms probably exist for differential control of adhesin gene expression. Lateral flagella genes are known to be transcribed in response to contact with, or growth on, a surface. Another regulatory strategy involves the variable or oscillatory expression of adhesin genes. This results in a heterogeneous population containing cells with many different adhesive phenotypes, some of which may be appropriate for the particular conditions in the habitat. Genetic technology is now being used to dissect these regulatory systems into their basic components (Silverman et al., this volume).

THE CONTROL OF ADHESION

One of the major objectives of research into microbial adhesion is to gain enough knowledge of the processes involved to be able to control them. The advantages, for example, of being able to prevent microbial adhesion to inanimate surfaces such as heat exchangers and ship hulls are obvious, as is that of the control of the adhesive interaction involved in the microbial infection of various animal and plant cells. It may also be desirable to increase microbial adhesion, particularly in bioreactor systems.

Microbial attachment to inanimate surfaces might be prevented by the use of protective molecular layers. The mechanism for this can be either a) charge repulsion, i.e., anionic polyelectrolytes can increase the negative charge of a surface, thus helping to repel negatively charged cells, b) phase separation between the protective polymer and microbial cell surface macromolecules. Usually when two polymers are dissolved together, they separate into two phases, each rich in one of the polymers. Thus, the adhesion of bacteria to a surface might be prevented by using a protective layer of polymer which will phase separate with the extracellular polymer of the bacteria. This protective layer can be anionic or neutral and it should extend to a significant distance from the protected surface (say, 10 nm) (Rutter and Vincent, this volume), c) ablating polymers that at regular intervals "unzip" the surface layer, thereby removing any adhering microbial film.

A variety of approaches are being investigated in attempts to control or prevent infections by interfering with bacterial attachment to host tissues. One of the most common is to use bacterial adhesins as immunogens in vaccines (Jones, this volume). This approach is probably closest to becoming practical, and, indeed, some success has already been realized in the case of enteric infections of farm animals. In some systems immunization with bacterial surface antigens other than the adhesins can also prevent adhesion by mechanisms such as clumping, steric hindrance, etc. Some investigators have attempted to remove adherent bacteria through use of enzymes which degrade their bridging polymers. An example is the application of glucan hydrolases which attack the extracellular glucans which promote accumulation of Streptococcus mutans cells on teeth. It has also been possible to modify glucan synthesis by S. mutans. Thus, mixing of low M.W. dextran fragments with sucrose leads to the synthesis of smaller glucan polymers which are less effective in promoting S. mutans accumulation. Both glucan hydrolases and low M.W. dextrans are effective in reducing streptococcal accumulation in in vitro models and when administered to S. mutans-infected animals. They also reduce the development of dental caries in animal models, but their practicality for use in humans is questionable. (See (5) for a review of the adhesion of oral organisms.)

Administration of receptor analogues to prevent attachment or to promote desorption is another approach. Thus, instillation of a methyl mannoside into the bladder of mice can prevent the attachment and colonization of Escherichia coli strains which adhere by means of mannose-sensitive fimbriae. It has also been shown that certain lectins will bind to tissue surfaces and inhibit bacterial attachment, presumably by masking receptors. These observations raise the possibility that dietary lectins may influence bacterial colonization in the mouth and elsewhere in the alimentary canal. The implantation of avirulent mutants which occupy binding sites on tissue surfaces is also being studied. For example, lactic dehydrogenase-defective mutants have been produced which form little acid during glycolysis and which can no longer induce dental decay. However, these mutants can colonize and compete effectively with wild-type strains. Their prior establishment on teeth will inhibit implantation of virulent strains.

Other avenues being studied include application of materials to modify tissue surfaces, such as enzymes which destroy tissue receptors, or amphipathic molecules (i.e., emulsan) which may block hydrophobic domains. Sublethal concentrations of certain antibiotics have been found to inhibit the synthesis of some bacterial adhesins, and, therefore, new chemotherapeutic agents may be uncovered which could control bacterial colonization in this manner.

Microbial association with substrata has also been increased in the case of both animate and inanimate surfaces. For example, the introduction of genes from Rhizobium trifolii required for receptor synthesis into related bacteria enhances the attachment of these cells (Dazzo, this volume). In some commercial processes using continuous fermentation, it is important to retain the microorganisms in the reactor vessel. This may be achieved using various organic polymers to trap the organisms within a gel matrix (Atkinson, this volume).

Although apparently very different methods are being used to control microbial adhesion, it will be interesting to see whether similarities emerge as continued work unravels more of the complexities of the interactions involved.

CONCLUSION

Perhaps one of the most important conclusions to arise from this workshop was that not only could physical chemists and microbiologists agree on certain definitions, but that their continued cooperation would lead to fruitful and exciting research in the future.

It was felt that the following suggestions might stimulate further work and lead to significant increases in the understanding of microbial attachment.

Questions

Given the possibility of using genetic manipulation to produce a 1 micron scientist, how can we decide whether he/she should be a microbiologist, a physical chemist, or a hybrid?

How do microbial cell surface and substratum surface macromolecules interact both with each other and with one another?

How do the composition and stereochemistry of a conditioning film adsorbed to a usually hard substratum surface influence subsequent microbial adhesion?

How can we stabilize surfaces in biological environments?

How can we obtain quantitative information regarding the changes in chemical potential of the components involved in the formation of an adhesive junction? (See (6).)

How can the force required to remove adherent microbial cells be measured? (This is often of more practical relevance than the force of adhesion.)

How can defined hydrodynamic conditions for microbial adhesion and removal be achieved in complex biological environments?

How do microbial cells sense contact with a substatum surface and respond to it?

How and why are adhesin and receptor genes regulated?

Technique Development

1. Methods of nondestructively characterizing surfaces in vivo are needed.

2. Extension of the use of whole cell n.m.r. should be considered.

Definitions

Adhesin. An adhesin is a surface macromolecule of a cell known to be responsible for specific adhesion.

Adhesion. Adhesion can be defined unambiguously only in terms of the energy involved in the formation of the adhesive junction. Thus, two surfaces may be said to have adhered when work is required to separate them to their original condition.

Nonspecific Adhesion. Microbial adhesion involves macromolecules on the surface of the microorganism concerned. These may interact with the surface of the substratum or the macromolecules present on that surface. A number of classes of interactions may be involved (ionic, dipolar, H-bonds, hydrophobic interactions). The number and strength of these interactions vary considerably from case to case, so that the level of adhesion of a given microbe to a variety of surfaces (or a range of microbes to a given surface) will also vary considerably. This, however, does not imply specificity.

Specific Adhesion. An additional requirement for specific adhesion is some form of stereochemical constraint which brings more than one (normally several) pair of neighboring, interacting groups on the microorganism and the substratum into contact. An example would be sugar residue-protein interactions on two complementary polymers, i.e., a lock-and-key mechanism is required. Monomers and oligomers which sufficiently mimic the combining sites of either interacting surface polymer can occupy the combining sites and block specific attachment, reducing it to a nonspecific level of adhesion characteristic for that system.

Adsorption. Adsorption is the accumulation of molecules at a fluid interface at a concentration exceeding that in the bulk fluid, brought about as a result of random Brownian motion.

Association. This is a term which is useful in describing the interaction between a microorganism and a surface, without specifying a mechanism for this interaction. The initial indication of a reaction between microorganisms and a complex biological surface is usually the observation of an increased retention time of the bacterium at the surface (compared to that of a random nonsticky collision). Such increased retention time may be due to adhesion (as defined here) or to nonadhesive mechanisms (e.g., trapping in mucus gel). Until the precise mechanism of the interaction has been elucidated, it may be properly described as microbial association with a surface.

Deposition. Deposition is normally used to describe the accumulation of particles at a fluid interface brought about by the application of an external force, e.g., gravity.

Hydrophobicity and Hydrophilicity. The structure of water in the region near any surface is perturbed over distances of maybe up to several tens of molecular layers. Near a so-called hydrophobic surface the water is less structured in terms of intermolecular hydrogen bonding between the water molecules, whilst near a hydrophilic surface water is more structured. If two similar surfaces come together such that the two perturbed water layers overlap, then there must be a displacement of perturbed water molecules into bulk water. This will lead to a decrease in free energy (i.e., an attraction) in the case of a hydrophobic surface, but to an increase in free energy (i.e., a repulsion) in the case of a hydrophilic surface. These structural interactions (which clearly vary with separation between the surfaces) are an addition to the Van der Waals interactions between the particles concerned. If one brings a hydrophobic surface up to a hydrophilic surface, one cannot predict a priori whether there will be a net repulsion or attraction.

Receptor. A receptor is a component on the surface which is bound by the active site of an adhesin during the process of specific microbial attachment.

Selectivity. This implies the ability of microorganisms to achieve different degrees of adhesion to a variety of surfaces. This could be achieved, for example, by means of a series of substrata with different Van der Waals attractive properties. Alternatively, selectivity may result from quantitative differences in the numbers of specific adhesive molecules on the surfaces. Consequently, it is not possible to decide whether adhesion has a specific molecular basis merely by observing relative numbers of cells which attach to surfaces or by measuring their strength of attachment. Specificity requires some demonstration of molecular stereochemical complementarity.

REFERENCES

(1) Absolom, D.R.; Lamberti, F.V.; Policova, Z.; Zingg, W.; van Oss, C.J.; and Neuman, A.W. 1983. Surface thermodynamics of bacterial adhesion. Appl. Envir. Microbiol. 46: 90–97.

(2) Adamczyk, Z.; Dabros, T.; Czarnecki, J.; and van de Ven, T.G.M. 1983. Particle transfer to solid surfaces. Adv. Coll. Interface Sci. 19: 183–252.

(3) Fattom, A., and Shilo, M. 1984. Hydrophobicity as an adhesion mechanism of benthic Cyanobacteria. Appl. Envir. Microbiol. 47: 135–143.

(4) Fisher, L.R.; Israelachvili, J.N.; Parker, N.S.; and Sharples, F. 1980. Adhesion measurement in microbial adhesion to surfaces. In Microbial

Adhesion to Surfaces, eds. R.C.W. Berkeley et al., pp. 515-517. Chichester: Ellis Horwood.

(5) Gibbons, R.J. 1980. Adhesion of bacteria to the surfaces of the mouth. In Microbial Adhesion to Surfaces, eds. R.C.W. Berkeley et al., pp. 351-388. Chichester: Ellis Horwood.

(6) Pethica, B.A. 1980. Microbial and cell adhesion. In Microbial Adhesion to Surfaces, eds. R.C.W. Berkeley et al., pp. 19-45. Chichester: Ellis Horwood.

Microbial Adhesion and Aggregation, ed. K.C. Marshall, pp. 21-38. Dahlem Konferenzen 1984. Berlin, Heidelberg, New York, Tokyo: Springer-Verlag.

Physicochemical Interactions of the Substratum, Microorganisms, and the Fluid Phase

P. R. Rutter* and B. Vincent**
*Minerals Processing Branch, BP Research Center, Sunbury, Middlesex
**Physical Chemistry Dept., Bristol University, England

Abstract. The adhesion of microorganisms to surfaces is influenced by long-range, short-range, and hydrodynamic forces. In the study of particle adhesion using well-defined, nonliving systems, long-range forces are adequately described by DLVO theory (due to Derjaguin and Landau, and Verwey and Overbeek), and hydrodynamic forces can be controlled. The quantitative description of short-range forces, however, remains a problem. In biological systems the application of DLVO theory as well as the quantitative description of short-range forces becomes difficult. Consequently, carefully designed experiments are required.

The authors suggest that in order to advance the study of microbial adhesion, it is necessary for physical chemists and microbiologists to collaborate in a detailed study of adhesion in a well characterized microbiological system.

INTRODUCTION

Any discussion of the physical chemistry of microbial adhesion is complicated by the nature of the particles and substrata concerned. Microorganisms are far from being "ideal" particles: they have neither a simple geometry, nor a simple, uniform molecular composition. They may be deformable, and internal chemical reactions can lead to changes in molecular composition both in the interior and at the surface of the bacteria; indeed, molecules and ions may cross the interface.

A further problem in applying physicochemical ideas in this area lies in identifying the nature of the strongly specific effects that frequently occur. Only certain microorganisms may adsorb on a given surface and then only under specific conditions, or at a specific stage in their growth. Clearly, the short-range interactions are very complex in biosystems.

An even more fundamental problem concerns the legitimacy of applying thermodynamic principles to microbial adhesion. Considerations of whether the process of adhesion of a microorganism to a surface takes place under equilibrium conditions, and whether, after adhesion, the equilibrium state is achieved, underlie the application of "DLVO-type theories" of particle/substratum interactions and the "wetting theories" of adhesion, as enunciated, for example, by Pethica (18). In practical terms, the establishment of the relevant equilibrium conditions requires detailed knowledge of the relative time scales over which the possible processes which could occur actually take place. For example, the magnitude of the electrostatic interaction between a particle and a surface depends on whether the charges on the two surfaces are effectively "locked-in," or whether the surface charges can adapt (to maintain constant potential) on a time scale comparable to the time scale of the adhesion process itself.

Similarly, if, in a microorganism which becomes attached to a surface, further chemical processes occur in the interior or at the surface of the cell, then that microorganism cannot be in complete physicochemical equilibrium. Moreover, equilibrium may only be achieved in a (thermodynamically) closed system (i.e., where no exchange of matter or energy with the surroundings occurs); in an open system, which is the case for many biological systems, equilibrium may never be reached (although a temporary steady state condition may be realized). In these cases, the much more complex theory of nonequilibrium thermodynamics needs to be considered.

How far, then, can physical chemists go in spanning the bridge across to biological systems, in particular, in the case of microbial adhesion? In our opinion, there is tremendous insight to be gained from considerations of physicochemical phenomena at the qualitative, conceptual level, but caution is required in applying full, quantitative theories. Before even semiquantitative applications are possible, physical chemists would require the microbiologists to come halfway across the bridge to meet them, by working with as well-defined and as characterized systems as possible. This may even require, in the first instance, the modelling of microorganisms as has been attempted recently, for example, by Imperial Chemical Industries (unpublished data). They have managed to imitate the surface of bacteria used in the "Pruteen" process for protein production by grafting the extracellular protein extracts to the surface of polymer latex particles. Alternatively, physical chemists and microbiologists should collaborate on a detailed examination of a well-defined microbiological system. The adhesion of Escherichia coli to epithelial cells, for example, might prove to be suitable, although the final choice should emerge from discussion.

The emphasis in this review, therefore, will be in stressing the nature of the physicochemical phenomena pertinent to microbial adhesion. We start with a review of the long- and short-range interactions between particles and a macroscopic surface, first in the absence and then in the presence of macromolecules (adsorbed or in solution). We then consider particle adsorption isotherms and kinetics including the effects of shear. In some cases equations are given, but this is simply because this is frequently a useful way of summarizing a concept, rather than expressing it in words. It is not necessarily implied that the equation is directly applicable in the biological context.

INTERACTIONS BETWEEN CHARGED PARTICLES AND MACROSCOPIC BODIES

One may discuss the net interaction between a charged particle and a macroscopic body in terms of the mutual force (f_i) or free energy (G_i) as a function of separation (h) (23). The form of $G_i(h)$ was originally formulated in the now classical DLVO theory, developed in the context of colloidal dispersions, the basic premise of which is that the total interaction between two particles is comprised of two additive terms: one (G_A) due to the van der Waals forces, and one (G_E) due to the overlap of the electrical double layer associated with charge groups present on the particle and macroscopic surfaces. The situation is schematically illustrated in Fig. 1.

One approximate form for G_A is (31):

$$G_A = -\frac{A}{6}\left[\frac{2a(h+a)}{h(h+2a)} - \ln\left(\frac{h+2a}{h}\right)\right], \qquad (1)$$

where A is the net Hamaker coefficient for the system.

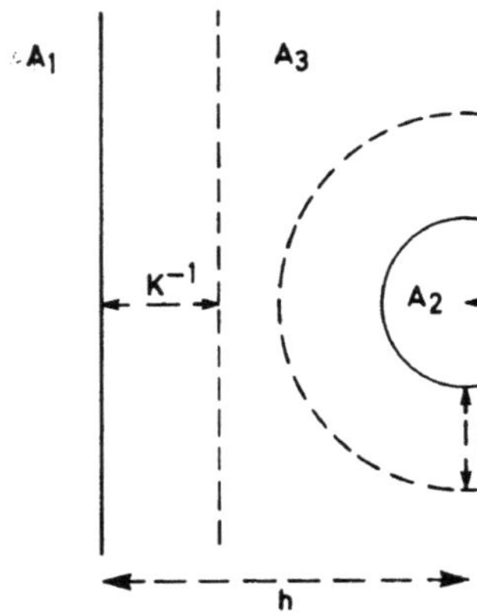

FIG. 1 - Schematic representation of a particle approaching a macroscopic surface.

In order to evaluate the electrical double layer interaction, we need to define the electrical potential at each surface, i.e., ψ_{13} and ψ_{23}. For order-of-magnitude calculations, the electrokinetic (or zeta) potential (ζ) suffices. This may be determined, in the case of the particles, by electrophoresis measurements. Typical values for ζ lie in the range 0 to ±100 mV, depending primarily on the net number of surface charge groups and the bulk electrolyte concentration; an increase in the latter generally reduces ζ.

As mentioned in the introduction, in evaluating G_E one has to make an assumption as to whether the interaction takes place at constant charge (i.e., the charge on each surface remains unchanged during the interaction) or at constant potential, although in practice this only leads to a significant difference at small values of h. For the case of constant potential, G_E is given by Eq. 2, for small values of ψ (<50 mv)(12).

$$G_E = \Pi \epsilon \epsilon_o a(\psi_{13}^2 + \psi_{23}^2)\left[\frac{2\psi_{13}\psi_{23}}{\psi_{13}^2+\psi_{23}^2}\ln\left(\frac{1+e^{-\kappa h}}{1-e^{-\kappa h}}\right) + \ln(1-e^{-2\kappa h})\right], \qquad (2)$$

where ϵ is the dielectric constant of the medium (78.2 for water at 25°C) and ϵ_o is the permittivity of free space ($8.85 \times 10^{-12} J^{-1} C^2 m^{-1}$).

κ^{-1} is an estimate of the effective thickness of the electrical double layer and depends on the total ion concentration in the bulk solution.

The forms of G_A, G_E, and G_i (= $G_A + G_E$) are shown schematically in Fig. 2, for those cases where the particle and macroscopic surface have the same sign.

The exact shape of the curves depends critically on the values of a, ψ_{13} and ψ_{23} (or ζ_{13} and ζ_{23}), and κ; the forms indicated in Fig. 2a–c are only intended to be schematic. It can be seen that at low electrolyte concentrations (Fig. 2a), there is a large free energy barrier to overcome if a particle is to come into close contact with the surface; thus, effectively, the particle is repelled from the surface. At high electrolyte concentrations (Fig. 2c), this energy barrier is eliminated, and there is a strong net attraction between the surfaces and the particles. At intermediate electrolyte concentrations (Fig. 2b), the free energy barrier is still present but much smaller. Therefore, a certain fraction of the particle collisions with the surface will result in (effectively) permanent contacts being made, i.e., the particles will "sit" in the primary minimum. To a first approximation, we can say that the fraction, f, of such collisions resulting in primary minimum contact being established will be proportional

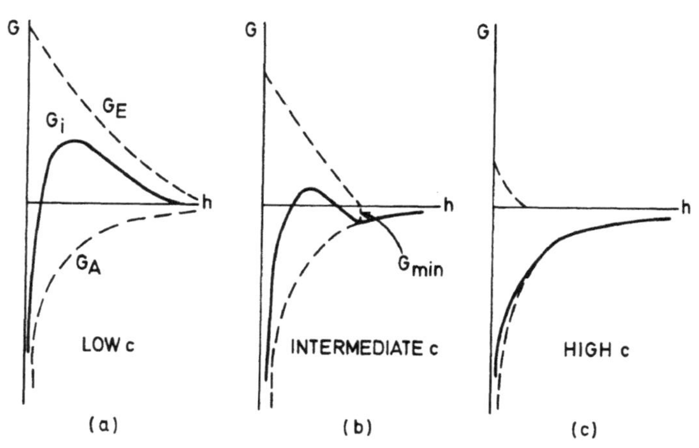

FIG. 2 - Interaction free energy (G) versus separation (h) curves as a function of electrolyte concentration (c) for a charged particle approaching a macroscopic surface of the same sign.

to $\exp(-G_{max}/kT)$, (i.e., when $G_{max} \rightarrow \infty$, $f \rightarrow 0$, but when $G_{max} \rightarrow 0$, $f \rightarrow 1$). It should be noted, however, that for intermediate electrolyte concentrations a secondary minimum may also exist. Because this is generally much shallower than the primary minimum, particles will reside in this minimum temporarily. We can surmise that the residence time, t, for a particle in the secondary minimum will be proportional to $\exp(|G_{min}|/kT)$, so that as $G_{min} \rightarrow \infty$, $t \rightarrow \infty$.

In those cases where the surface and the particle are of opposite sign, the net effect is a strong attraction, decreasing with increasing bulk electrolyte concentration, but in all cases leading to strong adhesion.

The most obvious parameter to emerge from a study of the long-range interactions which might influence adhesion is surface charge. However, although the adhesion of Streptococcus mutans to glass is influenced by ionic strength (2), two detailed studies using various oral organisms (1, 16) failed to find any simple correlation between zeta potential and adhesion either to glass or dentine. It is also interesting to note that certain species of trypanasome are positively charged at physiological pH yet circulate freely in the bloodstream (18).

These examples suggest that whilst long-range forces may control the time which a bacterial cell spends in close proximity to a surface, other forces determine whether adhesion will occur or not (see Rutter et al., this volume).

SHORT-RANGE FORCES

In the above discussion, only long-range forces (i.e., the dispersion contribution to the van der Waals forces and the electrical double layer interactions) were considered. At short separations between the surface and the particles, other interactions may have to be taken into account, e.g., dipole-dipole (Keesom) interactions, dipole-induced dipole (Debye) interactions, ion-dipole interactions, H-bonds, etc. These types of interactions include many which have been identified as important to biological systems, for example, the effect of divalent cations on the aggregation and fusion of phosphatidyl-serine vesicles (4) and the implication of Ca^{2+} in bacterial adhesion in the mouth (20). Short-range effects are particularly important in aqueous systems and may be repulsive or attractive, depending on the nature of the surfaces involved. We may discuss this in terms of the displacement of water molecules from the interaction zone as a particle approaches the macroscopic surface.

If both surfaces are "hydrophilic" there is a net increase in free energy of the displaced water molecules. This is because the local ordered water structure near the surfaces has to be broken down. This leads to a short-range repulsion force which may be sufficient to prevent the particles coming close to the macroscopic surface. An example of this effect is observed with aqueous dispersions of precipitated silica particles which do not readily coagulate with the addition of electrolytes (11). On the other hand, when both the surfaces involved are hydrophobic, the short-range interaction is a net attraction, i.e., over and above the dispersion force term. In effect this leads to a deepening of the primary minimum. This arises because the water molecules displaced into bulk solution now decrease their free energy since in bulk solutions there is a net increase in their H-bonding. This is the basis of the so-called "hydrophobic interaction."

In the case of hydrophilic microorganisms, the above analysis predicts that their adsorption to a hydrophobic surface (e.g., teflon) will be stronger than to a hydrophilic surface (e.g., clean glass).

AFFECTS ARISING FROM THE PRESENCE OF MACROMOLECULES

Macromolecules (neutral polymers or polyelectrolytes), present in solution or located at an interface, may lead to significant changes in the interaction between particles and macroscopic surfaces. Most microorganisms have a thick layer of macromolecules associated with their surface. In certain circumstances these may be released into the solution surrounding the microorganism.

One effect of macromolecules present in the continuous phase is a purely hydrodynamic one; the rate constant for adsorption is reduced as a result of the increase in solution viscosity. However, at high concentrations of free polymer, it has been observed (7, 27, 32) that weak reversible flocculation can occur in particulate dispersions. The mechanism is ascribed to a "depletion layer" of polymer near the surface. Overlap of the depletion layers on two particles leads to an attraction between them. The significance of this effect in biological systems could be very important. For example, the flocculation of erythrocytes by dextrans (6) can probably be ascribed to this effect. Even when the particle surfaces carry an attached layer of polymers, the addition of high concentrations of free polymer may lead to destabilization (7).

When macromolecules adsorb at an interface to form a layer of thickness

δ, there are additional effects on particle interactions (see, e.g., reviews by Vincent (25, 28)). First of all, there are indirect effects, in the sense that both G_A and G_E may be modified. The general situation is depicted in Fig. 3 (for simplicity, we assume here that the same polymer is adsorbed on both surfaces). In terms of G_A, strictly speaking, a fourth Hamaker coefficient (A_4), i.e., that of the adsorbed layer, has to be considered, but in many practical cases $A_4 \simeq A_3$, so that Eq. 1 suffices. Also with regard to G_E, Eq. 2 applies, provided the electrokinetic potentials corresponding now to the peripheries of the two adsorbed polymer layers (i.e., ψ'_{13} and ψ'_{23}) are used and h is replaced by (h - 2δ). In this way, G_i (= G_E + G_A) for h $\geq 2\delta$ may be evaluated.

When h < 2δ, there is a direct effect due to the "interference" of the two adsorbed layers i.e., the so-called "steric interaction," G_S, comes into play (15, 27) The significant point regarding this interaction is that provided the adsorbed polymer chains are strongly attached to the surface, at full coverage, and are in a good solvent environment, a steep repulsion starts at h = 2δ. A typical G_i(h) curve for neutral (uncharged) systems is illustrated in Fig. 4. It is characterized by an interaction free energy minimum, G_{min}, which itself depends primarily on δ, the thickness of the adsorbed polymer layers, and the G_A curve: G_{min} G_A, at h = 2δ.

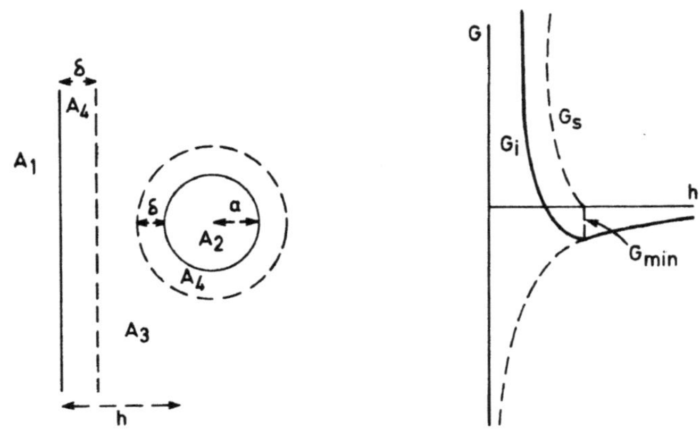

FIGS. 3 and 4 - The interaction between a particle and macroscopic surface in a sterically stabilized system.

If the surfaces are only partially covered by adsorbed polymer molecules, then it is possible for "polymer bridging" to occur, i.e., the co-adsorption of polymer molecules onto both surfaces. Even if the surfaces are fully covered, then it may be possible for a second polymeric species to "bridge" the two adsorbed polymer layers, provided specific interactions (Coulombic, acid/base) occur between the added polymer and the attached polymer. Lipoteichoic acid (LTA) can cross-link the two protein layers on a substratum and bacterium in this way (3).

Polymers play an important part in determining the behavior of nonbiological dispersions, and in many cases interesting parallels between "living" and "nonliving" systems emerge. For example, Feigin and Napper (9) have shown that selective flocculation can occur in systems containing particles stabilized by different water soluble polymers. This implies that polymer-stabilized particles can distinguish "self" from "not-self" under suitable conditions. The efficiency of filtration has been shown to improve if polymer is added (24). The polymer molecules increase the chance of particle capture by bridging between the particles and filter media. It is also well-known that the flocculation of small mineral particles by anionic polyacrylamides, important in many mineral processing operations, is strongly influenced by the presence of small quantities of divalent cations, e.g., Ca^{2+} (13). This is thought to be due to changes in the adsorption characteristics of the polymer brought about by the formation of complexes with calcium.

Investigation of the similarities of the effect of polymers on living and nonliving systems could provide a fruitful field of research leading to a greater understanding of the interactions involved in both areas. Clearly this will involve detailed study of microbial polymer adsorption, composition, conformation, gelation, solubility, degradation, etc.

PARTICLE ADSORPTION ISOTHERMS

The discussion so far has only considered the interaction between isolated particles and a macroscopic surface. As the coverage of the surface increases, we have to take into account lateral interactions between particles. We first consider the case of charged particles and surfaces in the absence of macromolecules. In general, if the adsorbing particles are all of the same type, then the lateral interactions will be repulsive at low bulk electrolyte concentrations but attractive at high bulk electrolyte concentrations, i.e., when the effects of the electrical double layers are sufficiently reduced.

We can now speculate as to the shape of the adsorption isotherms for particles adsorbed at a macroscopic surface in those cases where: (a) these are of the same sign, or (b) these are of opposite sign.

In Figs. 5a and b, the isotherm is sketched in the form of the coverage, θ , where $\theta = \Gamma / \Gamma_{h.c.p.}$ (Γ is the adsorption in particles/area and $\Gamma_{h.c.p.}$ the maximum value corresponding to hexagonal close packing in a two-dimensional array), versus the equilibrium particle concentrations. "High," "intermediate," and "low" electrolyte concentrations are somewhat arbitrary designations, but, as a rough guide, one infers by "low" $<\sim 10^{-4}$ mol dm^{-3}, and "high" $>\sim 10^{-1}$ mol dm^{-3} 1:1 electrolyte. At high electrolyte concentrations, electrical double layer effects are minimal, and a high affinity adsorption isotherm results in both cases a and b corresponding to irreversible adsorption into the deep primary minimum (Fig. 2). Maximum coverage ($\theta = 1$) may not be attained, however, in practice since the particles will be attracting laterally and are unlikely, therefore, to form a close-packed, two-dimensional array. At low electrolyte concentrations one would not expect to observe any adsorption because of strong normal and lateral repulsions. At intermediate concentrations there will be a weaker, but still effective lateral repulsion between the particles, but the role of the secondary minimum now has to be considered in the case of the normal (i.e., surface particle) interaction. The results will be weak, reversible (i.e., responsive to changes in the bulk particle concentration) adsorption of the particles

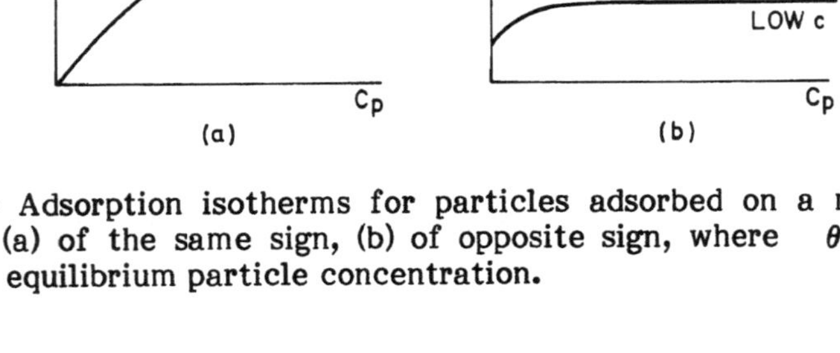

FIG. 5 - Adsorption isotherms for particles adsorbed on a macroscopic surface: (a) of the same sign, (b) of opposite sign, where θ = coverage and C_p = equilibrium particle concentration.

into the secondary minimum, i.e., most of the particles will not be in primary contact but rather held at some distance (typically ~10-100 nm) from the surface, although the occasional particle, as discussed above, may surmount the free energy barrier and come into primary contact. If no contacts of this sort are made, then the (reversible) adsorption isotherm will be Langmuiran in form. Note that this would not represent the equilibrium isotherm, however, since the "true" equilibrium situation (i.e., the minimum free energy state) is when all the adsorbing particles are in close contact with the surface; this would then correspond to a high affinity, irreversible adsorption isotherm.

Isotherms similar to those illustrated in Fig. 5 have been obtained for the adsorption of microorganisms onto polystyrene (10) and glass (17).

With regard to the case where the particles and the adsorbing surface are of opposite sign, the adsorption is high affinity at all electrolyte concentrations; what differs is the maximum coverage attainable.

When the particles/macroscopic surface are charged and carry an adsorbed polymer layer, the situation is quite complex. The equilibrium aspects of this situation have been discussed by Vincent, Young, and Tadros (29, 30) for the case of small, positive polystyrene particles adsorbing onto larger negative particles, in which both sets of particles carried a pre-adsorbed layer of poly(vinyl alcohol). Again a subtle balance between the primary, normal adsorbing forces and the lateral (repulsive or attractive) forces has to be considered. Both reversible and irreversible adsorption can be achieved depending on the thickness of the polymer layer, δ , and the bulk electrolyte concentration. For a given value of δ , there exists a critical electrolyte concentration c^*, below which the adsorption is high affinity and irreversible, but above which it is low affinity and reversible. c^* decreases as δ increases.

Figure 6 shows a Scanning Electron Microscope (SEM) picture representing high affinity particle adsorption in these mixed latex particle systems. It may be seen that, in the main, the small particles are adsorbed as discrete individuals, rather than in clusters, illustrating their mutual lateral repulsion on the surface.

In conclusion, one can state that the adsorption/desorption behavior of particles at a macroscopic surface involves a subtle balance of forces. The adsorption may be weak and reversible or strong and (effectively) irreversible; the particles may be in close contact with the surface or

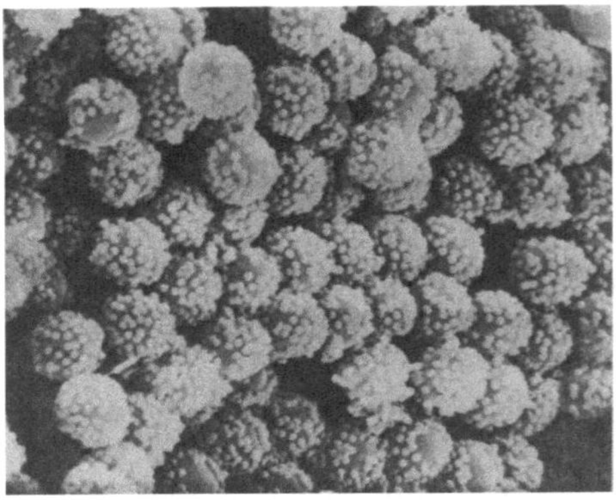

FIG. 6 - Small latex particles adsorbed to large latex particles from 10^{-4}M NaCl solution.

held at some distance from it; it may be possible to achieve high levels of coverage or only low levels. All these features are relevant to the discussion of the adsorption/adhesion behavior of microorganisms to macroscopic surfaces. Some examples of the influence of polymers on the attachment of bacteria to surfaces are given in a review by Rutter (21).

ADSORPTION KINETICS

The sequence of events which leads to the eventual adhesion of a particle to a surface is strongly influenced by the hydrodynamic conditions which exist at the point of contact. Particles captured by a surface in a primary minimum have a finite probability of escape due to thermal motion and local shear forces provided by the suspension medium which is usually in some state of flow. Under these conditions it becomes very difficult to quantify the principal forces involved, especially if microroughness effects become important.

Ideally, studies concerned with bacterial adhesion should be carried out under conditions of defined steady shear, using apparatus which allow direct observation of the adhering cells. Cell multiplication due to growth should also be eliminated or minimized. Techniques using small glass capillaries have been developed by a number of authors (19, 22). Generally, the deposition rate of cells to the glass walls of the capillary is much less than the collision rate. It has been suggested (22) that in order for attachment to occur, the cell must be captured in a primary or secondary minimum long enough for a link to be made between the two surfaces via polymer adsorption. Furthermore, this should be strong enough to withstand the continuous shearing by the flowing suspension. This explanation is supported by the observed oscillation of microorganisms attached to glass surfaces (2), which is difficult to explain in the absence of polymer bridging.

A number of similarities between the adhesion of bacterial cells and polystyrene latex particles are illustrated in a recent study by Dabros and van de Ven (8). In common with many cells they found that negatively charged latex particles adhered to negatively charged glass (both with zeta potentials of -50 to -60 mV), oscillated about a given position on the glass surface with displacements of up to a few microns, and excluded the adhesion of further particles from an area of 20 to 30 times their geometrical cross section. It should be noted here that the observation of a low number of bacterial cells on a host surface is often attributed to specific site interactions!

Results of studies on the adhesion of both microorganisms and latex particles demonstrate the importance of desorption as well as adsorption. Whether the adsorption of a particle to a surface is reversible or not will depend upon the depth of the free energy minimum (G_{min}) involved in the normal (particle/surface) interaction.

If we can assume that the adsorption is irreversible (G_{min} > 10 kT, for example), then the kinetic analysis proposed by Langmuir (14), in the context of gas adsorption, may be adopted; one may then write for the number concentration of small particles as a function of time n(t) (26):

$$\frac{dn(t)}{dt} = - k\, n(t)\ [1 - \theta(t)], \tag{3}$$

where k is the rate constant, and $\theta(t)$ is the coverage of the macroscopic surfaces, as given by

$$\theta(t) = \frac{n_o - n(t)}{n_o - n_f}, \tag{4}$$

where n_o and n_f are, respectively, the initial and final number concentrations of the particles in suspension.

Substitution of Eq. 4 into Eq. 3 and subsequent integration yields

$$\ln\left[\frac{n}{n - n_f}\right] = \ln\left[\frac{n_o}{n_o - n_f}\right] + \left[\frac{n_f}{n_o - n_f}\right] kt. \qquad (5)$$

Thus, a plot of $\ln[n/(n - n_f)]$ versus t should give a straight line, from which k can be determined, provided k is independent of θ; this may only apply at low coverage. The interpretation of k, even at low coverages, depends on the hydrodynamic conditions operating.

The case of reversible adsorption has been considered by Boughley et al. (5) and Vincent et al. (26). We now have to include a term for the desorption, i.e., Eq. 3 becomes

$$dn/dt = -\overrightarrow{k}n(1 - \theta) + \overleftarrow{k}(n_o - n), \qquad (6)$$

which leads to the following integral expression:

$$-\ln\left[\left(\frac{n - n_f}{n_o - n_f}\right) \Big/ \left(1 - (1 - \frac{n}{n_o})\ \theta_f\right)\right] = \overrightarrow{k}t\left(\frac{n_o}{n_o - n_f} - \theta_f\right), \qquad (7)$$

where θ_f is the equilibrium coverage. Note that in Eq. 7 $\overrightarrow{k}$ is the adsorption rate constant, and $\overleftarrow{k}$ is the desorption rate constant; $\overleftarrow{k} \propto \exp(-|G_{min}|/kT)$, at low coverages; at high coverage $\overleftarrow{k}$ will also depend on the lateral interactions.

CONCLUSIONS

A consideration of the long-range forces operating between microorganisms and surfaces can provide a useful first approach to explaining adhesive phenomena. The subtle differences between the adhesiveness of different microorganisms and their specificities for particular surfaces cannot, however, be explained in terms of long-range interactions.

In order to arrive at a satisfactory explanation of the events leading up to and during the formation of an adhesive junction between a cell and substratum, all the interactions between the surfaces and environment must be considered. This means that it is no longer sufficient to ignore or at best qualitatively discuss short-range forces. These must be described and quantified. To borrow a phrase from Pethica (18): "Further advance in the study of the physical chemistry of microbial adhesion requires an understanding of the stereochemical 'grain' of the interacting surfaces." Collaboration between physical chemists and microbiologists should therefore be directed towards identifying all the variables involved in the adhesive process ideally by beginning with a system which is well understood biologically. The relative importance of, for example, specific binding molecules, extracellular polymers, specific and nonspecific electrolytes, temperature, pH, cell surface polymer structure and solvation, shape, metabolic state, motility, shear, etc., could then be identified. Although necessarily complex, such a study would provide valuable insights into the adhesive mechanisms involving microorganisms in vivo and perhaps lead to methods of control as subtle as those used by nature itself.

REFERENCES

(1) Abbott, A.; Berkeley, R.C.W.; and Rutter, P.R. 1980. Sucrose and deposition of Streptococcus mutans at solid liquid interfaces. In Microbial Adhesion to Surfaces, eds. R.C.W. Berkeley, J.M. Lynch, J. Melling, P.R. Rutter, and B. Vincent, pp. 117-142. Chichester: Ellis Horwood.

(2) Abbott, A.; Rutter, P.R.; and Berkeley, R.C.W. 1983. The influence of ionic strength, pH and a protein layer on the interaction between Streptococcus mutans and glass surfaces. J. Gen. Microbiol. 129: 439-445.

(3) Beachey, E.M.; Simpson, W.A.; and Ofek, I. 1980. Interaction of surface polymers of Streptococcus pyogenes with animal cells. In Microbial Adhesion to Surfaces, eds. R.C.W. Berkeley, J.M. Lynch, J. Melling, P.R. Rutter, and B. Vincent, pp. 389-406. Chichester: Ellis Horwood.

(4) Bentz, J.; Nir, S.; and Wilschut, J. 1983. Mass action kinetics of vesicle aggregation and fusion. Coll. Surf. 6(4): 333-364.

(5) Boughley, M.T.; Duckworth, M.M.; Lips, A.; and Smith, A.L. 1978. The observation of weak primary minima in the interaction of polystyrene with nylon. J. Chem. Soc. Faraday Trans. I 74: 2200-2208.

(6) Brooks, D.E. 1973. The effects of neutral polymers on the electrokinetic potential of cells and other charged particles III. The experimental studies on the dextran/erythrocyte system. Coll. Interface Sci. 43: 700-713.

(7) Cowell, C., and Vincent, B. 1982. The stability of polystyrene latices in the presence of poly(ethylene oxide). In The Effects of Polymers on Dispersion Properties, ed. T.F. Tadros, pp. 263-284. London:

Academic Press.

(8) Dabros, T., and van de Ven, T.G.M. 1983. A direct method for studying particle deposition onto solid surfaces. Coll. Polym. Sci. 261: 694-707.

(9) Feigin, R.I., and Napper, D.H. 1978. Heterosteric stabilisation and selective flocculation. Coll. Interface Sci. 67(1): 127-139.

(10) Fletcher, M. 1977. The effects of culture concentration and age, time and temperature on bacterial attachment to polystyrene. Can. J. Micro. 23: 1-6.

(11) Harding, D. 1971. Stability of silica dispersions. J. Coll. Interface Sci. 35: 172-174.

(12) Hogg, R.; Healy, T.W.; and Fuerstenau, D.W. 1960. Mutual coagulation of colloidal dispersions. Trans. Faraday Soc. 62: 1638-1651.

(13) Hunter, R.J. 1982. Some effects of dissolved salt on flocculant performance. Filtra. Separ. July/August: 289-291.

(14) Langmuir, I. 1918. Adsorption of gases onto plane surfaces of glass, mica and platinum. J. Am. Chem. Soc. 40: 1361-1403.

(15) Napper, D.H. 1977. Steric stabilisation. J. Coll. Interface Sci. 58: 390-407.

(16) Olsson, J.; Glantz, P.O.; and Krasse, B. 1976. Surface potentials and adherence of oral streptococci to solid surfaces. Scand. J. Dent. Res. 84: 240-242.

(17) Ørstavik, D. 1977. Sorption of Streptococcus faecium to glass. Acta Path. Microbiol. Scand. B 85: 38-46.

(18) Pethica, B.A. 1980. Microbial and cell adhesion. In Microbial Adhesion to Surfaces, eds. R.C.W. Berkeley, J.M. Lynch, J. Melling, P.R. Rutter, and B. Vincent, pp. 19-46. Chichester: Ellis Horwood.

(19) Powell, M.S., and Slater, N.K.H. 1983. The deposition of bacterial cells from laminar flows onto solid surfaces. Biotech. Bioeng. 25: 891-900.

(20) Rolla, G. 1980. On the chemistry of the matrix of dental plaque. In Microbial Adhesion to Surfaces, eds. R.C.W. Berkeley, J.M. Lynch, J. Melling, P.R. Rutter, and B. Vincent, pp. 425-440. Chichester:

Ellis Horwood.

(21) Rutter, P.R. 1980. The physical chemistry of the adhesion of bacteria and other cells. In Cell Adhesion and Motility, eds. A.S.G. Curtis and J.D. Pitt, pp. 103-135. Cambridge University Press.

(22) Rutter, P.R., and Leech, R. 1980. The deposition of Streptococcus sanguis NCTC 7868 from a flowing suspension. J. Gen. Microbiol. 120: 301-307.

(23) Rutter, P.R., and Vincent, B. 1980. The adhesion of microorganisms to surfaces: physico-chemical aspects. In Microbial Adhesion to Surfaces, eds. R.C.W. Berkeley, J.M. Lynch, J. Melling, P.R. Rutter, and B. Vincent, pp. 79-92. Chichester: Ellis Horwood.

(24) Sehn, P., and Gimbel, R. 1983. Effect of polymers on particle adhesion mechanisms in deep bed filtration. In Advances in Solid-Liquid Separation, ed. J. Gregory. London: Academic Press, in press.

(25) Vincent, B. 1974. The effect of adsorbed polymers on dispersion stability. Adv. Coll. Interface Sci. 4: 193-277.

(26) Vincent, B.; Jafellici, M.; Luckham, P.F.; and Tadros, T.F.T. 1980. Adsorption of small positive particles onto large negative particles in the presence of polymer. II. Adsorption equilibrium and kinetics as a function of temperature. J. Chem. Soc. Faraday Trans. F I 76(3): 674-682.

(27) Vincent, B.; Luckham, P.F.; and Waite, F.A. 1980. The effect of free polymer on the stability of sterically stabilised dispersions. J. Coll. Interface Sci., in press.

(28) Vincent, B., and Whittington, S. 1983. Polymers at interfaces and in disperse systems. In Surface and Colloid Science, ed. E. Matijevic, vol. 12, pp. 1-118. New York: Plenum.

(29) Vincent, B.; Young, C.A.; and Tadros, T.F.T. 1978. Equilibrium aspects of heteroflocculation in mixed sterically stabilised dispersions. Faraday Disc. Chem. Soc. 65: 296-305.

(30) Vincent, B.; Young, C.A.; and Tadros, T.F.T. 1980. Adsorption of small positive particles onto large negative particles in the presence of polymer. I adsorption Isotherms. J. Chem. Soc. Faraday Trans. F I 76(3): 665-673.

(31) Visser, J. 1976. Adhesion of colloidal particles. In Surface and

Colloid Science, ed. E. Matijevic, vol. 8, pp. 3-84. New York: Plenum Press.

(32) Vrij, A. 1976. Polymers at interfaces and the interactions in colloidal dispersion. Pure Appl. Chem. 48: 471-483.

Microbial Adhesion and Aggregation, ed. K.C. Marshall, pp. 39-49. Dahlem Konferenzen 1984. Berlin, Heidelberg, New York, Tokyo: Springer-Verlag.

Stereo-biochemistry and Function of Polymers

I.D. Robb
Unilever Research Laboratory
Port Sunlight, Merseyside, L63 3JW, England

Abstract. The role of exocellular polymers in the adhesion of bacteria to surfaces has been considered by comparing it with the adsorption of free polymers on surfaces. In this, the roles of the various interactions between the polymer segment, solvent, surface, and solution components have been discussed. The importance of the multisegment nature of a polymer has been emphasized, together with the special properties and behavior of polyelectrolytes. Factors influencing the time dependence of the adhesion of (inert) particles to surfaces and their applicability to bacterial adhesion are discussed.

INTRODUCTION

A simple view of a bacterium is that it consists of a roughly spherical core, often surrounded by macromolecules that are generally, though not always, securely attached to the core (9). The size of the central core is often about 1µm, so that superficially, at least, a bacterium can be modelled by a colloidal particle having an adsorbed polymer layer. In a bacterium, the core (cell wall and contents) is living, i.e., it is able to use chemical energy to produce new molecular structures, whereas the morphology of an inert colloidal particle is essentially independent of time. The interactions of inert colloidal particles are governed by certain laws of nature that must also apply to bacteria. The important difference, of course, is that with bacteria the effects arising from the energy (hence molecules, shapes, or structures) of metabolism is superimposed on the behavior of inert colloidal particles and may have a dominant or insignificant influence.

Macromolecules adsorbed on the surface of inert particles have a significant effect on the interactions and behavior of the latter. Some of these effects, such as the stabilization of colloidal dispersions, have been utilized empirically for centuries, though a sound, fundamental study of them has been undertaken only in the last 25 years (20, 24). The purpose of this article is to outline what is known of these effects and to what extent they apply to the behavior of bacteria.

POLYMER ADSORPTION

To understand the possible roles of the exocellular polymers in the adhesion of bacteria to surfaces, it is necessary to understand those factors controlling the behavior of model, isolated macromolecules on surfaces. A model of such an adsorbed molecule is shown in Fig. 1.

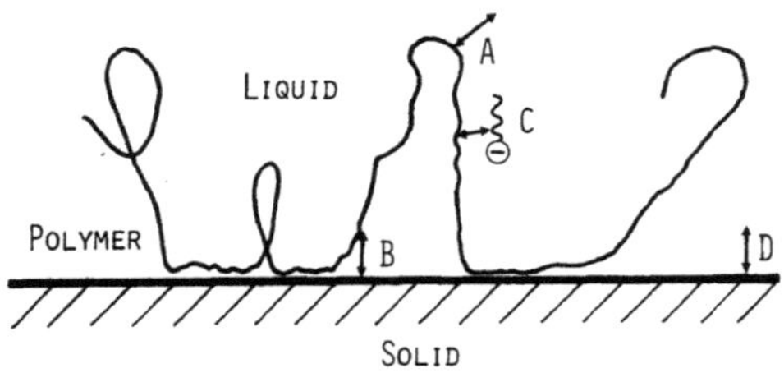

FIG. 1 - Representation of a polymer adsorbed at a solid-liquid interface. Important interactions are: A. segment-solvent; B. segment-surface; C. segment-other solvent components; D. surface-solvent.

There are four important constituents in the model, i.e., a) polymer segments, b) the solid surface, c) the solvent molecules, and d) other components such as phosphate, proteins, or lipids present in the solvent. Each of these constituents can interact with themselves and the other ones, and the behavior of the adsorbed polymer is the net balance of these interactions. The nature of these interactions, together with their likely consequences for bacterial adhesion, will be considered in the next sections.

Polymer Segment-surface Interactions

This is probably the most important of the many interactions. As polymers may have as many as 10^5 units in their chain, it is in principle possible for one molecule to have this number of bonds to the surface. The probability of a polymer having all its segments in contact with the surface at any one time is small, though it is commonly (3, 6, 7) found experimentally that a polymer will have between 30% and 60% of its segments in contact with the surface. Thus a random coil polymer may commonly have 10^4 contacts with a surface. This number helps explain the common observation (8) that once a macromolecule is attached to a surface, it is unlikely to be desorbed, since for this to occur, each of these bonds must be broken at the same time without new ones forming. Thus, even if the bond energy between each polymer segment and the surface is very small, say 0.01kT, then the total adsorption energy for the polymer with only 10^3 contacts with the surface is 10kT. The equilibrium distribution of particles between two sites (1 and 2) having a difference in energy of ΔE is governed by the Boltzmann distribution, i.e., $N_2 = N_1 e^{-\Delta E/kT}$, where N_1 and N_2 are the numbers of particles in each site. For ΔE = 10kT, $N_2/N_1 \sim 5 \times 10^{-5}$ so that at equilibrium only $\sim 10^{-2}$% of the particles would be in state 2.

The net energy of interaction between polymer segments and surfaces is largely unknown for any real systems, though the nature of the bonding involved may give some order of magnitude. Hydrogen bonds (17) have energies of about 4kT. If the system being considered is the adhesion between bacteria and glass, the net energy of adsorption for a segment of exocellular polysaccharide (assuming that this is the bonding mechanism) on glass would be the energy difference between the polysaccharide-water bond and the polysaccharide-SiOH bond. As each of these are 4-5kT, a net energy of adsorption of 0.1kT is quite feasible. Thus, this simple treatment illustrates that very weak bonding between macromolecules and solid surfaces can lead to very strong, apparently irreversible adsorption of the macromolecule on the surface.

The above ignores one point worthy of consideration. A molecule prefers to have the maximum possible number of degrees of freedom, i.e., its entropy tends to increase. In order to transfer from one state to another (having fewer degrees of freedom), the loss in entropy will have to be compensated by energy from some other part of the system. When a polymer adsorbs from solution (where it has three degrees of spatial freedom) to a surface (where it only has two), there has to be some other energetic incentive for that to occur. This other energy is that released from the polymer-surface bond. For free individual polymers adsorbing from solution to a solid surface, this energy is about 0.5kT per segment (13, 19). Polymers already attached to bacteria in effect are already constrained (by the cell surface) in their ability to move through the whole of the solution, so that less energy (than the 0.5kT per segment for free polymers) is needed for adsorption to begin.

Summary point. Polymers can adsorb to solid surfaces in an apparently irreversible fashion, as a result of very weak polymer-surface bonds. This is essentially a consequence of the large number of contacts there can be between polymers and the substratum.

An important related point is the following. Particles such as bacteria may have a net negative charge and yet be able to adsorb onto negatively charged surfaces without any (biologically) specific mechanisms being involved. The negative charge on the bacteria may originate on the cell wall and/or the exocellular polymer layer. The repulsion between the negative surface and bacterium may well be overcome by the polymer-surface interaction, i.e., many weak polymer-surface contacts may provide sufficient energy to overcome an electrical repulsion in the adsorbed state. An example (14) of this is the adsorption of sodium poly(styrene sulphonate) on silica particles (negatively charged) from water. An electrostatic repulsion can produce a kinetic barrier to adsorption, giving an impression of preventing adsorption at equilibrium. This is analogous to the (kinetic) stability of charge-stabilized latex particles. The height of the kinetic barrier to adsorption (or the extent into solution of the electrical field from the bacterium) will depend on the distribution of charges near the surface of the bacterium. If they occur entirely at the cell wall, the extension of the exocellular polymer will often exceed the significant extension of the electrical field, particularly at ionic strengths of about 0.1 mol dm^{-3}. If the exocellular polymers are also charged, their interaction with the surface will be more characteristic of polyelectrolyte adsorption, discussed below.

Summary point. Negatively charged bacteria can adsorb (via polymers) onto negatively charged solid surfaces without necessarily invoking any specific biological mechanisms. However, rates of adsorption may be reduced considerably if both surface and bacteria are negatively charged.

An important contribution to adhesion may come from the exocellular enzymes often associated with the bacteria. Enzymes contain both anionic and cationic groups, and the latter in particular can provide strong bonding to certain sites, especially on negatively charged surfaces. Synthetic amphoteric polyelectrolytes (i.e., polymers containing both positive and negative groups) having an overall negative charge have been found (26) to bind strongly to the anionic surface of silica. The strength of the bond between the cationic group and the SiO^- on the surface was sufficient to overcome the repulsion between the surface and the polymer. For the adsorption of proteins, this repulsion would be a minimum at their point of zero charge, at which pH their adsorption is often a maximum (10).

Summary point. In addition to the polysaccharides, the exocellular enzymes and proteins may provide important binding/adsorption sites, since although they may have a net negative charge, the positive groups can form strong bonds to negative sites on the surface.

Polymer Segment-solvent Interactions

The relative affinity of polymer segments for solvent molecules (compared to other polymer segments) largely governs whether polymers are soluble. For soluble polymers, the greater the relative affinity of the polymer segments for the solvent, the more expanded is the size of the polymer coil and the greater is the repulsion between adjacent polymer chains. This repulsion is important when single polymer molecules adsorb from solution (at concentrations ~0.1% w/v) onto surfaces where their concentration may be as much as 10% w/v. The better the solvent (i.e., the greater the affinity between polymer and solvent), the less polymer adsorbs (2) onto a surface, simply because of the greater lateral repulsion between the chains.

In the case of bacterial adhesion, the exocellular polymers are more concentrated than if they were separate, free molecules, so that solvent quality effects will be less than for the adsorption of free polymers. Although this (smaller) solvent effect will apply to the first monolayer of bacteria, one would expect a larger solvent quality effect on multilayer adsorption. For multilayer adsorption, there will be an overlap of the exocellular polysaccharides, as shown in Fig. 2.

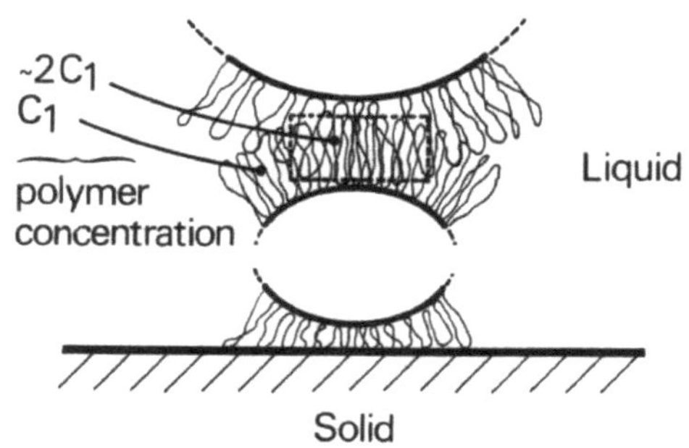

FIG. 2 - Two bacteria at a close separation showing the increased concentration of polymer in the dotted overlap region.

In the overlap region the concentration of polymer will be approximately double the concentration on the surface of a single bacterium. As water is a good solvent for these polymers, there will be a tendency for solvent to dilute the concentrated overlap region, i.e., force the particles to separate. This is one important mechanism in preventing the indiscriminate aggregation of cells both in solution and in the adsorbed state. For cells to aggregate, some attractive interaction, usually provided by a specific mechanism, must be applied to overcome this polymer-polymer repulsion. It is well-known that multilayers of bacteria can form on surfaces, implying that bacteria are able to adhere to one another. However, if bacteria do not aggregate with one another in solution, it is unlikely that they will form multilayers on surfaces by addition of one bacterium to another that is already on a surface. Of course, multilayer formation by bacteria reproducing is not influenced by these considerations.

Summary point. The affinity of the exocellular polymers for the solvent (water) can provide a mechanism for maintaining bacteria as separate individual cells, both in solution and in the adsorbed state.

Polyelectrolyte Adsorption

The adsorption of polyelectrolytes are especially influenced by both segment-surface and segment-solvent interactions. This arises from the charged nature of the segments. Anionic polyelectrolytes can (7, 25) adsorb onto crystals (e.g., Ca^{++} or Ba^{++} salts), with the adsorption often being accompanied by desorption (7) of the anions from the crystal surface. On surfaces where this anion desorption is not possible (e.g., glass, plastic, or metal), the level of adsorption of anionic polyelectrolytes increases markedly with ionic strength (21). Thus, to the extent that bacterial adhesion is determined by the polyelectrolyte nature of its exocellular polymers, it will be strongly influenced by the ionic strength of the solution.

Summary point. Where the exocellular polymers are charged and largely control the adhesion of bacteria, the ionic strength of the media will have an important influence on the level of adhesion.

Prevention of Polymer Adsorption by Other Molecules

The adsorption of polymers onto surfaces can obviously be influenced by the presence of other molecules that can compete for the surface sites that the bacteria would otherwise occupy. This has been demonstrated with the adsorption of both charged and uncharged polymers on solids and is illustrated in Fig. 3.

In the case of uncharged polymers (5, 23), the simultaneous adsorption of poly(methyl methacrylate) and polystyrene onto silica was studied. The former adsorbs more strongly than the latter, to an extent that polystyrene was excluded from the silica surface by the more strongly adsorbed macromolecule, until all the poly(methyl methacrylate) had been removed from solution. Similarly (1), for the simultaneous adsorption of poly(styrene sulphonate) PSS, and sodium poly(acrylate) NaPA onto $CaCO_3$ crystals from water, the more strongly adsorbed NaPA was adsorbed in preference to the PSS, even where the PSS was of lower molecular weight and able to reach the surface before NaPA. This lack of adsorption of the PSS is a result of its weak bonding to $CaCO_3$ being insufficient to overcome the electrostatic repulsion of the adsorbed NaPA. In the presence of 0.1 mol dm^{-3} NaCl, both PSS and NaPA were able to absorb simultaneously, as the electrolyte lowered this electrostatic repulsion.

Small molecules can also reduce the adsorption of polymers, provided they are more strongly adsorbed than the groups of the polymer. Thus, small phosphate molecules such as pentasodium triphosphate can prevent (7) the adsorption of sodium polyacrylate onto calcium salts and even cause the desorption of those polyelectrolytes that may have been pre-adsorbed.

For the adhesion of bacteria in living systems to solid surfaces, there will be many molecules competing for surface sites. It is reasonable to expect that the most strongly bound molecule will be preferentially adsorbed initially, i.e., the type of molecules adsorbed may be determined by thermodynamics rather than kinetics (unusual in systems that appear to involve irreversible adsorption). An example of this is shown in the

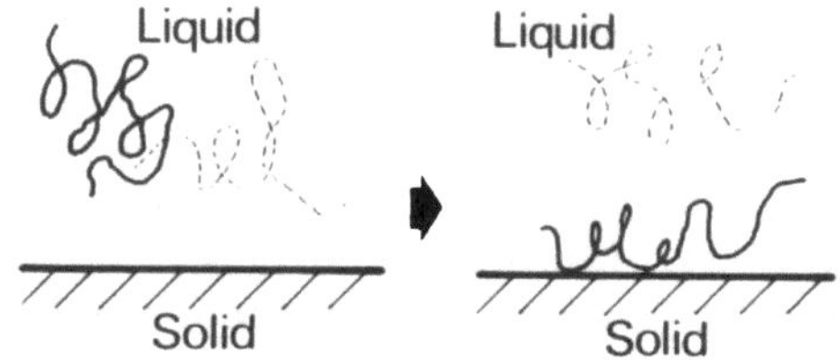

FIG. 3 - Preferential adsorption of the more strongly adsorbed polymer ——————.

establishment (12) of pellicle on the hydroxyapatite surfaces of teeth. The first molecules to adsorb are proteins rich in phosphate groups, as might be expected from the strong interaction between phosphates and calcium surfaces. Bacterial colonization of the tooth surfaces follows this establishment of the pellicle.

Summary point. It is well established that other solution components, including both small molecules and polymers, can prevent the adsorption of macromolecules onto surfaces. These "interfering" molecules may well have a significant influence on the ability of a bacterium to adsorb on a surface via its exocellular macromolecules.

Extension of Adsorbed Layers

Naturally occurring random coil macromolecules normally have molecular weights no greater than $\sim 10^{-6}$, as larger molecules are particularly susceptible to degradation by shearing. The dimensions of such molecules in solution will obviously depend on the particular molecule and nature of the solvent. For highly charged polyelectrolytes in pure water, the dimensions (as measured by the radius of gyration, Rg, of the molecule) will be much greater than in solution of high ionic strength. Nevertheless, polyelectrolytes having a molecular weight of about 5×10^5 have radii of gyration (16, 22) of 15-100 nm in electrolyte solutions of < 0.1 mol dm^{-3}. In general, these dimensions will be greater than the extension of the repulsive electrical double layer. The extension of this, which decreases with increasing ionic strength, is only about 10 nm at 10^{-3} mol dm^{-3} ionic strength. Thus, at quite modest ionic strengths, the extension of exocellular polymers is likely to be greater than any electrical repulsion arising from the cell wall (i.e., all the counter-ions for the charges on the cell wall probably lie within the extension of the exocellular polymer).

Summary point. The extension of exocellular polymer on a bacterium is probably sufficient to contain all the counter-ions from the charges on the cell wall. In this case, the net charge on a bacterium would arise only from groups on the polymer itself.

Influence of Residence Time on Strength of Adhesion

It has been observed (11, 15) that the strength of adhesion of bacteria to surfaces increases with time. A commonly accepted explanation for this is that the adsorbed bacteria continue to produce exocellular polymers, forming additional bonds to the surface and thus increasing the adhesion of the bacteria. Electron microscopy of thin sections at different times provides evidence for this and shows strands thickening to a more solid matrix. While such an explanation is perfectly reasonable, though difficult to prove conclusively, an additional observation should be borne in mind.

In studies of adhesion of plastics or rubbers to solid surfaces such as glass, it has been noted that the adhesion between the surfaces increases with the time they are in contact (4, 18). Clearly this cannot be a result of the production of new material at the interface. If deformable spherical particles are brought into contact with solid surfaces, they deform under the attractive van der Waals' forces, giving a profile as shown in Fig. 4.

Most real surfaces such as glass, metals, or plastic are, at an atomic level, rough, and the adsorbed or contacted surface can slowly deform to the shape of these asperities. This deformation can be accompanied by a viscous flow of the deformable surface, and the energy lost in the flow corresponds to the increase in work of adhesion with time of contact of the surfaces (energy involved in elastic flow on deformation would, of course, be recoverable).

Application of this observation to the adhesion of bacteria is not straightforward, as it is not clear what forms of viscous flow would take place within the cell. As the core of the cell is mainly an aqueous medium, viscous changes following adsorption are most likely in the cell wall.

Summary point. The increase with time in adhesive strength of rubbers on solid surfaces is probably a result of viscous flow under the forces of attraction between the two contacted bodies. It is possible that this factor could also contribute to the observed increase in strength of adhesion with time of contact of bacteria on surfaces.

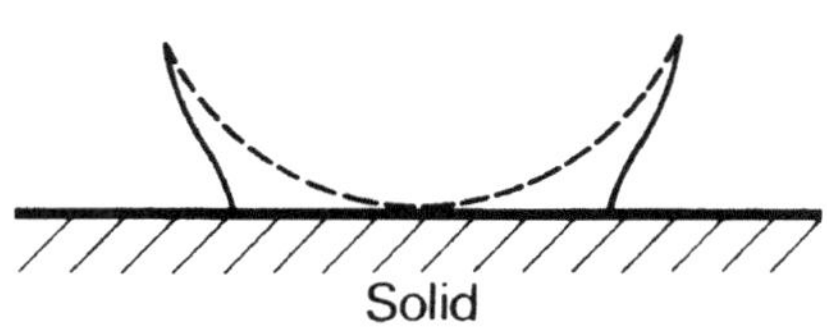

FIG. 4 - Deformation of a plastic surface in contact with a solid surface such as glass or metal. Dotted line - undeformed surface; full line - after deformation.

REFERENCES

(1) Adam, U., and Robb, I.D. 1983. Adsorption and exchange of polyelectrodes on crystal-surfaces. J.C.S. Faraday I 79: 2745-2753.

(2) Ashmead, B.V., and Owen, M.J. 1971. Adsorption of polydimethyl siloxanes from solution on glass. J. Poly. Sci. A2(9): 331.

(3) Barnett, K.G.; Cosgrove, T.; Vincent, B.; Sissons, D.S.; and Cohen-Stuart, M. 1981. Measurement of the polymer-bound fraction at the solid-liquid interface by pulsed nuclear magnetic-resonance. Macromolecules 14: 1018-1020.

(4) Barquins, M. 1982. Influence of dwell time on the adherence of elastomers. J. Adhes. 14: 63-82.

(5) Botham, R.A.; Shank, C.P.; and Thies, C. 1970. Colloid morphological behaviour block graft copolymer. Proc. Am. Chem. Soc. Symp. 247.

(6) Botham, R.A., and Thies, C. 1970. The adsorption behavior of polymer mixtures. J. Poly. Sci. C 30: 369-380.

(7) Cafe, M.C., and Robb, I.D. 1982. The adsorption of polyelectrolytes on Barium-sulfate crystals. J. Coll. Interface Sci. 86: 411-421.

(8) Cohen-Stuart, M.A.; Scheutjens, J.M.H.M.; and Fleer, G.J. 1980. Polydispersity effects and the interpretation of polymer adsorption isotherms. J. Poly. Sci. Phys. 18: 559-573.

(9) Davies, B.D.; Dulbecco, R.; Eisen, H.N.; Ginsberg, H.S.; and Wood, W.B., Jr. 1973. Microbiology. New York: Harper and Row.

(10) van Dulm, P.; Norde, W.; and Lyklema, J. 1981. Ion participation in protein adsorption at solid-surface. Coll. Interface Sci. 82: 77-82.

(11) Jones, G.W.; Richardson, L.A.; and vanden Bosch, J.L. 1980. Microbial Adhesion to Surfaces. Ch. 11. Chichester: Ellis Horwood.

(12) Juriaanse, A.C. 1980. Ph. D. Thesis. Rijksuniversiteit de Groningen, The Netherlands.

(13) Lal, M.; Turpin, M.A.; Richardson, K.A.; and Spencer, D. 1975. Configurational behavior of isolated chain molecules adsorbed from athermal solutions. Adv Chem Se 8: 16-27.

(14) Marra, J.; van der Schee, H.A.; Fleer, G.J.; and Lyklema, J. 1982. Adsorption from Solution. New York: Academic Press.

(15) Marshall, K.C.; Stout, R.; and Mitchell, R. 1971. Mechanism of initial events in sorption of marine bacteria to surfaces. J. Gen. Microbiol. 68: 337.

(16) Miyamoto, S.; Ishii, Y.; and Ohnuma, H. 1981. Intrinsic-viscosity of polyelectrodes having hydrophobic side-residues. Macromol. Chem. 182: 483-500.

(17) Pimental, G.C., and McClellan, A.L. 1960. The Hydrogen Bond. San Francisco: W.H. Freeman and Co.

(18) Roberts, A.D., and Othman, A.B. 1977. Rubber adhesion and dwell time effect. Wear 42: 119-133.

(19) Scheutjens, J.M.H.M., and Fleer, G.J. 1979. Statistical theory of the adsorption of interacting chain molecules. I. Partition-function, segment density distribution, and adsorption-isotherms. J. Phys. Chem. 83: 1619-1635.

(20) Tadros, T.F. 1981. The Effect of Polymers on Dispersion Properties, vol. 1. New York: Academic Press.

(21) Takahashi, A.; Kawaguchi, M.; and Kato, T. 1980. Adhesion and Adsorption of Polymers. In Polymer Science and Technology, ed. L.H. Lee, vol. 12B, p. 729. New York: Plenum.

(22) Tanford, C. 1961. Physical Chemistry of Macromolecules, pp. 495-500. New York: John Wiley and Sons.

(23) Thies, C. 1966. The adsorption of polystyrene-poly(methyl methacrylate) mixtures at a solid-liquid interface. J. Phys. Chem. 70: 3783-3790.

(24) Vincent, B. 1974. The effect of adsorbed polymers on dispersion stability. Adv. Coll. Interface Sci. 4: 193-277.

(25) Williams, P.A.; Harrop, R.; Phillips, G.O.; Pass, G.; and Robb, I.D. 1982. Effect of electrolyte and pH on the interaction of sodium carboxymethyl cellulose on Barium-sulfate particles. J.C.S. Faraday I 78: 1733-1740.

(26) Williams, P.A.; Harrop, R.; and Robb, I.D. 1984. The effect of an amphoberic polymer on the stability of silica dispersions. J.C.S. Faraday I, in press.

Microbial Adhesion and Aggregation, ed. K.C. Marshall, pp. 51-70. Dahlem Konferenzen 1984. Berlin, Heidelberg, New York, Tokyo: Springer-Verlag.

Adhesion to Inanimate Surfaces

S. Kjelleberg
Dept. of Marine Microbiology, University of Göteborg
413 19 Göteborg, Sweden

INTRODUCTION

Two experimental rationales applied to understanding interfaces in microbial ecology are: a) the exchange of bacteria between interfaces and the bulk phase, and b) the strong influence of environmental conditions on the extent and degree of adhesion. Recent studies (16, 20) very clearly show that surface active molecules, immobilized at both a solid- and an air-water interface, are scavenged and by this process transported into the bulk phase by both rod-shaped hydrophobic and hydrophilic bacteria and reversibly adhering "crawling" cells of Leptospira. With respect to the second point, it was shown that flux of nutrients in normally low nutrient, oligotrophic, aquatic environments (29) led to correspondingly rapid changes in bacterial cell physiology, morphology, and surface characteristics. Small starved cells in aquatic ecosystems appear to have an increased tendency for firm adhesion (7, 21), and growth under C- or N-limitation as compared to growth without nutrient limitation leads to drastic changes in cell surface properties and the adhesion pattern (5).

These variables complicate considerably a discussion on mechanisms of bacterial adhesion to inanimate surfaces, but they a) bridge the gap between traditional microbial ecology and strict adhesion studies; b) improve our understanding of some complex adhesion phenomena, such as variations in polymer production and physicochemical characteristics of the cell surface; and c) lead to the possibility of designing meaningful, defined model system experiments. A study of bacterial adhesion without these aspects in mind is potentially simplistic and misleading.

In view of these statements, I will elaborate on: a) the relative proportions of attached and free-living bacteria in aquatic systems and its ecological significance, b) the significance of structure- and force-mediated adhesion, c) the influence of nutrient availability on bacterial adhesion, and d) a thermodynamic model for describing bacterial adhesion. The final section considers model system measurements of hydrophobicity and bacterial adhesion at the gas-liquid interface.

This paper puts some emphasis on the possible role of hydrophobic interaction. This might at first appear too simplistic. Cells certainly utilize different mechanisms to adhere to different surfaces, and it is difficult to decide whether the interactions involved are predominantly hydrophobic or dependent on charge or other bonds. Surfaces are also variably coated with complex macromolecules with structural arrangements interacting to determine the resulting adhesion mechanism. In view of available data related to nonspecific adhesion at inanimate surfaces, and bearing in mind the limited space available for this article, I have chosen to present certain aspects rather than a comprehensive review of the "state of the art" of the topic. It is also a difficult task to combine well characterized experiments, in order to determine adhesion mechanisms to inanimate surfaces, with studies that are relevant to microbial ecology!

Definitions of basic physicochemical concepts and expressions, used throughout the text, are given in Rutter et al. and in Rutter and Vincent (both this volume).

ATTACHED AND FREE-LIVING BACTERIA IN AQUATIC ECOSYSTEMS

It has been suggested that suspended particles in the marine environment have a large fraction of associated heterotrophic bacteria and microflagellates (15). This forms the basis for the aggregate-spinning wheel concept for food chain dynamics (15). Contrary to classical ecological theory, heterotrophic bacteria, in addition to or instead of being decomposers, seem to conserve particulate matter by the uptake of dissolved organic matter. This bacterial production, as well as predation by microflagellates and zooplankton and release of nutrients, is hypothesized to take place within the same milieu (aggregate) (15). On the basis of measured growth efficiencies of natural bacteria, Williams (38) calculated that as much as 40% of the initially fixed organic material (photosynthesis) could be passed on by bacteria to the next trophic level. Heterotrophic microflagellates, preyed upon by microzooplankton, appear to control the numbers of free bacteria. The flagellates and microzooplankton are grazed upon by zooplankton and therefore provide a means for returning energy to the classical food chain. A rapid cycling of energy and material in this way takes place in association with microaggregates. "Interfaces in microbial ecology" seem to play a far more important role than hitherto anticipated.

It is possible that many bacteria in natural habitats lack the ability to adhere irreversibly to surfaces but are still capable of reversible adhesion to such surfaces. Williams (37) stated than an extensive investigation on size distribution of marine microheterotrophic populations led to the conclusion that should bacteria be associated with particles, the attachment would be very loose or the particles friable. The shear forces involved in most isolation or observation techniques would certainly remove any reversibly-adhering bacteria. Roper and Marshall (32) furthermore showed that the natural bacterial populations of sediments can be removed by lowering the electrolyte concentration. This indicates reversible adhesion at the distance of long-range attractive and repulsive forces.

Indirect experimental evidence for the association between bacteria and microzones of increased nutrient levels is, in fact, accumulating. Kinetic uptake studies of heterotrophic activity and substrate turnover of aquatic bacteria reveal: a) higher substrate levels (Sn) as determined by chemical analysis than the sum of the transport constant (K_t) and Sn obtained from the microbiological kinetic approach, and b) nonlinear plots of turnover time versus substrate concentrations which indicate a wide range of different K_t values and maximum uptake rates in bacterial populations. Azam and Hodson (3) suggest that part of the substrate is not equally available (adsorbed) for uptake (by nonattached bacteria), and/or bacteria with high K_t and V_{max} values exist within microzones (particles) and become overlooked due to lower levels of added substrate.

The suggested turnover of carbon associated with microaggregates and the large proportion of possibly reversibly adsorbed heterotrophic bacteria imply that we must consider the dynamics of carbon at an interface, including its interchange with the bulk phase (B. Dahlbäck, personal communication). Often stated problems in trying to separately measure attached versus free-living bacteria should not be regarded as experimental drawbacks, rather this reflects the key to our understanding of microbial ecosystems.

STRUCTURE- AND FORCE-MEDIATED ADHESION

Adhesion to inanimate surfaces is obviously of enormous ecological significance, not only an easy-to-observe phenomenon in model system experiments or at man-made structures in the environment. Is it possible to perform relevant experiments in the study of these events? It is clear that many bacterial species have characteristics that enable them to adhere to surfaces with which they rarely come into contact. An excellent example is the adhesion of Acinetobacter calcoaceticus RAG-1 to a variety of surfaces; plastics, buccal epithelial cells, the tooth surface, as well as the oil-water interface (36). A display of such a presumably general mechanism indicates that it is relevant to study interactions at interfaces in more detail.

Most aquatic bacteria appear to adhere to surfaces by means of surface polymers. These include, for example, lipopolysaccharides, extracellular polymers and capsules, pili, fimbriae, flagella, and more specialized structures such as appendages and prosthecae (see (13) for review). Although these surface components play a role in the initial reversible adhesion, they often serve to anchor more firmly the bacterium at an interface via polymeric bridging, at the point of effective short-range forces.

The composition and quantity of bacterial cell surface polymers vary considerably and are furthermore strongly influenced by growth or environmental conditions. Although extracellular polysaccharides have often been reported as responsible for irreversible adhesion, this is not always the case. For example, Brown et al. (5) demonstrated widespread adhesion from mixed populations in a carbon-limited culture, despite no obvious evidence of extracellular polymer production. A nitrogen-limited culture, however, resulted in poor adhesion, although large extracellular polymer production was observed. Although large quantitative differences were reported, one should bear in mind the possibility that polymers present between the cell and the substratum, not observed in light or scanning electron microscopy, could be responsible for the adhesion. Pringle et al. (31) found, by growing a freshwater strain of Pseudomonas fluorescens in a fermentor, that: a) a mucoid exopolysaccharide-producing mutant dominated the aqueous phase; b) a mutant with crenated colony morphology that produced little exopolymer dominated the walls; and c) the wild type was foam fractionated by the adsorptive bubble separation process. A further example of direct experimental evidence relates the capsular polysaccharides of Acinetobacter calcoaceticus RAG-1 (emulsan) and A. calcoaceticus BD4 (rhamnose-glucose polymer) to adhesion to hydrocarbon and to cell surface hydrophobicity (34). Both of these exopolymers interfere with adhesion to hydrocarbon and decrease cell surface hydrophobicity. Figure 1 shows the adhesion of capsulated wild-type BD4, a minicapsulated mutant BD413, and enzymatically decapsulated BD4 cells. Rosenberg et al. (34) suggest that starved cells producing little or no capsular material adhere more firmly to hydrocarbon substrate. As growth proceeds and the interface becomes densely colonized, desorption, in addition to adhesion of bulk phase cells and daughter cells released from irreversibly adhering bacteria, may be a necessary first step in the colonization of fresh interfaces. Desorption could be brought about by the synthesis of a capsule. In fact, the amphipathic polysaccharide emulsan prevents and reverses adhesion of hydrophobic bacteria and was recently found to desorb over 70% of the normal biota from over 500 epithelial cells examined (33). Mechanisms for detachment of cyanobacteria by production of hydrophilic rather than hydrophobic surface structures and layers are reported by Breznak et al. (this volume).

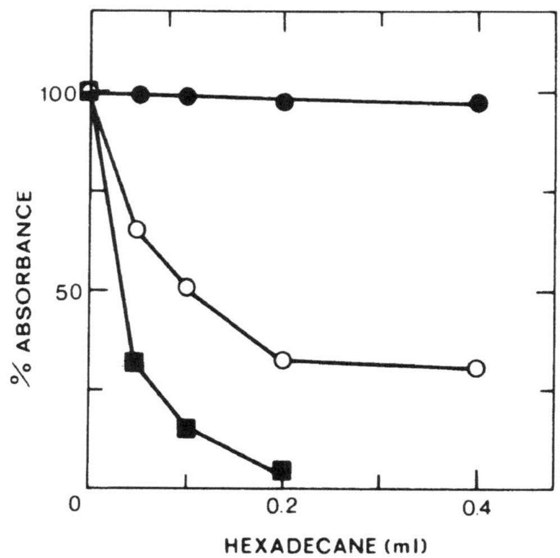

FIG. 1 - Adhesion of Acinetobacter calcoaceticus BD4 (●), BD413 (○), and enzymatically decapsulated BD4 (■) to hexadecane. The amounts of capsule material in percent of cell dry weight were 27, 11, and 4, respectively (from (34)).

TABLE 1 - Enrichment, E, of rough (Rc) and smooth (S), fimbriated (fim^+) and nonfimbriated (fim^-) cells of Salmonella typhimurium SH 6749 and fim^+ and fim^- Serratia marcescens at an air-water interface. E is the ratio between the number of cells in the surface film relative to that in the bulk phase (from (17)).

Bacteria		$E \times 10^{-3}$
S. typhimurium	$Rcfim^+$	1.43 ± 0.71
S. typhimurium	$Sfim^+$	0.69 ± 0.34
S. typhimurium	$Rcfim^-$	0.36 ± 0.15
S. typhimurium	$Sfim^-$	0.10 ± 0.04
S. marcescens	fim^+	2.01 ± 0.15
S. marcescens	fim^-	5.59 ± 1.50

It appears that with respect to inanimate surfaces, there is a subtle balance between cell surface components which may reduce (extracellular polysaccharides, lipopolysaccharides (LPS)) or promote (fimbriae) adhesion to inanimate surfaces. Jonsson and Wadström (19) recently showed that an encapsulated Staphylococcus aureus strain did not bind to hydrophobic Octyl-Sepharose gel, whereas a noncapsulated variant showed binding properties similar to protein A-producing (hydrophobic) strains. In fact, polyanionic extracellular carbohydrate material may not be primarily concerned with the initial adhesion process, but rather with development of subsequent bacterial film (31). The presence of LPS reduces cell surface hydrophobicity and decreases adhesion to the air-water interface of Salmonella typhimurium (Table 1) (17). Fimbriae often appear to confer increased hydrophobicity and degree of interaction at air-, solid-, and oil-water interfaces (17, 35, 40). The subtle balance between surface components is, however, reinforced by the findings that the highly hydrophobic Serratia marcescens exhibited a decreased adhesion to the air-water interface for fimbriae bearing cells. Surface localized pigments are to a greater extent exposed on nonfimbriated cells, and this results in increased hydrophobicity and adhesion to an interface (Table 1).

In addition to these structures, lectin-like, receptor-mediated interactions have been considered important even for bacterial adhesion to inert surfaces such as hydroxylapatite. Furthermore, Ofek et al. (28) consider the possibility of a specific hydrophobic interaction involving a specific recognition between a hydrophobic moiety on the bacterial cell surface and a substratum receptor. These examples may be relevant for inert surfaces in nature and are interesting to bear in mind for discussions on short-range forces and the general correlation between degree of hydrophobicity and adhesion that exists in a wide range of systems (see (36) for review).

At short distances several forces are important (ionic and dipolar, H-bonds, and hydrophobic interactions). It is therefore possible that the obvious variability in bacterial surfaces leads to various types of interactions that take place simultaneously with a single bacterial type. In many biological systems, hydrophobic interactions adjacent to ionic or hydrogen bonds can stabilize an otherwise energetically weak binding complex (9). A similar situation could occur for receptor-mediated interactions. Doyle et al. (9) propose that the adhesion of Streptococcus sanguis to saliva-coated hydroxylapatite beads can be described in terms of positive cooperativity, i.e., once the initial bacterium is bound, compounds on the substratum may change conformation, creating new receptor sites for other cells. The formation of new sites could be the result of the influence of hydrophobic interactions, but enzymatic action exposing saccharide receptors or cell-cell binding at the high bacterial concentrations used should also be considered.

Theories on specificity in adhesion at inanimate surfaces and the ecological significance thereof must also be discussed in view of seemingly opposing experimental data. The oil-degrading strain Acinetobacter calcoaceticus RAG-1 adheres avidly to both teeth and epithelial cells (36). It was suggested that the hydrophobic interactions per se can mediate adhesion in the absence of stereo-specific recognition mechanisms. In this strain, the surface hydrophobicity is mediated by several surface components, and the degree of adhesion is the sum of proportions and localization of these. Adhesion to liquid hydrocarbon is not limited to oil-degrading bacteria. Actually, several oil-degrading strains have low affinities for hydrocarbon, and the possibility of a substrate-specific adhesion mechanism was not considered tenable (36). A study on a large number of bacterial isolates as well as bacteria obtained directly from the human tooth surface showed that the vast majority exhibited pronounced cell surface hydrophobicity (36). Moreover, hydrophobic interaction was shown to mediate attachment of a nonoral species to the tooth surface as seen by comparisons of the wild-type and an isogenic non-hydrophobic mutant. An extensive survey of bacterial isolates from both the air-water interface and the bulk in marine environments showed that, of the broad range of different bacterial types tested, those residing in the surface microlayers exhibited the highest degree of hydrophobicity (Fig. 2) (6).

It must be emphasized that extensive, preferably in situ surveys, as in two of the examples given above, are much needed in contemporary studies of bacterial adhesion. This helps to avoid confusing results and misleading interpretations from studies with a few selected bacterial strains. The possibility of hydrophobic interaction being involved in adhesion of bacteria at solid surfaces was first presented by Marshall and Cruickshank (23). The general application of this finding has been stressed above, and it might at present be possible to suggest that hydrophobic bacteria in many systems tend to adhere to a higher extent to solid surfaces than do hydrophilic bacteria (20, 21). This approach naturally restricts the discussion on binding capacity and correlations

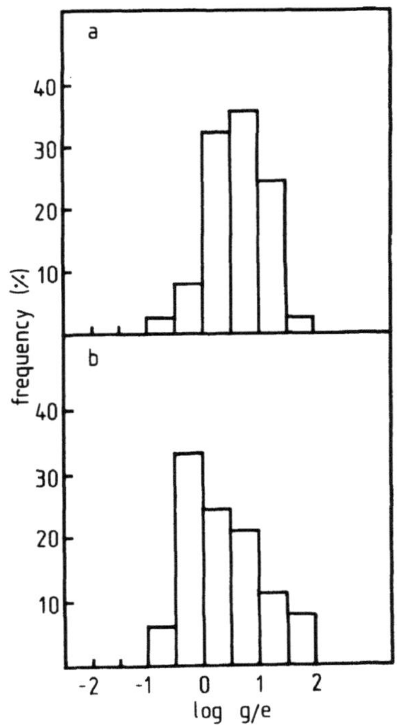

FIG. 2 - Distribution of degree of hydrophobicity (gel/eluate = g/e) among 109 bacterial isolates from a) air-water interface and b) subsurface water at one station off the Swedish west coast as measured by hydrophobic interaction chromatography (from (6)).

between adhesion and cell surface characteristics to consideration of the ecological significance. A distinction between the relative importance of, for example, lectin-receptor binding and less specific general hydrophobic interaction cannot be made.

INFLUENCE OF NUTRIENT AVAILABILITY

An understanding of the flexibility of bacterial responses to environmental changes will cast more light on the attachment process. The previous section included several examples of environmental changes influencing bacterial adhesion. This can be illustrated further. Phosphate limitation, often reported in aquatic environments, will, for example, induce elongation of the prosthecae of Caulobacter (29). This structure mediates adhesion. Fletcher (10) gave ample evidence for how the presence of inorganic ions and dissolved organic matter in the medium can inhibit as well as promote bacterial attachment. A recent study by McEldowney and Fletcher (unpublished data) readily illustrates the complexity of changes in environmental and growth conditions on subsequent bacterial attachment. Variations in culture conditions, such as different carbon sources, limiting nutrients, and growth rates, induced large changes in the surface characteristics of four freshwater bacterial species. Although it is difficult to make general statements about correlations between the hydrophobicity of the cell surface and the degree of attachment based on only four strains, it is very interesting to note that attachment to hydrophobic and hydrophilic solid surfaces, respectively, was independently affected by alterations in bacterial surface characteristics.

Growth under low nutrient concentrations, 2 and 10% of a conventional marine growth medium that contains approximately 1 g organic C l^{-1}, gave significantly higher adhesion values for both hydrophilic and hydrophobic surfaces than did growth in the full strength medium. This study was limited to four freshly isolated marine bacteria (K. Pedersen, personal communication). Marshall et al. (24) reported that glucose concentrations above 7 mg l^{-1} markedly inhibited firm adhesion of a Pseudomonas to glass surfaces in an artificial seawater medium.

The appreciation of very different nutritional types of bacteria, for convenience defined as oligotrophs and copiotrophs (29), in low nutrient environments adds to our understanding of bacteria at inanimate surfaces. Copiotrophic, and to a certain extent facultative oligotrophic bacteria, respond to the flux of nutrients, typical of microhabitats in low nutrient ecosystems, by rapid changes in cell volume, morphology, endogeneous metabolism, and intracellular material ((7, 21), see (25) for review). These copiotrophic bacteria, which are capable of growth under high nutrient concentrations, undergo a series of processes in low nutrient environments which lead to the formation of small, starved, but metabolically competent cells. Oligotrophic bacteria are capable of growth at much lower concentrations of nutrients and, in case of obligate oligotrophs, do not respond to microzones of higher substrate concentrations. A characteristic feature of oligotrophic bacteria is a high affinity and low specificity system for nutrient uptake. These different types have been considered to be responsible for the multiphasic kinetics revealed in substrate uptake experiments (3). With respect to distribution, it is worth mentioning that particle-associated bacteria are easily grown on high (laboratory) nutrient media and affected by short-term phenomena, while planktobacteria are more uniform in distribution and not markedly influenced by short-term events such as nutrient availability (39). Ishida et al. (18) reported that freshwater obligate oligotrophs were free-living while facultative oligotrophs were found surface-associated to a greater extent. Based on in situ adhesion of marine bacteria to glass surfaces, small, dwarf-like forms were found to be the primary colonizers. The subsequent appearance of larger rods has recently been interpreted as growth and conversion of small, starved copiotrophic bacteria into larger forms at the nutrient-enriched interface (22).

Starvation of seven, marine Gram-negative copiotrophic bacteria in the laboratory shows that visible, small, starved cells had altered surface topography, possibly due to folding of outer membrane material, and, in five strains, increased degree of hydrophobic character, as seen in hydrophobic partitioning experiments, and increased degree of irreversible binding to glass surfaces (Kjelleberg and Hermansson, in preparation).

Finally, a note of caution should be raised in relation to the findings that some small, starved bacteria are more efficient at binding soluble carbon substrates than non-starved cells (25). This has been suggested as acting as a survival strategy for small, starved bacteria. Surface localized substances are, however, often surface active and exhibit low water solubility. For example, the presence of long chain fatty acids in surface microlayers, phytoplankton production zones, and sediments is well documented. Such molecules at the solid- and air-water interfaces can be efficiently scavenged by bacteria of different types, degree of irreversible binding, and mode of motility (16, 20). It is obvious that there would be no attraction gradient of such surface active molecules, and chemotaxis as an important prerequisite for adhesion may not always be valid. A gradient of some other surface-associated compound that is not firmly bound could possibly serve as the chemotactic attractant, leading to subsequent scavenging of the surface active molecules. Chemotaxis would not, however, be an important factor in turbulent or flowing systems (Breznak et al., this volume).

A THERMODYNAMIC MODEL FOR BACTERIAL ADHESION

Experimental data have given an apparent lack of correlation between numbers of attached bacteria and substratum wettability. The substratum wettability can be measured, for example, in terms of surface free energy, (γ_S), approximated by determining the critical surface tension, (γ_C), by determining γ_S from a single contact angle measurement using the equation of state approach (26), or indirectly by the water contact angle (12), and the work of adhesion between water and the solid surface (30). To summarize briefly some of the conflicting results reported in the literature: the lowest number of attached bacteria in aquatic habitats existed for a narrow "minimally bioadhesive" range of γ_C values, 25 to 35 mNm^{-1} (8); there was a direct relationship between increasing adhesion and increasing water contact angle (increasing hydrophobicity) (12, 14), and a majority of freshwater isolates showed the highest attachment to more hydrophobic surfaces (30); and a given substratum type could not be proven generally favorable or unfavorable for attachment of all of a series of bacteria in a given environment (10).

Although there are several possible explanations for these discrepancies, such as variable effects of conditioning films, different bacterial characteristics, and bacterial competition for binding sites in field but not in defined laboratory studies (K.C. Marshall, personal communication), this section shall be restricted to an experimentally verified thermodynamic model that accounts for the interfacial tensions between all three interacting components of the system: the cell surface, the solid substratum, and the liquid phase. Experimental problems involved in this approach are mentioned by Rutter et al. (this volume).

The basic concept in the thermodynamic model is that the free energy will be minimized at equilibrium. Thus, the process under consideration, i.e., bacterial adhesion, will be favored if the process itself causes the thermodynamic function to decrease. Assuming that electric charges and specific receptor bindings can be neglected, the change in the free energy function is (per unit surface area):

$$\Delta F^{adh} = \gamma_{BS} - \gamma_{BL} - \gamma_{SL} \qquad (1)$$

where ΔF^{adh} is the free energy of adhesion, γ_{BS} is the bacterium-substratum interfacial tension, and γ_{BL} and γ_{SL} the interfacial tensions between the bacteria and the liquid and the substratum and the liquid, respectively. These interfacial tensions have been derived by the use of Young's equation:

$$\gamma_{SV} - \gamma_{SL} = \gamma_{LV} \cos \theta \ , \tag{2}$$

where γ_{SV}, γ_{SL}, and γ_{LV} are the substratum-vapor, substratum-liquid, and liquid-vapor interfacial tensions, respectively, and θ is the contact angle of the liquid of the substratum. Only γ_{LV} and θ are readily determined experimentally, but in conjunction with the equation-of-state relation that exists between interfacial tensions in a two-component three-phase system (26): γ_{SV} and γ_{SL} as well as all of the interfacial tensions in Eq. 1 can be calculated from a computer program. Absolom et al. (1) should be consulted for a more detailed theoretical outline as well as for the experimental procedures behind the results presented in the following.

Two sets of experimental results, examining the effects of the surface tension of the substratum (γ_{SV}) and of the liquid medium (γ_{LV}) on the extent of bacterial adhesion to a wide range of polymer surfaces, are given in Fig. 3. The surface tension of the liquid can be lowered by addition of, for example, dimethyl sulfoxide (DMSO). At the lowest concentration of dimethyl sulphoxide used (highest γ_{LV}), bacterial adhesion decreased with increasing γ_{SV}. As γ_{LV} was lowered (increasing DMSO concentrations), the change in bacterial adhesion was less pronounced. At intermediate γ_{LV} values, specific for each strain, adhesion became independent of γ_{SV}, and at even lower values of γ_{LV}, adhesion increased with increasing γ_{SV}. These results were in excellent agreement with theoretical calculations of ΔF^{adh}. The thermodynamic concept also predicts that ΔF^{adh} is independent of γ_{SV} and equals zero in the case of $\gamma_{LV} = \gamma_{BV}$ (means that $\gamma_{BL} = 0$). This is seen from the curves in Fig. 3. Slopes that are equal to zero correspond to a γ_{LV} that is characteristic for each bacterial species and equal to the surface tension of the bacterium.

As predicted by the model, adhesion to hydrophilic surfaces is more pronounced than to hydrophobic surfaces when the bacterial surface tension is larger than that of the bulk liquid. A higher surface tension of the liquid than of the bacterial surface, the normal situation in natural waters, leads to a higher degree of adhesion to the hydrophobic than

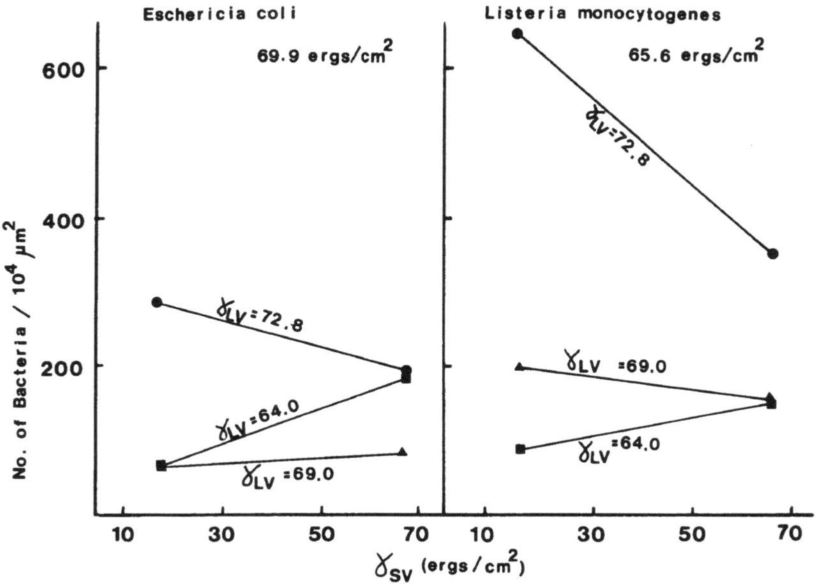

FIG. 3 - Bacterial adhesion vs. substratum surface tension, γ_{SV}, for various surface tensions of medium, γ_{LV} (from (1)).

to the hydrophilic surface. In all experimental situations, the relative order of bacterial adhesion was maintained, the most extensive adhesion being exhibited by the most hydrophobic bacteria for all γ_{LV} values examined.

The significance of each of the phases involved in adhesion, the solid and bacterial surfaces and the medium, and therefore the three interfacial tensions between these, is supported also by other workers. Fletcher (11), in a study on the effects of alcohols on the attachment of a marine pseudomonad to hydrophilic and hydrophobic interfaces, found a consistent minimum attachment at medium γ_{LV} values that could reflect the surface free energy of the bacteria. Gerson and Scheer (14) found a basically straight line relationship between ΔF^{adh} and the logarithm of the number of cells attached in a study of three bacteria to each of five substrata.

It should be emphasized that a thermodynamic model, as outlined above, should only apply to surfaces in contact where there are no interfacial bonds. The good correlations observed are noteworthy, considering that in aqueous or polar systems it is likely that there are also interfacial bonds between the two interacting bodies.

Attraction forces responsible for a favorable free energy difference upon adhesion, without involving bonds between interacting phases, are van der Waals forces and hydrophobic interaction. Although molecular solubility data have not been proven to form experimentally the basis for hydrophobic interaction or to predict free energy changes of hydrophobic contacts at the surface of, e.g., macromolecules at the cell surface, the essay on the hydrophobic effect by, e.g., Arakawa et al. (2) is worth mentioning for the purpose of understanding the process. Briefly, the nature of solute-water and solute-solute interactions in aqueous solutions of inert gases and hydrocarbons is considered in comparison with such interactions in nonaqueous solutions. Such studies led in one case to the following: the free energy for hydrophobic interaction (ΔG^{HI}) between two methane molecules in aqueous environments is expressed as a sum of several terms. Of these, the largest negative contribution to ΔG^{HI} comes from the so-called cavity effect, i.e., the arrangements of water molecules around nonpolar entities, and therefore the entropy effect of reducing the hydrocarbon-water interface, from transferring water molecules to the bulk phase. Contributions from the solute-water interactions and from the solute-solute interactions are basically negligible. A definition of hydrophobic interaction is given by Rutter et al. (this volume).

MODEL SYSTEM MEASUREMENTS OF HYDROPHOBICITY AND ADHESION

This final section considers some methods for the determination of bacterial cell surface hydrophobicity and the suitability of the gas-liquid interface for defined model system studies of bacterial adhesion.

Methods to Measure Hydrophobicity

Ofek et al. (28) suggest that the various methods used to measure hydrophobicity detect the availability of quantitatively and qualitatively different lipophilic residues on the bacterial surface. These may not necessarily influence the bacteria-substrata interaction in identical ways. That is, methods that measure degree of hydrophobicity only detect one type of interaction at a time, whereas some bacteria most likely employ various types of binding, displaying a complex nature of hydrophobic and polar regions. By specific blocking or removing of surface structures and repeated measurements of partitioning in hydrophobic assays and binding capacity, however, much useful information can be obtained. I agree with the authors' notion and would like to add that the partitioning methods commonly used, of great value in surveys, display a wide range of sensitivity for degree of hydrophobicity of the bacterial cell surface, and incorrectly interpreted results are easy to encounter. Partitioning in aqueous biphasic polymer mixtures (polyethylene-glycol and Dextran) is sensitive to small differences in the hydrophobicity of relatively hydrophilic bacteria. The interfacial tension between the phases is thus low (similar to that of aqueous high surface tension diluents and Octyl-Sepharose beads in hydrophobic interaction chromatography), compared to, e.g., aqueous-hydrocarbon interfaces. The use of more than one method has therefore often been found to be vital. It should be noted that partitioning in aqueous biphasic polymer mixtures would lead to the separation of three cell types, including bacteria that accumulate at the interface due to the different surface free energy values of the two phases. Furthermore, hydrophobic bacteria in two-phase aqueous-hydrocarbon systems would reside at the oil-water interface. Cells found in the oil phase must not be considered as hydrophobic, but rather as displaying surface groups that are able to react with the small amount of aqueous liquid dissolved or mixed with the hydrocarbon.

It has often been assumed that bacteria sticking to oil deform the interface. It would be of primary importance to measure the extent to which bacteria deform or stick through the interfaces of, e.g., oil- and air-water. This main difference between non-deformable solids and other interfaces would provide direct information on the energy involved in adhesion.

Methods like those mentioned above are important components in adhesion studies. The fact that hydrophobicity as measured by various techniques correlates reasonably well with bacterial adhesion in many systems supports the idea that hydrophobic residues with affinity for low energy surfaces contribute to the attraction forces operating between bacteria and substrata. In fact, the proportions of strongly hydrophobic bacteria in, e.g., the oral cavity is noteworthy. Are most bacteria in nature hydrophobic? Do we, after screening fresh isolates, discarding aggregating strains, and passing them through cultivation procedures, end up with disproportionally high numbers of hydrophilic bacteria (E. Rosenberg, personal communication)?

Gas-liquid Interfaces

Bacterial adhesion to inanimate surfaces has generally been studied using solid-water interfaces. From the experiments referred to in this paper it is obvious that a series of confusing and contradicting results have been part of the information obtained. Many factors such as substratum rigidity and the dependence of adhesion on the adsorption of organic substances to form a conditioning film (22) are difficult to control at solid surfaces. Nonspecific adhesion should ideally be studied using substrata previously coated with defined molecules, while being able to continuously monitor the properties of the substratum during the adhesion experiment. Some of these problems could be overcome by the use of gas-liquid interfaces. For example, experiments designed to distinguish between receptor-mediated and force-dependent binding must involve the use of a defined system. A range of experimental designs, including both simple static containers with the surface film overlying a bacterial suspension and maneuverable insoluble monolayers in surface balance studies (27), as well as gas bubbles rising through a water column or bacterial suspension, have yielded valuable information on the mechanisms of bacterial adhesion. Furthermore, a series of samplers can be used for collecting the surface microlayers with adhering microorganisms, and jet and film drops from bursting bubbles are sampled for numbers and types of test organisms.

Air-water interfaces can easily be modified, and the amounts of applied surface film lead to defined studies of substrate uptake at a surface and allow us to follow the exchange of material and organisms between the boundary and the bulk phase. Physicochemical parameters such as surface pressure and surface potential can be controlled, which leads to an understanding of the importance of the architecture of the film (surface topography) and allows bacteria and bacterial surface groups to interact in various, controlled ways with molecules residing at the boundary. In addition to changes in surface free energy, the charge characteristics of the interface can be manipulated by spreading of appropriate substances.

Acknowledgements. I am grateful to the following colleagues for stimulating discussions and constructive criticisms: B. Dahlbäck, K.C. Marshall, B. Norkrans, and M. Rosenberg. Work in this review done by the author was supported by the Swedish Natural Science Research Council and the Australian Research Grants Committee.

REFERENCES

(1) Absolom, D.R.; Lamberti, F.V.; Policova, Z.; Zingg, W.; van Oss, C.J.; and Neumann, A.W. 1983. Surface thermodynamics of bacterial adhesion. Appl. Envir. Microbiol. 46: 90-97.

(2) Arakawa, K.; Tokiwano, K.; Ohtomo, N.; and Kedaira, H. 1979. A note in aqueous solutions. Bull. Chem. Soc. Jpn. 52: 2483-2488.

(3) Azam, F., and Hodson, R.E. 1981. Multiphasic kinetics for D-glucose uptake by assemblages of natural marine bacteria. Mar. Ecol. - Progr. Ser. 6: 213-222.

(4) Blanchard, D.C., and Syzdek, L.D. 1978. Seven problems in bubble and jet drop researches. Limn. Ocean. 23: 389-400.

(5) Brown, C.M.; Ellwood, D.C.; and Hunter, J.R. 1977. Growth of bacteria at surfaces: Influence of nutrient limitation. FEMS Microbiol. Lett. 1: 163-166.

(6) Dahlbäck, B.; Hermansson, M.; Kjelleberg, S.; and Norkrans, B. 1981. The hydrophobicity of bacteria - an important factor in their initial adhesion at the air-water interface. Arch. Microbiol. 128: 267-270.

(7) Dawson, M.P.; Humphrey, B.A.; and Marshall, K.C. 1981. Adhesion, a tactic in the survival strategy of a marine vibrio during starvation. Curr. Microbiol. 6: 195-198.

(8) Dexter, S.C.; Sullivan, J.D., Jr.; Williams, J. III; and Watson, S.W. 1975. Influence of substrate wettability on the attachment of marine bacteria to various surfaces. Appl. Microbiol. 30: 298-308.

(9) Doyle, R.J.; Nesbitt, W.E.; and Taylor, K.G. 1982. On the mechanism of adherence of Streptococcus sanguis to hydroxylapatite. FEMS Microbiol. Lett. 15: 1-5.

(10) Fletcher, M. 1980. The question of passive versus active attachment mechanisms in non-specific bacterial adhesion. In Microbial Adhesion to Surfaces, eds. R.C.W. Berkeley, J.M. Lynch, J. Mellling, P.R. Rutter, and B. Vincent, pp. 197-210. Chichester: Ellis Horwood.

(11) Fletcher, M. 1983. The effects of methanol, ethanol, propanol and butanol on bacterial attachment to surfaces. J. Gen. Microbiol. 129: 633-641.

(12) Fletcher, M., and Loeb, G.I. 1979. Influence of substratum characteristics on the attachment of a marine pseudomonad to solid surfaces. Appl. Envir. Microbiol. 37: 67-72.

(13) Fletcher, M., and Marshall, K.C. 1982. Are solid surfaces of ecological significance to aquatic bacteria? In Advances in Microbial Ecology, ed. K.C. Marshall, vol. 6, pp. 199-236. New York: Plenum

Press.

(14) Gerson, D.F., and Scheer, D. 1980. Cell surface energy, contact angles and phase partition. III. Adhesion of bacterial cells to hydrophobic surfaces. Biochim. Biophys. Acta 602: 506-510.

(15) Goldman, J.C. 1983. Oceanic nutrient cycles. In Flow of Energy and Materials in Marine Ecosystems, ed. M.J. Fasham. New York: Plenum Press, in press.

(16) Hermansson, M., and Dahlbäck, B. 1983. Bacterial activity at the air/water interface. Microbial Ecol. 9: 317-328.

(17) Hermansson, M.; Kjelleberg, S.; Korhonen, T.K.; and Stenström, T.A. 1982. Hydrophobic and electrostatic characterization of surface structures of bacteria and its relationship to adhesion at an air-water surface. Arch. Microbiol. 131: 308-312.

(18) Ishida, Y.; Shibahara, K.; Uchida, H.; and Kadota, H. 1980. Distribution of obligately oligotrophic bacteria in Lake Biwa. Bull. Jap. Soc. Sci. Fish. 46: 1151-1158.

(19) Jonsson, P., and Wadström, T. 1983. High surface hydrophobicity of Staphylococus aureus as revealed by hydrophobic interaction chromatography. Curr. Microbiol. 8: 347-353.

(20) Kefford, B.; Kjelleberg, S.; and Marshall, K.C. 1982. Bacterial scavenging: Utilization of fatty acids localized at a solid-liquid interface. Arch. Microbiol. 133: 257-360.

(21) Kjelleberg, S.; Humphrey, B.A.; and Marshall, K.C. 1983. Initial phases of starvation and activity of bacteria at surfaces. Appl. Envir. Microbiol. 46: 978-984.

(22) Marshall, K.C. 1979. Growth at interfaces. In Strategies of Microbial Life in Extreme Environments, ed. M. Shilo, pp. 281-290. Dahlem Konferenzen. Weinheim: Verlag Chemie.

(23) Marshall, K.C., and Cruickshank, R.H. 1973. Cell surface hydrophobicity and the orientation of certain bacteria at interfaces. Arch. Mikrobiol. 91: 29-40.

(24) Marshall, K.C.; Stout, R.; and Mitchell, R. 1971. Mechanism of the initial events in the sorption of marine bacteria to surfaces. J. Gen. Microbiol. 68: 337-348.

(25) Morita, R.Y. 1982. Starvation survival of heterotrophs in the marine

environment. In Advances in Microbial Ecology, ed. K.C. Marshall, vol. 6, pp. 171-198. New York: Plenum Press.

(26) Neumann, A.W.; Good, R.J.; Hope, C.J.; and Sejpal, M. 1974. An equation-of-state approach to determine surface tensions of low-energy solids from contact angles. J. Coll. Interface Sci. 49: 291-304.

(27) Norkrans, B. 1980. Surface microlayers in aquatic environments. In Advances in Microbial Ecology, ed. M. Alexander, vol. 4, pp. 51-85. New York: Plenum Press.

(28) Ofek, I.; Whitnack, E.; and Beachey, E.H. 1983. Hydrophobic interactions of group A streptococci with hexadecane droplets. J. Bacteriol. 154: 139-145.

(29) Poindexter, J.S. 1981. Oligotrophy: Feast and famine existence. In Advances in Microbial Ecology, ed. M. Alexander, vol. 5, pp. 63-89. New York: Plenum Press.

(30) Pringle, J.H., and Fletcher, M. 1983. Influence of substratum wettability on attachment of freshwater bacteria to solid surfaces. Appl. Envir. Microbiol. 45: 811-817.

(31) Pringle, J.H.; Fletcher, M.; and Ellwood, D.C. 1983. Selection of attachment mutants during the continuous culture of Pseudomonas fluorescens and relationship between attachment ability and surface composition. J. Gen. Microbiol. 129: 2557-2569.

(32) Roper, M.M., and Marshall, K.C. 1974. Modification of the interaction between Escherichia coli and bacteriophage in saline sediment. Microbial Ecol. 1: 1-13.

(33) Rosenberg, E.; Gottlieb, A.; and Rosenberg, M. 1983. Inhibition of bacterial adherence to epithelial cells and hydrocarbons by emulsan. Infec. Immun. 39: 1024-1028.

(34) Rosenberg, E.; Kaplan, N.; Pines, O.; Rosenberg, M.; and Gutnik, D. 1983. Capsular polysaccharides interfere with adherence of Acinetobacter calcoaceticus to hydrocarbon. FEMS Microbiol. Lett. 17: 157-160.

(35) Rosenberg, M.; Bayer, E.A.; Delarea, J.; and Rosenberg, E. 1982. Role of thin fimbriae in adherence and growth of Acinetobacter calcoaceticus on hexadecane. Appl. Envir. Microbiol. 44: 929-937.

(36) Rosenberg, M.; Rosenberg, E.; Judes, H.; and Weiss, E. 1983.

Hypothesis. Bacterial adherence to hydrocarbons and to surfaces in the oral cavity. FEMS Microbiol. Lett. 20: 1-5.

(37) Williams. P.J. leB. 1970. Heterotrophic utilization of dissolved organic compounds in the sea. I. Size distribution between respiration and incorporation of growth substrates. J. Marine Biol. Ass. UK 50: 859-870.

(38) Williams, P.J. leB. 1981. Incorporation of microheterotrophic processes into the classical paradigm of the planktonic food web. Kiel. Meeresforsch. Spec. Publ. 5: 1-28.

(39) Wilson, C.A., and Stevenson, L.H. 1980. The dynamics of the bacterial population associated with a salt marsh. J. Exp. Mar. Biol. Ecol. 48: 123-138.

(40) Young, L.Y. 1978. Bacterioneuston examined with critical point drying and transmission electron microscopy. Microbial Ecol. 4: 267-277.

Microbial Adhesion and Aggregation, ed. K.C. Marshall, pp. 71-84. Dahlem Konferenzen 1984. Berlin, Heidelberg, New York, Tokyo: Springer-Verlag.

Adhesion to Animal Surfaces

G.W. Jones
Dept. of Microbiology and Immunology
University of Michigan Medical School
Ann Arbor, MI 48109, USA

Abstract. Bacterial adhesins are most commonly proteinaceous in nature, although there is evidence that lipoteichoic acids (LTA) can function as the adhesin of Streptococcus pyogenes and that the deposition of insoluble polysaccharides facilitates the accumulation of Streptococcus mutans on the tooth surface. The specificity of attachment to the animal cell appears to result from the binding of proteinaceous adhesins to carbohydrate receptors of the animal cell glycoconjugates, or perhaps from unique hydrophobic interactions between the acyl groups of LTA and particular hydrophobic domains of the surface. The filamentous form and reduced negative charge of the adhesive appendage probably allows attachment to occur when the distances of separation between the bacterium and the animal cell surface is such that the mutual repulsion between the surfaces is minimized.

INTRODUCTION

Surfaces of the animal body composed of keratinized or non-keratinized epithelial cells, hydroxyapatite, or salivary pellicles adsorbed to tooth surfaces, are colonized by characteristic populations of bacteria which appear to be intimately attached to the surface. Other bacteria are found lodged in the mucus gel layer of mucosae (3), and yet others attach to the surfaces of bacteria of established adhesive populations. Many species and genera are known to associate with these surfaces, but a given species, or even strain of a species, is often restricted to a particular site of the body (1, 7, 11). Such selectivity is initially due to the unique attachment mechanisms responsible for adhesion and, subsequently, to the ability of the organism to grow on the surface. For the purpose of discussion, these many adhesive activities are assigned to one of three categories according to the chemical composition of the component believed to be responsible for the adhesive activity; it should be noted, however, that the chemical compositions of many other adhesive activities are unknown (13). In the first category may be placed the adhesive mechanisms of most Gram-negative and some Gram-positive bacteria which result from the synthesis of unique adhesive proteins (adhesins). These adhesin proteins, which differ markedly from one another in molecular weight, amino acid composition, and quaternary structure, appear to recognize and bind to carbohydrate receptors of the animal cell surface glycocalyx. The second category consists of the adhesive activities such as those of Streptococcus pyogenes which are thought to attach to the host cell membranes via the fatty acid residues of lipoteichoic acids (LTA). In the third category are placed those attachment mechanisms which result from the deposition of insoluble polysaccharide matrices by bacteria, followed by the binding of the extracellular polysaccharide by the bacterial surface components. This mechanism, which has been studied principally in S. mutans (8), is probably responsible for bacterial accumulation on a surface rather than for initial attachment (4). Thus, not only do the surfaces to which bacteria attach differ both in physical, chemical, and biological properties, but the microbial mechanisms involved in attachment are equally diverse, differing not only between species, but also between strains of a species. In consequence, no model adhesive mechanism, generally applicable to all forms of microbial adhesion, can be cited, although a general similarity of the adhesins within each of the categories given above is apparent.

Microbial activities other than adhesion can influence the efficiency of bacterial attachment to a surface in vitro and sometimes in vivo (3). The influence of bacterial motility, chemotaxis, and hydrophobicity (for example, due to cell surface roughness) has been discussed, as has surface modification and the influence of the fluid phase. Bacterial adhesion in vitro, it should be noted, is seldom expressed in terms of affinity constants and never in terms of binding energy. Rather, adhesion is usually measured in terms of the number, or proportion, of bacteria which attach, and the results are evaluated on the relative basis of strain comparison. Consequently, microbial activities which modify adhesion can distort such results and lead to their erroneous interpretation. These modifying activities, however, may constitute real and important events in a sequence leading to adhesion (13) and therefore must be taken into account if a comprehensive picture of microbial adhesion is to be drawn (3).

The terminology to be employed has been discussed (10, 13). The term fimbriae is used for discrete nonflagella, filamentous structures of uniform diameter which probably function as adhesive organelles. The frequently used synonym, pili, is better reserved for those organelles involved in conjugation and DNA transfer between cells. The terms fibrillae or fimbrillar material describe the more amorphous surface adhesive appendages which appear to lack the regular filamentous forms of fimbriae. The functional term adhesin will be used for all surface structures known to be adhesive organelles, and receptor for a known or putative surface component to which an adhesin binds.

Adhesion is envisaged as a two-component system depending as much on the number and distribution of surface receptors as it does on the extent of adhesin production by the bacteria. Presumably, a minimum number of adhesin-receptor interactions must be achieved before adhesion occurs. Consequently, a binding site for an adhesive bacterium exists on a surface where the receptor concentration is sufficient to allow attachment. Such would account for the differing receptivities of surfaces in terms of the numbers of bacteria of a strain which can attach, and also for the occasional uneven distribution of attached bacteria on some surfaces.

ADHESINS

Bacterial adhesion can be monitored in vitro in several ways (3). The animal cell used in such tests need not be the one normally colonized by the bacteria (e.g., erythrocytes). It is assumed that interactions with cells such as erythrocytes result from bacterial adhesins reacting with receptors that approximate the structure of the real receptor on the surface to which the bacteria normally attach. The kinetics of adhesion, with the exceptions of the attachment of Streptococcus sanguis to saliva-coated hydroxyapatite surfaces (6), for example, have not been explored in detail. Whether or not such in vitro measurements are realistic reflections of adhesion in vivo, or of adhesin-receptor interactions when adhesion is influenced by other bacterial activities noted above, is open to question. Such studies, based on Langmuiran adsorption, however, have provided information on the relative affinity constants, the numbers of binding sites involved, and the existence of multiple binding activities in a culture (6). The reversibility of attachment, which needs to be the case for Langmuiran adsorption to hold true, is assumed; it should be noted, however, that detachment can result from loss of adhesiveness by the bacteria rather than by the dissociation of the adhesin from its receptor (3).

Morphology

The proteinaceous type 1 fimbriae of Escherichia coli are the most thoroughly studied adhesins. One may speculate that form imposes constraints on the general organization of these organelles in such a way that much of what can be stated about E. coli type 1 fimbriae is true of other fimbriae of similar morphology. Fimbriae have an obvious ordered alignment of subunits which gives the organelle a uniform cross-sectional dimension and a defined lateral edge. Type 1 fimbriae have a diameter of about 7 nm (although others may be as narrow as 2 nm or as wide as 10 nm) and may reach 4 μm in length. X-ray diffraction studies have demonstrated that type 1 fimbriae are composed of repeating subunits arranged in a right-handed helix about an axial hole; the subunit pitch is 2.32 nm and there are 3-1/8 subunits per turn (13).

In contrast to fimbriae are the proteinaceous fibrillae such as the E. coli K88 and K99 adhesins. These filaments, which are about 2 nm in diameter, appear to be composed of an ordered linear alignment of subunits, possibly spatially arranged as open helices (13). Their apparent flexibility, however, causes filaments of this type to become entangled and to appear at relatively low magnification as somewhat amorphous masses on the cell surface.

Some fimbriae are inserted through the cell wall at discrete points and have a basal body analogous to the basal body of flagella (10). Others, such as the K99 adhesin, are thought to be assembled from their constituent subunits in the outer membrane of the cell and, if attached to the cell, must be so at some outer membrane site (9).

Adhesins composed of LTA, or involving the binding of deposited insoluble carbohydrate polymers, have a much less well-defined structure. Those of the Gram-positive cocci are associated with a surface layer of short, chemically heterogenous fibrillae composed of several proteins, LTA, and carbohydrates. Individual LTA molecules of the fibrillae may be joined into chains by proteins (2). It is assumed that linkage between the LTA-protein complexes and the cell results from the normal intercalation of the acyl groups of so-called intracellular LTA (LTA_i) into the membrane of the bacterium (15, 20).

A filamentous structure, therefore, is often highly characteristic of the proteinaceous adhesins and, perhaps to a lesser degree, typical of the LTA and other adhesins. A filamentous form is probably critical to effective adhesin function (10, 13). The mathematical treatment of the interaction between a charged planar surface, such as that of formalized animal cells and particles of like charge the size of bacteria, suggest that the long-range forces which generate weak attraction at the secondary minimum of potential energy hold the bacterium adjacent to the animal cell surface, whereas the forces of repulsion generated between the surfaces prevent the bacterium from coming into more intimate physical contact with the animal cell (10, 13). However, such intimate contact is necessary for the presumed stereochemical fit of adhesin with receptor. It seems reasonable to suppose, therefore, that adhesins have evolved as filaments because this form enables them to bridge the space between the bacterium and the surface which then allows their active adhesive site to interact with the surface receptors and thereby to bind the bacteria to the substratum. Adhesins are able to carry out this bridging function because they are long appendages (up to 4 µm) of small radius of curvature (3.5 nm or less compared with 250 nm for E. coli) which carry a reduced negative charge compared to the bacterial cell wall/outer membrane. The two latter properties would minimize repulsion between the adhesin and the surface and their length would enable them to interact with their receptor in the absence of any interaction between the bacterial cell wall and the animal cell surface (13). The filamentous form of the glycocalyx on the surface of the animal cell, components of which act as receptors, would further enhance this interaction (13). Although a general increase in hydrophobicity, or the presence of an overall positive charge, would facilitate adhesion through hydrophobic interactions and electrostatic bonding, they alone would not account for the specificity of adhesive reactions.

Not all known adhesive activities, however, can be identified with a particular filamentous surface structure (13). Such adhesins, therefore, may lack the bridging distances of, for example, type 1 fimbriae. Suggestive of this is the observation that divalent cations of appropriate ionic size are necessary for the function of a Vibrio cholerae adhesin; presumably these ions act as a zipper between the surfaces enabling the adhesin to bind to its putative L-fucose receptors (12). The adhesive activities of other adhesins are impeded by reductions in ionic strength, suggesting that their effective bridging distance is less than 15 nm (14).

Composition

The proteinaceous adhesins which have been analyzed are composed of an array of identical subunits, each of which is composed of a single polypeptide (13). Subunit molecular weights range from 8,000 to 64,000 daltons (Table 1). Non-amino acid substitutes may be remnants of envelope attachment sites rather than integral components of the adhesin. Most of these adhesins are negatively charged at physiological pH, with isoelectric points between pH 3.7 and 5.6. The notable exception is the K99 adhesin, with an isoelectric point of pH 10.1 (Table 1). The length of each filament and its molecular weight is variable because it is determined by the number of subunits present; the cross-sectional dimension, however, is constant because it is defined either by the number of subunits per helical turn, or by the size of the subunits when these are in a linear array. Adhesins are readily dissociated by detergents and weak acids, suggesting that the subunits are held together by hydrogen and hydrophobic bonds; disulfide linkages, when these exist, are probably intramolecular. Intersubunit bonding of these types is consistent with the apparent ability of the subunits of some adhesins to undergo reorganization within the filament and thereby to bring about changes in filament morphology or to depolymerize and cause the filament to contract (13). The possibility that contraction of an attached adhesin could bring the bacterium into closer contact with a surface to facilitate additional binding of the same or other adhesins remains unexplored.

TABLE 1 - Properties of selected fimbrial proteins of Gram-positive and Gram-negative bacteria. See (13) for original sources.

Bacteria	Subunit molecular weight	pI	Hydrophobicity (-cal $mole^{-1}$)
Gram-positive			
A. viscosus	64,000	-	537
C. renale	19,000	4.5	550
Gram-negative			
P. aeruginosa	17,800	3.9	635
E. coli Type 1	17,100	3.9	511
K88	26,200	4.2	609
K99	19,500	10.1	577
N. gonorrhoea	19,500	4.9	623

Adhesins, in general, exhibit hydrophobic properties which may reduce bacterial electrophoretic mobility (10), alter partitioning in a two-phase polyethylene glycol-dextran system, etc. Hydrophobicity calculated on the basis of amino acid composition yields values between -499 and -735 cal mol^{-1} (Table 1), which, with two exceptions, places these adhesins in the same order as that found by the direct measurement of hydrophobicity (13). Clearly the latter measurement reflects more accurately the cell surface hydrophobicity, although it provides no information on the possible situation of hydrophobic domains. Hydrophobic domains, at or near the subunit binding site, may be expected to facilitate attachment by water displacement, whereas hydrophobic domains on adjacent subunits may play a role in subunit interactions and hence adhesin filament formation.

The sequences of the first twenty or so N-terminal amino acids of the adhesines of several, sometimes unrelated, species are surprisingly similar (13). Those of Neisseria gonorrhoea, Moraxella nonliquifaciens, and Pseudomonas aeruginosa, for example, differ by only three amino acids in the first twenty residues; each amino acid change could have resulted from a single base pair change. Type 1 fimbriae and the F7 adhesin of E. coli and type 1 fimbriae of Klebsiella pneumoniae on the one hand and the K88 and K99 adhesins on the other hand also exhibit remarkable similarities in sequence (13). Sequences which make up receptor binding sites are unknown, but are presumably unique and dependent on the tertiary structure for stereochemical alignment.

Since all subunits of an adhesin are identical, a receptor binding site is probably present on each subunit. This would account for lateral binding of adhesins at several points along their lateral margin (10, 13). Adhesins exhibit high degrees of specificity and hence the receptor binding site must be highly conserved, as must intersubunit polymerization sites if a consistent morphology is to be maintained. Indeed, cyanogen bromide fragments of N. gonorrhoea fimbriae show two highly conserved regions, probably representing the receptor binding and the intersubunit polymerization sites and a highly variable region (13). The latter probably is the region responsible for the existence of several antigenic types of this and other adhesin species. Such antigenic diversity probably reflects a highly evolved mechanism of antigenic drift localized at this immunodominant, hypervariable region, which allows the organism to avoid the consequences of anti-adhesive antibody binding to the adhesin and preventing bacterial attachment.

Far less is known about the composition of the fibrillar adhesins of such organisms as streptococci. The adhesive LTA component is a relatively simple amphiphilic molecule consisting of 20 to 30 repeating glycerol phosphate units with alanine and carbohydrate substitutions at the secondary glycerol hydroxyls (15, 20). Fatty acid, glycolipid, or phosphatidylglycolipid substitution which occur at one terminal glycerol of surface LTA (15, 20) mediate adhesion (2). Surface LTA molecules are probably anchored in position by cross-linking proteins interacting with the highly charged glycerol phosphate residues of LTA_i and of intermediate LTA molecules. Cross-linking may be achieved with M protein (one of the surface proteins of streptococcal cells) which forms insoluble complexes with LTA and deacylated LTA (2). Other surface proteins like R and T proteins which react less avidly with LTA may be involved in shielding surface fatty acids from the aqueous environment (10). The shielding protein could then be exchanged by the fatty acids for the animal cell receptor upon surface contact (10). In addition to the evidence that LTA functions in organisms such as S. pyogenes as the adhesin, information is accumulating which suggests that surface proteins may also be involved in the adhesion of other streptococci (e.g., (16)). Whether or not these function as part of the LTA complex, or independently, on the same or other surfaces, is unknown.

The S. mutans adhesive system is apparently more complex than most others (4). Initial adhesion probably results from the action of an adhesive protein (4). Subsequent bacterial aggregation and accumulation seem to result from the synthesis and deposition of the insoluble glucose polymer, mutan, which is bound by a cell surface glucan-binding protein and by a glucosyltransferase enzyme responsible for mutan synthesis. Only polysaccharides that are insoluble or form gels can function in this manner (19).

MULTIPLE ADHESIN SYNTHESIS AND ADHESIN PHASE VARIATION

It is important to stress that not all strains of a bacterial species produce the same adhesin. E. coli strains, for example, produce several adhesins that differ in morphology, composition, and activity (i.e., they have different receptors and/or exhibit differing selective attachment (3, 10, 13)) (Tables 1 and 2). conversely, morphologically and functionally identical adhesins, such as those of N. gonorrhoea, may vary considerably in other properties such as antigenicity (13). Of unknown significance is the ability of cultures of some strains of bacteria to produce up to three distinct adhesins, sometimes simultaneously on the same cell (3). Whether or not the multiple adhesins on a bacterial cell act in concert

TABLE 2 - Possible carbohydrate receptors of bacterial adhesins.

Possible Receptor Moieties	Examples of Inhibitors	Bacteria	Adhesive System	(Ref)
α-D-mannosyl	α-methyl-D-mannosides	E. coli	Type 1 fimbriae	(13)
β-D-mannosyl	β-D-mannosides	P. aeruginosa	Hemagglutinin	(13)
D-mannoside	D-mannose	V. cholerae	Adhesin/ hemagglutinin	(13)
α-D-galactosyl	α-D-galactosides	P. aeruginosa	Hemagglutinin	(13)
α-D-galactosyl	Chemically degraded glycoproteins	E. coli	K88 adhesin	(13)
β-D-galactosyl	Lactose	Actinomycetes	Fibrillae	(13)
Neuraminic acid	Neuraminic acid	Mycoplasma spp.	Adhesin	(13)
Neuraminic acid	Ganglioside GM_2	E. coli	CFA1 adhesin	(13)
	Ganglioside	S. sanguis	Adhesin	(5)
Neuraminic Acid	Ganglioside GM_2	E. coli	K99 adhesin	(13)
L-fucosyl	L-fucosides	V. cholerae	Adhesin/ hemagglutinin	
α-D-galactosyl	Globotetraosyl-ceramide	E. coli 3669	Hemagglutinin/ adhesin	(13)

or in sequence, represent possible alternative systems perhaps functioning in different habitats, or are aberrations of culture, is unknown (3).

The expression of most adhesins is a regulated process dependent on the phase of growth and conditions of culture (13). Qualitative phase variation is a reversible change from adhesin production to cessation of production. Quantitative phase variation is a phenomenon often regulated by different culture conditions than qualitative phase variation and results in greater or lesser amounts of adhesin synthesis. The conditions that promote expression differ for both the adhesin and the organisms. Conditions can be manipulated to accentuate the expression of one adhesin over that of another (3).

SURFACE RECEPTORS OF ADHESINS

Bacterial attachment to animal cells is a highly selective process. This is particularly evident in the oral cavity where it reflects in general the relative ability of the organism to colonize the various oral surfaces (7). Clearly, the latter selective colonization would also depend on the ability of the organism to grow on the surface. In another case, resistance to colonization and disease caused by a pathogenic E. coli is known to be due to the absence of specific components on the animal cell surface (13) and not on the inability of the organism to grow. It is believed that an active adhesive site on each adhesin subunit recognizes and then binds stereochemically to a specific receptor in a lock-and-key fashion. Receptor-bearing components of the animal cell surface are the glycoconjugates, or adsorbed films such as the saliva-derived pellicle of the tooth surface (10, 13). These are all likely to be very complex molecules. Experimental evidence, however, suggests that the receptor moiety which is bound by the adhesin is a relatively small molecule, perhaps no larger than a mono- or disaccharide (1, 10, 13). Evidence in support of the latter can be deduced from the fact that certain monosaccharides can specifically inhibit adhesion to normally receptive cells. As shown originally by Old (as described in (10)), inhibitory activity is very dependent on the stereochemistry of the inhibitor and is often greater with glycosides than with monosaccharides (10, 13). The marked specificity of inhibitors suggests that inhibitors are very similar, if not identical, to the natural receptor.

Lectins may also inhibit adhesion by competing with the adhesin for the adhesin receptor; in the absence of evidence that steric hindrance due to neighboring site attachment of the lectin is responsible for adhesion inhibition, the similarity of the carbohydrate bound by the lectin and the adhesin may be assumed. The inhibitory properties of an inhibitor or the receptivity of a surface can also be destroyed by enzymatic or chemical treatments known to cleave certain types of molecular bonds. The components removed or destroyed by treatment have been identified presumptuously with the receptor (1, 10, 13). The most direct demonstration of receptor function, however, is the attachment of an organism to a surface after an inhibitor of adhesin (i.e., a presumed receptor) has been linked to the surface (10, 13). Such studies support not only the contention that the inhibitor and natural receptor are chemically very similar, but also that the affinity of the adhesin for the inhibitor/receptor is sufficient to hold the bacterium on the surface (13).

The majority of known inhibitors/receptors detected in these test systems are simple sugars which, with the exception of L-fucose, are in the D-configuration (1, 10, 13) (Table 2). The increased inhibitory activities of glycosides compared to that of simple sugars indicate that such sugars are most probably the terminal residues of the larger oligosaccharide chains of glycoconjugates and that adhesins probably recognize at least the sugar residue and its glycosidic bond (1, 10, 13). Some bacterial adhesive activities which have no known inhibitors may recognize more complex carbohydrates, or only react with their receptor when the latter is in particular chemical environments (13). Much of this is somewhat speculative, however, in the absence of kinetic studies on the interaction of inhibitors/receptors with adhesins. Finally, it should be noted that a) those adhesins which are presumed to have carbohydrate receptors are proteins, b) other components of the animal cell surface may act as receptors, and c) some adhesive interactions may occur in the absence of any specific receptor.

A possible example of a noncarbohydrate receptor is fibronectin, which has recently been shown to bind to S. pyogenes (17, 18). Although there is some doubt that LTA is the streptococcal component which reacts with fibronectin LTA (18), specific LTA binding, if it occurs, would presumably be hydrophobic and occur at unique domains which accommodate fatty acids of a particular chain length and stereochemistry. Specificity in the binding of the lipopolysaccharide (LPS) of Gram-negative bacteria via particular fatty acid substituents of the LPS to recognizable proteins of the animal cell membrane has been described (10). The attachment of several species of oral streptococci to oral cavity cells is also reduced in the presence of animal cell membrane phospholipids, suggesting that these bacteria may not only bind to fibronectin, but also attach through the intercalation of their adhesins with the cell membrane (10).

LTA may attach bacteria to tooth enamel in a more direct fashion that does not involve any unique biological entity. Attachment to this surface may be caused by the formation of ionic bridges between the calcium ions of the hydroxyapatite hydration layer and ionized phosphate groups of the LTA backbone (15).

CONCLUSIONS

The need for bacteria to attach to the surfaces of the animal body has resulted in the evolution of a surprising number of adhesive mechanisms. Bacteria may rely on more than one specific mechanism to achieve this essential phase of colonization or may, in response to a particular environmental condition, synthesize a single adhesin to achieve these ends. However, the regulation of these adhesins at the genetic and at the environmental levels is little understood.

The adhesins have evolved as filamentous appendages with reduced negative charges compared with bacterial cells. It is probably because of these properties that the adhesive filament is able to approach closely the animal cell surface and act as a bridge between the bacteria and its substratum. Whether these properties alone would promote nonspecific adhesion (i.e., adhesion not involving a receptor) is unknown.

It can be readily imagined how unique proteins may have evolved to bind specific receptors such as the carbohydrates of the animal cell membrane glycoconjugates. What is less obvious is how fatty acid binding accounts for selective attachment. These is an obvious need to examine the ways adhesive proteins recognize receptor molecules and to clarify the precise composition and stereochemistry of these receptors. Indeed, there is no aspect of this subject which does not require considerable investigation.

REFERENCES

(1) Beachey, E.H., ed. 1980. Bacterial Adherence. London and New York: Chapman and Hall.

(2) Beachey, E.H., and Simpson, W.H. 1980. Interactions of surface polymers of Streptococcus pyogenes with animal cells. In Microbial Adhesion to Surfaces, eds. R.C.W. Berkeley, J.M. Lynch, J. Melling, P.R. Rutter, and B. Vincent, pp. 389-405. Chichester: Ellis Horwood Ltd.

(3) Freter, R., and Jones, G.W. 1983. Models for studying the role of bacterial attachment in virulence and pathogenesis. Rev. Infec. Dis. 5: 5647-5658.

(4) Gibbons, R.J. 1983. Importance of glycosyltransferase in the colonization of oral bacteria. In Glucosyltransferases, Glucans, Sucrose and Dental Caries, ed. R.F. Doyle and J. Ciardi, pp. 11-19. Washington, DC: IRL Press.

(5) Gibbons, R.J.; Etherden, I.; and Moreno, E.C. 1983. Association of neuraminidase-sensitive receptors and putative hydrophobic interactions with high-affinity binding sites for Streptococcus sanguis C5 in salivary pellicles. Infec. Immun. 42: 1006-1012.

(6) Gibbons, R.J.; Moreno, E.C.; and Etherden, I. 1983. Concentration-dependent multiple binding sites on saliva-treated hydroxyapatite for Streptococcus sanguis. Infec. Immun. 39: 280-289.

(7) Gibbons, R.J., and van Houte, J. 1975. Bacterial adherence in oral microbial ecology. Ann. Rev. Microbiol. 29: 19-44.

(8) Hamada, S., and Slade, H.D. 1980. Mechanisms of adherence of Streptococcus mutans to smooth surfaces in vitro. In Bacterial Adherence, ed. E.H. Beachey, pp. 105-135. London and New York: Chapman and Hall.

(9) Isaacson, R.E. 1983. Regulation of expression of Escherichia coli pilus K99. Infec. Immun. 40: 633-639.

(10) Jones, G.W. 1977. The attachment of bacteria to the surfaces of animal cells. In Microbial Interactions, ed. J.L. Reissig, pp. 139-176. London and New York: Chapman and Hall.

(11) Jones, G.W. 1980. Some aspects of the interaction of microbes with the human body. In Contemporary Microbial Ecology, eds. D.C. Ellwood, J.N. Hedger, M.J. Latham, J.M. Lynch, and J.H. Slater, pp. 253-282. London: Academic Press.

(12) Jones, G.W.; Abrams, G.D.; and Freter, R. 1976. Adhesive properties of Vibrio cholerae: Adhesion to isolated rabbit brush border membranes and hemagglutinating activity. Infec. Immun. 14: 232-239.

(13) Jones, G.W., and Isaacson, R.E. 1983. Proteinaceous bacterial adhesins and their receptors. Crit. Rev. Microbiol. 10: 229-260.

(14) Jones, G.W.; Richardson, L.A.; and Uhlman, D. 1981. The invasion

of HeLa cells by Salmonella typhimurium: Reversible and irreversible bacterial attachment and the role of bacterial motility. J. Gen. Microbiol. 127: 351-360.

(15) Kessler, R.E. 1982. Contribution of lipoteichoic acids to dental adhesion and pathogenesis of oral diseases. In Microbiology-1981, ed. D. Schlessinger, pp. 338-341. Washington, DC: American Society for Microbiology.

(16) Liljeviauk, W.F., and Bloomquist, C.G. 1981. Isolation of a protein-containing cell surface component from Streptococcus sanguis which affects its adherence to saliva-coated hydroxyapatite. Infec. Immun. 34: 428-434.

(17) Simpson, A.W., and Beachey, E.H. 1983. Adherence of group A streptococci to fibronectin on oral epithelial cells. Infec. Immun. 39: 275-279.

(18) Speziale, P.; Hook, M.; Switalsaki, L.M.; and Wadstrom, T. 1984. Fibronectin binding to a Streptococcus pyogenes strain. J. Bacteriol. 157: 420-427.

(19) Sutherland, I.W. 1980. Polysaccharides in the adhesion of marine and fresh water bacteria. In Microbial Adhesion to Surfaces, eds. R.C.W. Berkeley, J.M. Lynch, J. Melling, P.R. Rutter, and B. Vincent, pp. 329-338. Chichester: Ellis Horwood Ltd.

(20) Wicken, A.J. 1980. Structure and cell membrane-binding properties of bacterial lipoteichoic acids and their possible role in adhesion of streptococci to eukaryotic cells. In Bacterial Adherence, ed. E.H. Beachey, pp. 137-158. London and New York: Chapman and Hall.

Microbial Adhesion and Aggregation, ed. K.C. Marshall, pp. 85-93. Dahlem Konferenzen 1984. Berlin, Heidelberg, New York, Tokyo: Springer-Verlag.

Bacterial Adhesion to Plant Root Surfaces

F.B. Dazzo
Dept. of Microbiology and Public Health
Michigan State University
East Lansing, MI 48824, USA

Abstract. As plant roots grow, they come into constant contact with microorganisms. Following attachment, the microorganisms may influence plant morphogenesis, nutrition, symbiosis, and pathogenesis. Microbial attachment to plant roots is an important initial event of cellular recognition in infection processes of symbionts and pathogens. This review focuses on the selective attachment in Rhizobium-legume, Azospirillum and Klebsiella-grass, and Agrobacterium-dicot associations as models for understanding detailed biochemical mechanisms of bacterial attachment to roots. These studies emphasize the dynamic, multiphase nature of bacterial attachment to plant host cells and open new avenues for controlling these interactions for man's benefit.

INTRODUCTION

The process of attachment between microorganisms and higher plants is receiving considerable attention in light of its effect on plant morphogenesis, nutrition, symbiosis, and infectious disease. These positive cellular recognitions are believed to arise from a specific union, reversible or irreversible, between chemical receptors on the surface of interacting cells. This hypothesis implies that communication occurs when cells that recognize one another come into contact, and therefore the complementary components of the cell surfaces have naturally been the focus for most biochemical studies. In nature, a diverse collection of microorganisms attach to and colonize plant roots. The vast majority of these microorganisms are most likely attached by nonspecific mechanisms, but these have not been examined. The few cases which have been examined in detail have been shown to display a certain degree of specific adhesion (lock-and-key type) and are summarized here.

THE RHIZOBIUM-LEGUME SYMBIOSIS

Attachment of the rhizobial symbiont to root hairs of the host plant (mostly examined in hydroponic systems) can be viewed as an early recognition event of the infection process and has been recently reviewed (2). After attachment, the bacteria specifically infect these host cells, enter the root cortex, and incite the formation of nitrogen-fixing root nodules. Quantitative light microscopic studies indicate that attachment is accomplished by several mechanisms; some are nonspecific resulting in attachment at low background numbers, and others are host-specific, allowing selective attachment in significantly larger numbers under identical conditions. Host-specific attachment has been demonstrated in R. trifolii-clover, R. japonicum-soybean, and R. leguminosarum-pea root systems (2). However, specificity is not found when very high densities of radiolabeled rhizobia (10^9 cells per seedling) are used in root-binding studies, probably because many unattached bacteria are counted. When marble chips are used to dislodge bacteria attached to the root system, then quantitative plating assays will demonstrate host-specific "firm" attachment in R. trifolii-clover and R. meliloti-alfalfa systems (18).

Transmission electron microscopy discloses that bacterial attachment is initiated by contact between the fibrillar capsule of R. trifolii and electron-dense globular aggregates lying on the outer periphery of the clover root hair cell wall. This "docking" stage (Phase I Attachment) occurs within minutes after inoculation of encapsulated cells of R. trifolii on the host clover (2).

In the model system we have studied, nonspecific attachment accounts for 5-10% of the level of attachment of R. trifolii 0403 to white clover root hairs. The remaining level of attachment fits the definition of specific adhesion adopted by this workshop: where some form of stereochemical constraint brings more than one pair of neighboring interacting groups on the microorganism and the substratum into contact.

Our studies of the Rhizobium trifolii-clover symbiosis (reviewed in (2)) indicate that the specific, "Phase I" attachment process is initiated by a unique interaction between a multivalent lectin (called trifoliin A) associated with the root hair surface and carbohydrate receptors on the bacterial and plant cell walls. Immunochemical and genetic studies have demonstrated that the surfaces of R. trifolii and clover epidermal cells contain a cross-reactive carbohydrate antigen that binds to trifoliin A. Other studies have shown that lectins on pea, alfalfa, and soybean roots are accessible for binding to the appropriate rhizobia, and specific hapten-facilitated elution of lectin from pea, alfalfa, and soybean roots has also been found.

Specific Phase I attachment is hapten-reversible in the R. trifolii-clover system (2, 19) for approximately 12 hr. For instance, the specific hapten inhibitor of trifoliin A (2-deoxy-D-glucose) added with the inoculum blocks the attachment of the bacterial cells and their lectin-binding polysaccharides to the root hairs. This hapten is no longer effective in dislodging attached rhizobia from clover root hairs if they are preincubated for 12 hr before addition of the hapten (Dazzo et al., submitted).

Microscopic studies have disclosed that Phase I attachment is followed by an irreversible stage of Phase II adhesion, characterized by the firm anchoring of the bacterial cell to the root hair surface (Dazzo et al.,submitted). This step may be important in maintaining the close contact necessary for triggering the tight root hair curling (shepherd's crook formation) and successful penetration of the root hair cell wall by rhizobia. During Phase II adhesion, fibrillar materials are characteristically found associated with the adherent bacteria. It is not known whether these fibrils consist of cellulose, bundles of pili, or some other fibrillar polymers made by the attached bacteria. Future studies should be directed to isolate and characterize these fibrils and to identify what components of the system stimulate their synthesis in order to understand better the Phase II adhesion process. This is particularly important since the degree of host-specific firm attachment of rhizobial strains to the root shows a significant positive correlation with the degree of the strain's success in interstrain competition for nodule sites on the root (18).

The attachment process in the R. trifolii-clover symbiosis is regulated by several mechanisms. One involves a plant component, a second involves a bacterial component, and a third involves the interaction of both the bacterium and the plant host in situ in the root environment.

A mechanism regulated by the plant host seems to be a growth response of the root to an exogenous supply of nitrate (1). The specific attachment of R. trifolii to clover root hairs and the levels of trifoliin A on these epidermal cell walls decline in parallel when the roots are grown in a medium supplemented with nitrate. Recent studies indicate that this is not due to a direct interaction of nitrate with trifoliin A or to an inhibition of its de novo synthesis in roots, but rather to an inability of trifoliin A to bind to, and accumulate on, the root cell wall itself (2). These alterations of host receptors may reflect changes in wall chemistry associated with growth of the legume with combined nitrogen.

A mechanism regulated by the bacterium involves the growth phase-dependent accumulation of bacterial saccharide receptors for trifoliin A (2). These receptors are transient as a result of chemical modifications of the surface polysaccarides which occur at certain phases of culture growth. At discrete times in broth culture, the LPS is a dominant lectin receptor on the bacterium, and the lectin-binding sugar, quinovosamine (2-amino-2,6-dideoxyglucose) increases in the LPS as cells enter stationary phase (5). When grown on agar surfaces, the cells develop a well-defined capsule which contains an acidic heteropolysaccharide as the dominant lectin receptor. Cells grown on a defined agar medium for five days bind trifoliin A in greatest quantity uniformly around the capsule. With this distribution of lectin receptors, the bacteria attach in high numbers in random orientation to clover root hairs. However, lectin receptors are located predominantly at one cell pole before and after five days of growth on the agar surface, and the percentage of these cells which attach polarly to root hairs is proportionally increased (15). This indicates that the culture age-dependent distribution of trifoliin A receptors dictates the level and orientation of attachment of R. trifolii 0403 to clover root hairs. Studies are under way to define the biochemical basis for this transient appearance of lectin receptors on bacteria grown on solid media.

Similarly, the lectin receptors on R. japonicum and R. leguminosarum are transient. Methylation of galactose residues in the capsular polysaccharide of R. japonicum occurs as the culture enters stationary phase, reducing its affinity for the galactose-specific soybean lectin (12). The transient period when R. leguminosarum gains the ability to bind pea lectin occurs during arithmetic growth following exponential phase (17). These effects of culture age on the composition of lectin-binding polysaccharides of Rhizobium may reflect mechanisms which regulate cellular recognition in the Rhizobium-legume symbiosis.

The presence of multiple lectin receptors on rhizobia raises the question of whether each one has a different role in root hair infection. Infection studies by Kamberger (6) suggest that the lectin cross-bridging hypothesis needs to be modified. For example, the capsular polysaccharides could be responsible for attachment of high numbers of rhizobial cells to the target root hairs via cross-bridging lectins as an early recognition event. This would be followed by secondary recognition events requiring the host-range specific binding of lectin on localized sites of the root hair to LPS, which could trigger subsequent invasive steps. Studies are in progress to test this most intriguing hypothesis.

The ultimate level of regulation of attachment involves the interaction of the bacterial and plant symbionts in the rhizosphere. The clover root releases enzymes which alter the lectin-binding capsule of R. trifolii in a way which favors a preferred polar attachment (3). Based on kinetics of attachment (Dazzo et al., submitted) and this enzymatic alteration of the bacterial capsule in the rhizosphere, we have subdivided Phase I attachment into three sequential steps: a) encapsulated cells bind first in random orientation to root hair tips where trifoliin A accumulates, b) enzymes in root exudate alter the capsule of non-attached cells in the rhizosphere resulting in a polar capsule, and c) these cells then polarly attach along the sides of the root hair.

Genes in R. trifolii which are important to trifoliin A receptor synthesis and attachment to clover roots are encoded on the same large symbiotic plasmid ((14, 19), and Dazzo et al., submitted). Conjugal transfer of the nodulation plasmid from R. trifolii into pTi-cured Agrobacterium tumefaciens results in a hybrid which nodulates clover (4), synthesizes trifoliin A receptors, accomplishes the 3-step sequence of Phase I attachment described above, and infects clover root hairs (Dazzo et al., in preparation). It would be very interesting to know how these plasmid-encoded "root hair attachment" genes are regulated.

ATTACHMENT OF AZOSPIRILLUM AND KLEBSIELLA TO GRASS ROOTS

The attachment of Azospirillum brasilense Sp7 to root hairs of pearl millet in hydroponic culture is similar in some respects to attachment of R. trifolii 0403 to clover root hairs (16). Such attachment can be viewed as favoring the colonization of the rhizoplane by azospirilla to occupy niches which take advantage of root exudation. Attached azospirilla were associated with granular material on root hairs and fibrillar material on undifferentiated epidermal cells. Significantly fewer cells of azospirilla atached to millet root hairs when the roots were grown in culture medium containing 5 mM nitrate. Attachment to undifferentiated epidermal cells was unaffected. Root exudate from millet contained nondialyzable and protease-sensitive proteins which bound to azospirilla in the root environment and promoted their attachment to millet root hairs but not to undifferentiated epidermal cells. Millet root hairs adsorbed more cells of azospirilla than of R. trifolii, Pseudomonas sp., Azotobacter vinelandii, Klebsiella pneumoniae, or Escherichia coli. It is clear from these studies that attachment of azospirilla to root hairs of millet is accomplished by mechanisms which differ from those that attach azospirilla to undifferentiated epidermal cell surfaces.

It has recently been shown that the strains of Azospirillum which successfully occupy niches on the grass rhizoplane attach better to the grass roots than do other strains which occupy a more invasive niche and colonize the middle lamella of the cortex within the root (Jain and Patriquin, in preparation). Thus, the ability of Azospirillum to attach to the root surface may be very important in establishing which of these two habitats and niches can be occupied by the bacteria.

Korhonen et al. (7) have recently examined the in vitro adhesion of associative nitrogen-fixing Klebsiella spp. to grass roots. Klebsiellas were labeled with ^{3}H-amino acids and incubated with the grass roots. Their attachment was dependent on inoculum density, incubation time and temperature, pH and ionic strength of the incubation buffer, and the growth phase of the bacterial cells. Type 1 (mannose-sensitive) and type 3 fimbriae were isolated from the labeled bacteria and examined for their possible role in root attachment. The binding of type 1 fimbriae to roots was inhibited by the hapten α-methylmannoside and by Fab monovalent fragments directed against the purified fimbriae. However, non-fimbriated mutant strains attached very well to the whole grass roots. More work on the role of fimbriae in root attachment should include detailed microscopy to sort out the mechanisms of attachment of these nitrogen-fixing organisms to grass root hairs and other epidermal cells.

ATTACHMENT OF AGROBACTERIUM TUMEFACIENS TO PLANT CELL WALLS

Agrobacterium tumefaciens is the causal agent of the tumorous disease of higher plants called crown gall. Once introduced into a plant wound, the bacterium attaches to specific sites on the host cell wall and then transfers a part of its large Ti plasmid into the host cell where it integrates into the chromosome and causes the transformation of host cells into tumor cells.

The initial events of attachment have been examined both by measuring inhibition of tumor formation in wound sites on bean leaves and by measuring bacterial binding to tissue culture cells. Although the two systems generally give similar results, there are some differences depending on the system used. In wounded bean leaves, attachment appears to involve a bacterial polysaccharide: plant polysaccharide interaction of the O-antigen side chains of the bacterial LPS with non-methylesterified pectic material of the host cell wall (13). In the tissue culture system, bacterial binding may involve LPS, but pectic material from the plant has no effect on bacterial attachment (11). In both wounded plants and tissue cultured cells, this interaction shows considerable specificity and is complete after the 15-60 min after inoculation. Both chromosomal as well as Ti plasmid genes are capable of determining site attachment. In a way similar to the Rhizobium-clover system, most strains of Agrobacterium produce microfibrils after initial attachment which presumably add to the stability of the bacterium-host complex and can serve to entrap additional bacteria (10). These fibrils have been identified as cellulose. Mutants that have lost the ability to produce cellulose, however, are still virulent and can bind to host cells (8). However, these mutant bacteria, unlike the parent strain, can readily be removed from wound sites by water washing (8). Although agrobacteria produce tumors on a wide variety of dicotyledonous plants, monocotyledons as a group seem to be largely resistant. Inhibitory studies of agrobacterial attachment to wound sites which lead to tumor formation and studies of bacterial binding to tissue culture cells suggest that an initial factor contributing to the resistance of monocotyledonous plants would be the failure of these plants to support Agrobacterium attachment (9, 13).

FUTURE RESEARCH NEEDS

There continues to be a pressing need for definitive identification of the surface components on bacteria and plants which mediate both the specific and the nonspecific attachments to these hosts. A long-range goal of this effort would be to manipulate the interaction by controlling at will the initial attachment steps. This may make it possible to promote beneficial associations (e.g., broaden the host range for Rhizobium nitrogen-fixing symbioses) and suppress harmful ones which destroy major crops. The role of bacterial fimbriae and cellulose microfibrils in establishing firm attachments needs to be investigated in further detail, particularly in explaining those attachments which cannot be accounted for by plant lectins. Another most important aspect is to determine how the dynamic events of attachment integrate with subsequent steps within the "black box" to lead to successful infection in plant-microorganism interactions.

Acknowledgements. Supported in part by USDA-CRGO Competitive Grant No. 82-CRCR-1-1040, NSF Grant No. 80-21906, and Project No. 1314H from the Michigan Agricultural Experiment Station.

REFERENCES

(1) Dazzo, F.B., and Brill, W.J. 1978. Regulation by fixed nitrogen of host-symbiont recognition in the Rhizobium-clover symbiosis. Plant Physiol. 62: 18-21.

(2) Dazzo, F.B., and Truchet, G.L. 1983. Interactions of lectins and their saccharide receptors in the Rhizobium-legume symbiosis. J. Membr. Biol. 73: 1-16.

(3) Dazzo, F.B.; Truchet, G.L.; Sherwood, J.E.; Hrabak, E.H.; and Gardiol, A.E. 1982. Alteration of the trifoliin A-binding capsule of Rhizobium trifolii 0403 by enzymes released from clover roots. Appl. Envir. Microbiol. 44: 478-490.

(4) Hooykaas, P.J.; van Brussel, A.A.N.; den Hulk-Ras, H.; van Slogteren, G.M.; and Schilperoort, R.A. 1981. Sym plasmid of Rhizobium trifolii expressed in different rhizobia species and Agrobacterium tumefaciens. Nature 291: 351-353.

(5) Hrabak, E.M.; Urbano, M.R.; and Dazzo, F.B. 1981. Growth-phase dependent immunodeterminants of Rhizobium trifolii lipopolysaccharide which bind trifoliin A, a white clover lectin. J. Bacteriol. 148: 697-711.

(6) Kamberger, W. 1979. Role of cell surface polysaccharides in the Rhizobium-pea symbiosis. FEMS Microbiol. Lett. 6: 361-365.

(7) Korhonen, T.; Haahtela, K.; Ahonen, A.; Rehn, M.; Vaisanen, V.; Pere, A.; and Tarkka, E. 1982. Adhesion of nitrogen fixing Klebsiellas to plant roots. 2nd National Symposium on Biological Nitrogen Fixation. Helsinki, Finland, pp. 143-150. Helsinki: SITRA.

(8) Matthysse, A.G. 1983. The role of cellulose fibrils in Agrobacterium tumefaciens infection. J. Bacteriol. 154: 906-915.

(9) Matthyse, A.G., and Gurlitz, R.H.G. 1982. Plant cell range for attachment of Agrobacterium tumefaciens to tissue culture cells. Physiol. Plant Pathol. 21: 381-387.

(10) Matthysse, A.G.; Holmes, K.V.; and Gurlitz, R.H. 1981. Elaboration

of cellulose fibrils by Agrobacterium tumefaciens during attachment to carrot cells. J. Bacteriol. 145: 583-595.

(11) Matthysse, A.G.; Holmes, K.V.; and Gurlitz, R.H.G. 1982. Binding of Agrobacterium tumefaciens to carrot protoplasts. Physiol. Plant. Pathol. 20: 27-33.

(12) Mort, A.J., and Bauer, W.D. 1980. Composition of the capsular and extracellular polysaccharides of Rhizobium japonicum: changes with culture age and correlations with binding of soybean seed lectin to the bacteria. Plant Physiol. 66: 158-163.

(13) Rao, S.S.; Lippincott, B.B.; and Lippincott, J.A. 1982. Agrobacterium adherence involves the pectic portion of the host cell wall and is sensitive to the degree of pectin methylation. Physiol. Plant. 56: 374-380.

(14) Russa, R.; Urbanik, T.; Kowalczuk, E.; and Lorkiewicz, Z. 1982. Correlation between occurrence of plasmid pUCS202 and lipopolysaccharide alterations in Rhizobium. FEMS Microbiol. Lett. 13: 161-165.

(15) Sherwood, J.E.; Vasse, J.; Dazzo, F.B.; and Truchet, G.L. 1984. Development and trifoliin A-binding ability of the capsule of Rhizobium trifolii. J. Bacteriol. 159: 145-152.

(16) Umali-Garcia, M.; Hubbell, D.H.; Gaskins, M.; and Dazzo, F.B. 1980. Association of Azospirillum with grass roots. Appl. Envir. Microbiol. 39: 219-226.

(17) van der Schaal, I.A.M.; Logman, T.J.; Diaz, C.L.; and Kijne, J.W. 1983. Growth-phase dependent pea (Pisum sativum L.) lectin receptors of Rhizobium leguminosarum. In Advances in Nitrogen Fixation Research, eds. C. Veegers and W. Newton, p. 432. The Hague, Netherlands: Martinus Nijhoff.

(18) van Rensberg, H.J., and Strijdom, B. 1982. Root surface association in relation to nodulation of Medicago sativa. Appl. Envir. Microbiol. 44: 93-97.

(19) Zurkowski, W. 1980. Specific adsorption of bacteria to clover root hairs, related to the presence of the plasmid pWZ2 in cells of Rhizobium trifolii. Microbios 27: 27-32.

Microbial Adhesion and Aggregation, ed. K.C. Marshall, pp. 95-107. Dahlem Konferenzen 1984. Berlin, Heidelberg, New York, Tokyo: Springer-Verlag.

Genetic Control of Bacterial Adhesion

M. Silverman,* R. Belas,* and M. Simon**
*The Agouron Institute, La Jolla, CA 92037
**Div. of Biology, Caltech, Pasadena, CA 91125, USA

Abstract. Adhesive substances represent a solution to the problem of colonizing varied surfaces in the bacterial habitat. A given species may have genetic information for a large repertoire of adhesive substances, and the outcome of an encounter between a population of these bacteria and a particular surface may depend on how the bacteria regulate the expression of the adhesive substances which they encode. Synthesis of adhesive substances appears to be regulated to match the demands of the environmental circumstances, and we discuss several genetic strategies which bacteria could use to control expression of these functions.

INTRODUCTION

Bacteria adhere to a great variety of animate and inanimate surfaces, and it is clear that the adhesiveness of bacteria is a very important determinant in the colonization of a habitat in an ecosystem, whether that habitat is the surface of a stone in a stream, a mucosal cell in the intestinal tract, or the hull of a ship. The ability of a bacterium to interact with a given surface is now know to be governed by particular molecules or structures on the outer surface of the cell. We will refer to this chemically diverse group of molecules as adhesive substances. These may be proteins or lipopolysaccharides in the outer membrane, proteinaceous appendages such as fimbriae or flagella, or complex carbohydrates in the form of extracellular capsules. The association between a bacterial adhesive and a surface can be very specific and involve ligand-receptor interactions, or the adhesive binding can be quite nonspecific and depend upon the relative hydrophobicity of the bacterial surface (12). The process of adhesion is often temporally complex with initial attachment mediated by relatively weak reversible binding, followed later by the synthesis of adhesive substances which firmly cement the cell to the surface (10). Physiological conditions such as the ionic composition of the environment and the availability of nutrients affect the adhesion process. By adsorbing organic molecules, a solid surface can concentrate nutrients and encourage the growth of bacteria (18). In addition, some bacteria rely on motility and chemotaxis to assist them in finding a surface (4).

Considering the complexity of the adhesion process, the great variety of surfaces, and the vagaries of the bacterial environment, how does a bacterium produce adhesives appropriate to the particular circumstances? We will discuss several examples and possible mechanisms of genetic regulation of adhesive functions and describe the experimental system used to study adhesion in our laboratory. Although we will confine our discussion to genetic control, other important uses of genetic analysis should not be overlooked. Mutant studies often provide definitive evidence that a substance or structure is actually involved in adhesion. When several adhesive substances are present on an organism, the relative contributions to adhesiveness can be determined by comparing strains mutated in one of the several adhesive functions. Steps in a biosynthetic pathway for a complex adhesive such as lipopolysaccharide can be identified and ordered using mutants. Proteins from mutants are useful in understanding the biochemistry and stereochemistry of adhesive binding. This approach is particularly promising since it is now possible to clone and sequence a mutant gene and to identify specific amino acids or regions of a protein necessary for adhesion. By using genetics in concert with other disciplines, it is now possible to design genes which produce new adhesive substances.

STRATEGIES FOR GENETIC CONTROL

The genetic mechanism for control of expression of adhesive production may reflect the diversity of the environment in which the bacterium lives. For example, if the organism occupies several very similar habitats, constitutive expression of adhesive genes would meet the needs of the cell. However, since the bacterial environment is rarely homogeneous or unchanging, more complex genetic controls may be required to allow the organism to alter its adhesive composition. For the purpose of discussion, we will categorize genetic controls as "responsive" or "variable." If a population of bacteria changes its adhesive substance synthesis upon perception of an external signal, the control is "responsive." Alternatively, if a clonal population of bacteria is heterogeneous with respect to adhesive production, the control is "variable." Synthesis of host specific fimbriae (adhesins) by E. coli is affected by the composition of the growth medium and by temperature (5). Production of fimbrial adhesins such as K99 does not take place below 20°C. Expression of host-specific adhesins would only be advantageous when the bacterium grows in the gut, and the organism may turn off adhesin production when growing outside the host in response to nonphysiological temperature. Temperature could regulate the production of fimbriae by altering the fluidity of the membrane, which would affect the synthesis and transport of fimbrial subunits (7). A DNA fragment which encodes the K88 antigen has been isolated from a large plasmid, and using recombinant clones, the genes and most of the gene products involved in fimbriae formation have been identified (9). At least five genes are necessary for the syntheses of fimbriae, including adhD which specifies the fimbrial subunit. Further studies should reveal how expression of this adhesin gene system is controlled. Despite recent evidence that catabolite repression does not regulate transcription of the type 1 fimbriae genes of E. coli (3), the c-AMP control system is still an instructive model and a good candidate for regulation of other adhesins involved in specific adhesion reactions. A large number of bacterial characteristics are transcriptionally regulated simultaneously by the level of c-AMP in the cell. The c-AMP concentration in turn is controlled by the presence of catabolites in the culture medium. Control by c-AMP is mediated through binding to a positive regulator protein, the c-AMP receptor protein, which activates transcription of many operons, including ones for flagella and outer membrane protein.

The porin proteins of the outer envelope of E. coli are not considered to be adhesive substances, but the genetic mechanism which controls their expression is an interesting example of a "responsive" control system. OmpC and ompF encode major outer membrane porin proteins in E. coli, and the relative amount of the two proteins is regulated by the osmolarity of the culture medium. A third locus containing at least two genes, ompR and envZ, is responsible for regulation of the ompC and ompF genes (6). It is proposed that the ompR product acts as a bifunctional regulatory protein capable of turning on expression of ompC or ompF depending upon the multimeric nature of the protein; the monomeric form turns ompF expression on and the multimeric forms turns on ompC. The equilibrium of monomeric and multimeric forms is thought to be controlled by an envelope protein, the envZ product, which somehow responds to the concentration of solutes in the environment. Analogous systems may regulate other outer membrane proteins or capsular components which do act as adhesive substances.

Thus far, "responsive" systems which sense temperature, catabolites, and osmolarity have been mentioned, but this is only a partial list. The bioluminescence system of marine vibrios has a control mechanism which responds to cell density, and many gene systems respond to specific molecules such as amino acids and other metabolites. Microorganisms such as Agrobacterium and Rhizobium, which adhere to plant tissues, produce specific polysaccharide adhesins which mediate binding to plant surfaces. In Rhizobium, the accumulation and distribution of these adhesins is influenced by growth conditions. After initial attachment of Rhizobium and Agrobacterium, microfibrils are produced which firmly anchor the bacteria. In Agrobacterium these fibrils are composed of cellulose. Synthesis of these microfibrils appears to be a response to a signal resulting from the interaction between the bacteria and the plant surface (Dazzo, this volume). Some Gram-negative marine bacteria respond to nutrient starvation by becoming more adhesive (see Kjelleberg, this volume). Later, we will discuss a "responsive" control system in Vibrio parahaemolyticus.

The expression of type 1 fimbriae in E. coli is an example of "variable" genetic control. A population derived from a clone is heterogeneous and is composed of fimbriate and non-fimbriate bacteria. Individual cells switch from one form to the other at a frequency of about one cell in a thousand per generation. The switching has been shown to be under transcriptional control (2) and may be mechanistically similar to the phase variation system in Salmonella, which controls the oscillatory expression of flagella antigens. The genetic basis for phase variation is a DNA inversion which couples and uncouples a promoter element to a flagellar operon (14). The inversion is a site-specific recombination event mediated by a specific recombinase. This gene switch is closely related to the inversion systems of phages Mu and Pl which control host range types, and the recombinase function appears to have evolved from a mobile genetic element. The host range of these phages is determined by tail fibers which bind to lipopolysaccharide in the cell envelope, and the phage-bacterium interaction can be thought of as an adhesin-mediated process. Inversion switches could be quite common in nature and control the variable expression of adhesive substances in addition to type 1 fimbriae.

Neisseria gonorrhoeae displays an amazingly complex pattern of specific adhesins on its surface. Clonal populations of cells contain both piliated and non-piliated bacteria, and one form can convert to the other (16). In addition, a given strain can give rise to several pilus serotypes which vary in amino acid composition and subunit molecular weight. As many as fifty different pilus serotypes have been reported. Therefore, Neisseria turns pilin synthesis on and off and can also produce a great deal of variety in the nature of the pilin subunit. The opacity protein in the membrane of Neisseria also has adhesive properties and also shows an on-off variation. Like the pilus system, a large variety of opacity protein variants are observed (15). The gene for the pilin subunit has been cloned and used to probe for rearrangements in the Neisseria genome, and changes in the arrangement of DNA sequences in the pilin gene region have been found to correlate with the state of pilus expression (11). The pattern of change in DNA structure is complex, and no simple recombinational mechanism accounts for the observed DNA rearrangements. This phenomenon is in many ways similar to the variation of surface antigens known to occur in African trypanosomes and may be regulated by a similar recombinational mechanism.

The benefit derived from variable control is apparent if one considers a population of bacteria rather than individual cells. No single cell expresses the multiplicity of forms which are found in the population as a whole. The progeny of a single cell includes forms which have different adhesive properties, and the culture is in effect preadapted to changes in the ecosystem. Once a subpopulation of bacteria has colonized a surface, descendants arise with characteristics suited to a free-living life-style, leave the surface, and can colonize other habitats. With this mechanism of control the individual bacterium does not "know" where it is and cannot "decide" to synthesize an appropriate adhesive, but the solution to binding to a particular surface is nevertheless contained in the population. "Responsive" and "variable" controls are mechanistically different, but both methods of control could operate together to control complex adhesive phenotypes.

The expression of adhesive functions can be regulated in other ways. Genes for several fimbrial adhesins are encoded by conjugative plasmids or by plasmids which can be mobilized by other conjugative plasmids (5). In this way an adhesive property can be passed on to closely related bacteria. It is conceivable that adhesive functions could be transferred by mobile genetic elements, although no such transposon has been found. Developmental control of adhesive production is observed with Caulobacter

(13). With cell division there is a programmed cycle of change from attached stalk cell to free-living swarmer cell and back to stalked cell which then attaches to a surface. The mechanism responsible for this temporally programmed control is not known. This type of control is certainly common in multicellular organisms in which cell-cell association is mediated by a great variety of adhesive substances.

Experimental Paradigm

Bacterial adhesion to surfaces in the marine environment is an important component of a complex biofouling process. Organic molecules rapidly adsorb to clean surfaces immersed in seawater. Soon thereafter bacteria, some of which may be directed to the surface in response to chemoattractant gradients, adhere to the surface. Initially, attachment involves weak or reversible binding, and later adhesives are produced which result in tight association with the surface. We are interested in determining what substances bacteria use to attach to marine surfaces and how the expression of these adhesive substances is regulated. In the context of the previous discussion, are the regulatory controls "responsive" or "variable"? Research in our laboratory focuses on Vibrio parahaemolyticus, an organism common to many biofouling communities. This bacterium elaborates two very distinct kinds of flagellar appendage, the polar flagellum and the lateral flagella. Both organelles have been implicated in adhesion to marine surfaces (1). The polar flagellum is sheathed with a membrane similar in composition to the outer membrane, is synthesized constitutively under all conditions examined, and is necessary for motility in liquid culture. The polar flagellum could serve to bridge the repulsive barrier between the bacterial cell and a surface and appears to be important in initial reversible attachment. The second type of flagellar structure, the lateral flagella, is produced in the laboratory only when the bacteria are grown on agar (and other) surfaces and not when grown in liquid culture. Lateral flagella propel the bacteria over surfaces (swarming) and are responsible for a stronger adhesion to surfaces.

Since lateral flagella are synthesized only when the bacteria are grown on surfaces such as agar plates, do the bacteria actually respond to the presence of a surface? To explore the sensory aspect of surface perception a genetic dissection of the lateral flagella system has been initiated. Little genetic technology exists for examining Vibrio parahaemolyticus, but transposon mutagenesis and recombinant DNA methods are proving useful (Belas, Mileham, Simon, and Silverman, in preparation). Transposable elements were packaged in coliphage P1 and used to infect

V. parahaemolyticus. Since phage P1 could not replicate in the Vibrio host, the transposon, in this case mini-Mu, had to integrate into the chromosome to persist in the cell. Mutants were isolated by selecting for the transposon-encoded tetracycline resistance after infection with the transducing phage P1. From an extensive screening, several hundred non-swarming tetracycline-resistant mutants were collected, and most of these did not make lateral flagella. Transposon mini-Mu was designed to contain the coding region for the lacZ gene of E. coli. Upon integration into a target gene the mini-Mu, in addition to causing insertional inactivation of gene function, can integrate in an orientation which fuses the transcription of the target gene with the lacZ gene of the transposon. Consequently, many mini-Mu mutants were Lac^+ because of such fusion, and the synthesis of β-galactosidase was regulated as a function of transcription of the target gene (see Fig. 1). We thus have a sensitive, direct, and convenient means to measure expression of lateral flagella genes.

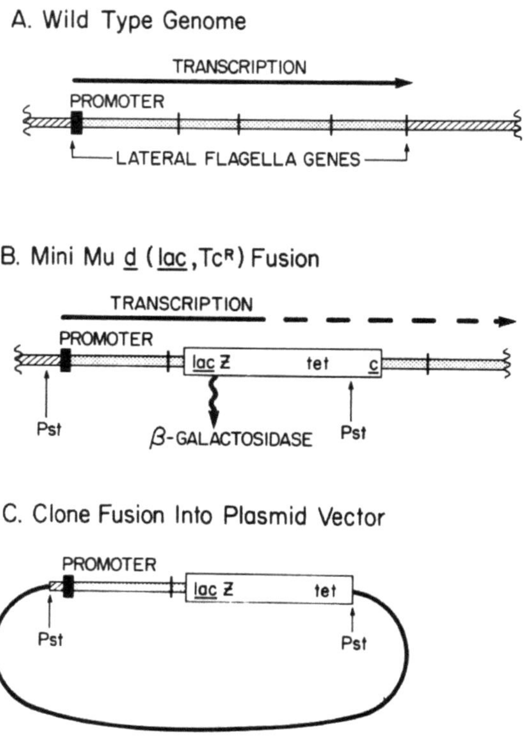

FIG. 1 - Mutagenesis and gene fusion using transposon mini-Mu.

Approximately fifty percent of the mini-Mu generated lateral flagella mutants were Lac^+ as was expected since one orientation of transposon insertion did not align lacZ with the direction of transcription of the target gene. Synthesis of β-galactosidase was initially monitored in vivo on agar plates containing the histochemical stain Xgal (5-bromo-4-chloro-3-indolyl-β-D-galactoside). When standard β-galactosidase assays were then performed using bacteria harvested from broth and agar plates, we found that almost half of the Lac^+ fusion strains did not synthesize β-galactosidase when grown in broth culture. Figure 2 shows the time course of surface-dependent turn-on of lateral flagella gene transcription. Turn-on did not occur immediately as might be expected if these bacteria were to have a tactile sense. Instead, sixty to ninety minutes elapsed before transcription began. We interpret this finding to mean that the cells must have time to synthesize and accumulate some molecule which is necessary to signal initiation of transcription of the lateral flagella genes. A variety of growth conditions are being examined to identify precisely what conditions signal the turn-on of these genes. For example, induction of transcription is being measured with fusion strains grown on a variety of membrane surfaces over a variety of substrata.

While work on the physiology of lateral flagella gene regulation is proceeding, we are using the transposon mutants to isolate lateral flagella genes. As shown in Fig. 1, the insertion of a transposon introduces a selectable marker into a target gene. It is then possible, using an appropriate restriction endonuclease enzyme, to clone a DNA fragment containing part of the target gene since this DNA is physically linked to DNA encoding the transposon drug resistance. In this way, several recombinant clones containing DNA for lateral flagella genes have been constructed. These clones have been used to probe a Vibrio parahaemolyticus cosmid bank to find large (>20kbp) cloned fragments containing lateral flagella genes. The particular cosmid (pLAFR1) is derived from a broad host range plasmid (RK2) and can be transferred from E. coli into V. parahaemolyticus where it will replicate. Therefore, DNA containing lateral flagella genes can be cloned from transposon mutants, and these recombinant clones can be used as probes to isolate larger DNA fragments on a cosmid which in turn can be reintroduced into V. parahaemolyticus. Using this kind of strategy, most if not all of the lateral flagella genes can be isolated and analyzed genetically. Of particular interest is the identification of those genes (and gene products) responsible for regulating lateral flagella production in response to growth on surfaces.

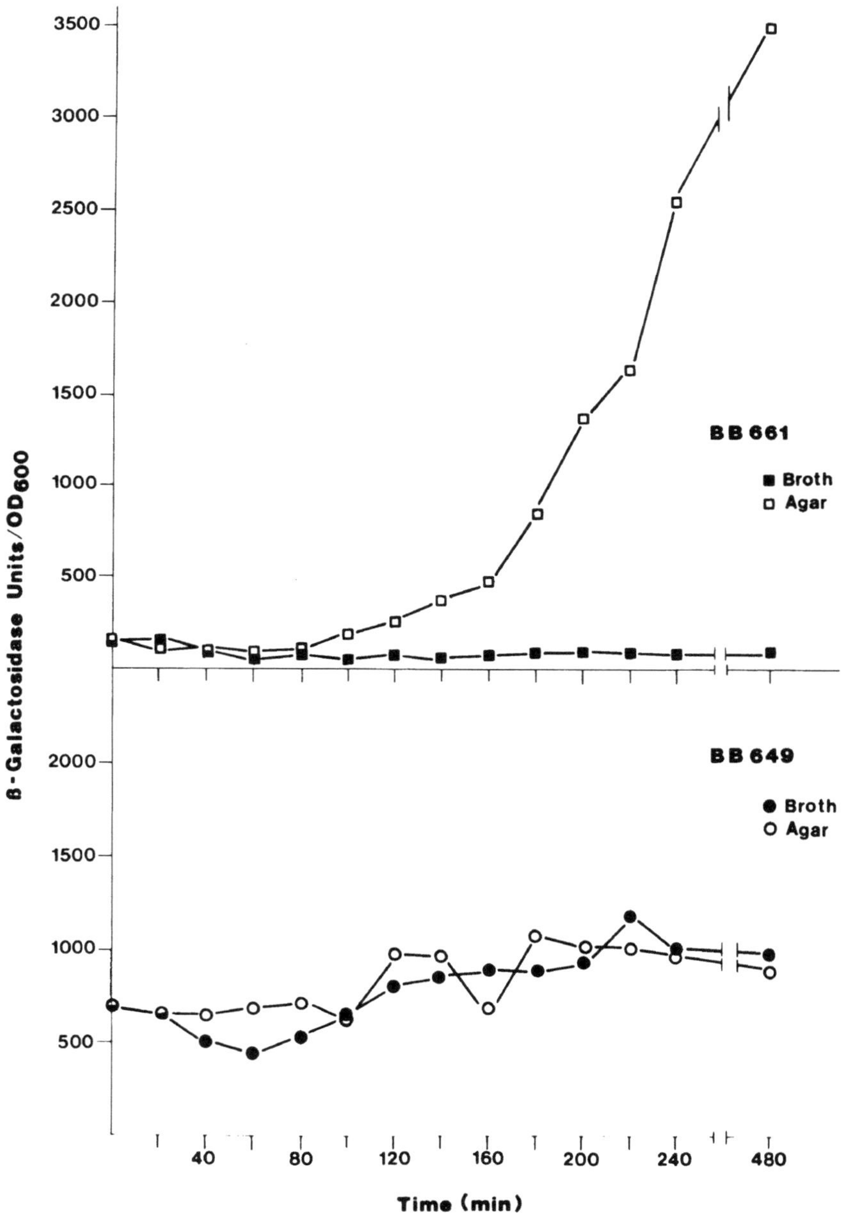

FIG. 2 - Surface-dependent expression of lateral flagella genes. β-galactosidase activity is shown for fusion strains BB661 (surface-dependent) and BB649 (surface-independent). β-galactosidase synthesis is a function of target gene transcription (see Fig. 1).

CONCLUSION

We have assumed that to maximize the benefit derived from the synthesis of adhesive substances bacteria must be capable of regulating expression of these substances, and that highly adaptable organisms "orchestrate" production in a manner appropriate to the changing conditions of the ecosystem. It remains to be seen how "smart" bacteria are, but it appears that at least some species have evolved complex genetic controls to regulate adhesive expression. To provoke discussion, control mechanisms have been characterized as "responsive" or "variable."

The essence of responsive control is information processing, whereby the bacterium senses some signal in the environment and responds by elaborating an adhesive substance. The bacterium can indirectly "know" where it is by sensing temperature, osmotic conditions, or metabolites, but can an organism sense a surface? With Vibrio parahaemolyticus it is clear that a new adhesive is produced after growth in contact with a surface (Fig. 3), but the bacterium may not actually possess a tactile sense. With higher organisms the direct interaction of cells with surfaces influences division, development, and adhesive properties (17), but it is not clear that microorganisms such as bacteria can sense and respond to direct contact between their cell envelope and a surface. A tactile response may occur with Myxococcus (8) in which the motility of different social types appears to be stimulated by cell-cell contact. At present we do not know what stimulates lateral flagella production in the marine Vibrio.

With "variable" control the appropriate adhesive substance is not produced in response to a signal but is produced by a subset of the population. A fraction of the cells are preadapted to adhere to a given surface, and individuals in the population are constantly interconverting to a variety of forms. In a population of bacteria growing on a surface, like fimbriated bacteria attached to epithelial cells, nonadherent variants arise, detach, and can move on to colonize other habitats (Fig. 3). More complex variations such as in Neisseria would give the organism more versatility. Of course, the variation itself could be subject to "responsive" control, so "responsive" and "variable" control mechanisms need not be thought to be mutually exclusive. The molecular mechanism which generates the variability of adhesive phenotypes is known for several gene systems and involves rearrangement of DNA structure. With Salmonella phase variation, the rearrangement is the inversion of DNA containing a transcriptional control element. The variation observed in Neisseria gonorrhoeae is controlled by DNA rearrangement, but the actual process is not understood. "Variable" control systems may be particularly advantageous where the generation of a large variety of related adhesive substances is required. For example, this regulatory mechanism could be used by pathogens to vary specific adhesin antigenicity to avoid immunological defenses or by other bacteria to generate diversity in the binding specificity of a particular adhesin.

The study of regulation of adhesive production is still in its infancy, but with the new genetic technologies progress should be rapid. Genetic tools are being developed which will be very useful for studying gene expression in organisms other than E. coli. As the mechanisms which regulate adhesive substances are more fully understood, new ways to intervene in biofouling or disease processes may be found, and the beneficial characteristics of bacteria may also be exploited.

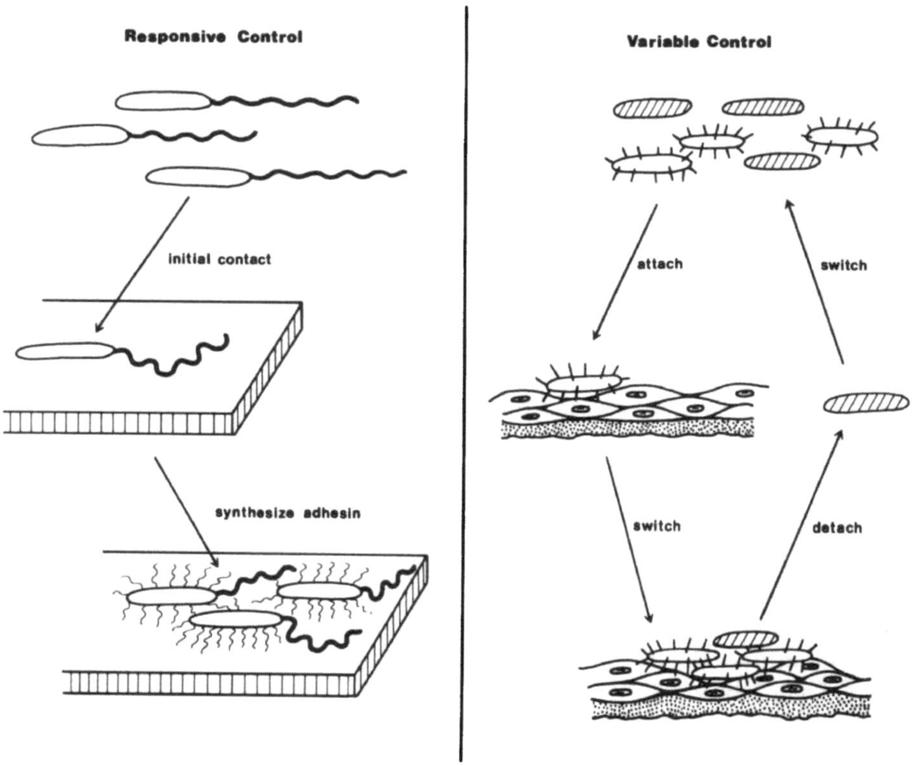

FIG. 3 - Strategies for control of adhesive substance expression.

Acknowledgement. Work on adhesion in our laboratory is supported by a contract from the Office of Naval Research.

REFERENCES

(1) Belas, M.R., and Colwell, R.R. 1982. Adsorption kinetics of laterally and polarly flagellated Vibrio. J. Bacteriol. 151: 1568-1580.

(2) Eisenstein, B.I. 1981. Phase variation of type 1 fimbriae in Escherichia coli is under transcriptional control. Science 214: 337-339.

(3) Eisenstein, B.I., and Dodd, D.C. 1982. Pseudocatabolite repression of type 1 fimbriae of Escherichia coli. J. Bacteriol. 151: 1560-1567.

(4) Freter, R. 1980. Prospects for preventing the association of harmful bacteria with host mucosal surfaces. In Bacterial Adherence, ed. E.H. Beachey, pp. 439-458. New York: Chapman and Hall.

(5) Gaastra, W., and DeGraaf, F.K. 1982. Host specific fimbrial adhesins of noninvasive enterotoxigenic Escherichia coli strains. Microbiol. Rev. 46: 129-161.

(6) Hall, M.N., and Silhavy, T.J. 1981. Genetic analysis of the ompB locus in Escherichia coli K-12. J. Mol. Biol. 151: 1-15.

(7) Jones, G.W., and Isaacson, R.E. 1983. Proteinaceous bacterial adhesins and their receptors. Critical Review Microbiol. 10(3): 229-260.

(8) Kaiser, D. 1979. Social gliding is correlated with the presence of pili in Myxococcus xanthus. Proc. Natl. Acad. Sci. USA 76: 5952-5956.

(9) Kehoe, M.; Winther, M.; and Dougan, G. 1983. Expression of a cloned K88ac adhesion antigen determinant: identification of a new adhesion cistron and role of a vector-encoded promotor. J. Bacteriol. 155: 1071-1077.

(10) Marshall, K.C.; Stout, R.; and Mitchell, R. 1971. Mechanisms of the initial events in the sorption of marine bacteria to surfaces. J. Gen. Microbiol. 68: 337-348.

(11) Meyer, T.F.; Mlawer, N.; and So, M. 1982. Pilus expression in Neisseria genorrhoeae involves chromosomal rearrangement. Cell 30: 45-52.

(12) Ofek, I., and Beachey, E.H. 1980. General concepts and principles of bacterial adhesion in animals and man. In Bacterial Adherence, ed. E.H. Beachey, pp. 1-29. New York: Chapman and Hall.

(13) Shapiro, L. 1976. Differentiation in the Caulobacter cell cycle. Ann. Rev. Microbiol. 30: 377-407.

(14) Silverman, M., and Simon, M. 1983. Phase variation and related systems. In Mobile Genetic Elements, ed. J.A. Shapiro, pp. 537-557. New York: Academic Press.

(15) Swanson, J. 1982. Colony opacity and protein II composition of gonococci. Infec. Immun. 37: 359-373.

(16) Swanson, J.; Kraus, S.J.; and Gotschlich, E.C. 1971. Studies on gonococcus infection. 1. Pili and zones of adhesion: Their relation to gonococcal growth patterns. J. Exp. Med. 134: 886-905.

(17) Yamada, K.M. 1983. Cell surface interactions with extracellular materials. Ann. Rev. Biochem. 52: 761-799.

(18) Zobell, C. 1943. The effect of solid surfaces upon bacterial activity. J. Bacteriol. 46: 39-56.

Standing, left to right:
Doug Caldwell, Gordon McFeters, James Bryers, Michael Bazin,
Ralph Schubert, Toyo Tanaka

Seated, left to right:
David Mirelman, Bill Characklis, Dave White, Ralph Mitchell, Daryl Lund

Microbial Adhesion and Aggregation, ed. K.C. Marshall, pp. 109-124. Dahlem Konferenzen 1984. Berlin, Heidelberg, New York, Tokyo: Springer-Verlag.

Biofilm Development and Its Consequences
Group Report

G.A. McFeters, Rapporteur
M.J. Bazin, D. Mirelman
J.D. Bryers, R. Mitchell
D.E. Caldwell, R.H.W. Schubert
W.G. Characklis, T. Tanaka
D.B. Lund, D.C. White

INTRODUCTION

The importance of cellular processes associated with interfaces is well established in many biological disciplines. In microbiology, however, this realization has been gained only recently for reasons discussed elsewhere (10). Biofilms are defined here as microorganisms and extracellular products associated with a substratum. Discussion of this group centered on the relevance of biofilm communities, the sequence of their development after cellular attachment, the properties and processes of biofilms along with methodological approaches to gain insights into these considerations, as well as techniques for controlling biofilms.

CONSEQUENCES OF BIOFILM FORMATION

The importance of biofilms is now acknowledged in a variety of settings (4, 6). The activity of microorganisms within these systems may result in a range of consequences. These include highly valuable processes and applications that have been exploited on the commercial scale and events of global consequence, as well as obscure interactions of unknown importance. On the other hand, microorganisms within some biofilms participate in processes that are usually recognized as non-beneficial or destructive. In each situation, however, the association of the participating organisms within the biofilm at an interface can result in reactions and processes that are not seen in unattached communities containing the same organisms. For example, experiments with adherent nitrifiers have indicated that attachment a) may result in the occurrence of multiple steady states, b) allows the microbial population to respond rapidly to increases in the nutrient concentration, and c) causes asymmetric transient behavior in response to perturbations of the system (1).

An exhaustive presentation of examples is not the purpose of this report. However, beneficial and detrimental consequences of biofilm activities from a number of systems are shown in Tables 1A and 1B. Published reports may be considered for a more comprehensive listing of specific examples (2-4, 6, 7). It must be recognized that new instances are being described with increasing frequency where biofilms are of significance in unanticipated ways, and many more undoubtedly await discovery.

TABLE 1A - Types of interactions in biofilms.

A. Nutrition and Development

1. Animal-animal interactions (e.g., rumen protozoa)

2. Animal-microbe interactions

   vertebrate (e.g., gut bacteria synthesize vitamin K, immunological effectors such as rheumatic fever)

   invertebrate
   - intracellular bacteria (legionella in amoeba)
   - protozoa and bacteria - termite gut
   - developmental cues (see Mitchell, this volume)

3. Plant-microbe interactions (e.g., mycorrhizas, rhizosphere
   - nodulation

4. Microbe-microbe interactions
   dental plaque

B. Pathology

1. Animal-animal (protozoan disease)
2. Animal-microbe (infectious disease)
3. Plant-microbe (blights)
4. Microbe-microbe (Bdellovibrio)

TABLE 1B - Consequences of biofilms from a process standpoint.

Process Influenced	Consequence	Application
Heat Transformations (energy)	Energy Losses	Power Industry, Chemical Processing Industry (CPI)
Fluid Flow (momentum)	Energy Losses	Pipelines Shipping Industry
Chemical Transformations (mass)	Material Deterioration	CPI, Power Industry, Shipping, Wastewater Treatment, etc.
	Potable Water Quality	Municipal Water Systems
	Natural Water Quality	Environmental Control
	Reduced Product Quality	Pulp and Paper, Steel, Electronics
	Product Formation	Biotechnology
	"Debris"	Sanitary/Hygienic

PROCESSES THAT DOMINATE IN BIOFILM DEVELOPMENT

Initiation and the subsequent development of a biofilm community follow the chemical conditioning of the surface and the adhesion of organisms or groups of organisms. Those events are described in detail elsewhere (see Characklis, Caldwell, Fletcher, Savage, and Rutter et al., all this volume). The processes important in biofilm development following adhesion, shown in Table 2, will be briefly discussed. All of these events and activities will fit into the framework of three sequential events: a) colonization, b) maturation, and c) detachment or sloughing.

Development of an active biofilm community is dependent on the metabolism and growth of attached cells upon the conditioned substratum. These processes are accompanied by the growth and reproduction of cells and the synthesis of exopolymers. This film may trap nutrients from the bulk phase, a process that is discussed elsewhere (2, 3). These activities lead to maturation of the film. One measurable property of this progression is an increase in the depth of many such communities. This results in alterations of microenvironments within the film, such as the development of anaerobic zones near the substratum due to a lack of oxygen diffusion through the film. Diffusion or other transport

TABLE 2 - Processes important in biofilm development.

Process	Mechanisms
Transport e.g., nutrients, cells	Chemotaxis Fluid dynamics Diffusion Surface forces
Attachment e.g., nutrients, cells, exopolymers	Induction of exopolymer formation Neutralization of surface change Adhesion (cell-substratum) Cohesion (cell-cell)
Transformation e.g., substrate, nutrients, cells	Interactions Substratum - biodeterioration - corrosion Lectin participation: Active transport Enzyme modulations; induction, repression Membrane changes (deformation) Microbe-microbe interactions Genetic exchange/nutrient exchange Microbial modification of local environment Redox potential, pH, osmotic, chelation, free radical formation, shift in polymer physical state and changed chemical structure. Predation Growth - reproduction Death Succession
Detachment e.g., cells, products	Change in physical structure of film Lysis Grazing Fluid dynamics

limitations may also restrict the availability of substrate to cells or accumulation of toxic substances in the deeper layers of the film. Other more subtle changes in the cells and exopolymers also occur in response to product accumulation and changes in the chemical and physical properties of the surrounding environment. Some of these physicochemical transformations in the exopolymer matrix can have profound implications for the community. These influences are discussed elsewhere (Hirokawa et al., this volume).

Accompanying the progressive development of the biofilm community are changes that promote ecological succession. Such alterations represent the evolution of new niches for the colonization by allochthonous organisms. Therefore, the development of biofilms may be considered from an ecological perspective. As in most ecosystems, there is an ecological succession on the colonized surface, leading to the formation of a climax or mature community (assuming that environmental conditions remain constant).

In many aquatic systems, the primary film is usually followed by the development of a population of prosthecate and filamentous bacteria within one or two weeks of active growth of the primary population (although prosthecate bacteria have been found as early colonizers of some substrata). This group of microorganisms is prevalent in nutrient-deficient bulk phases, where the stalks and filaments are utilized to concentrate nutrients. On wood substrata, a secondary population of fungi often develops, utilizing the substratum itself as a nutrient source.

During this secondary stage of biofilm development, exopolymer begins to accumulate and the biofilm takes on the macroscopic appearance of a gel. This gel acts as a trap for debris and other organic materials, sometimes including nutrients, from the bulk phase.

Ultimately, because of limitations of transfer through the gel, the inner layers of the biofilm may become anoxic. This new environment stimulates the development of an anaerobic microbial population in the reduced zone. Successions of this type have been documented (11).

In the final phase of biofilm development, protozoa may attach to the film. Very little is known about the specificity of protozoa living in biofilms. Questions concerning the existence of specific periphytic protozoa have not been addressed. Large populations of ciliate or flagellate protozoa are found in many mature biofilms. They graze on the bacterial population and appear to be important participants in the maintenance of a stable climax microbial community.

Interactions between species within biofilm communities have been described. Epilithic biofilms composed of both green algae and copiotrophic bacteria were seen in an extreme oligotrophic alpine system (9). The heterotrophs utilized extracellular products from the phototrophs for extensive reproduction. In this way, cross-feeding between two very different microorganisms within a biofilm allowed the growth of a copiotroph under oligotrophic nutrient conditions that were restrictive in the overlying water column. Interactions between organisms in biofilms might also involve genetic exchange and other forms of communication. However, little information exists in this area.

The mature biofilm community may exist in a relatively stable state. Sloughing of small amounts of the film occurs resulting in a balance between recruitment from reproduction, attachment, and biomass release. This loss of both cellular and exopolymer constituents, together and in patches, ranges in size from a few cells to larger macroscopic quantities of the film. Parts of the substratum may also be released in this process, e.g., intestinal mucosa and biodegradable materials. Eventually the biofilm may experience a catastrophic release where extensive portions are sloughed leading to a nearly denuded substratum. Such an event might be caused by excessive exopolymer or gas production in the anaerobic zone yielding pressure gradients and subsequent detachment of the film. Following this kind of event, only a relatively thin exopolymer layer may remain attached to the substratum. Subsequent recolonization may be enhanced on this type of substratum. These processes are discussed in more detail elsewhere (Characklis, this volume). Some of the events and processes may be described to some extent by the mathematical models of the type described by Characklis (see his Table 3) and Caldwell (both this volume).

ENVIRONMENTAL FACTORS INFLUENCING BIOFILM DEVELOPMENT

Biofilm formation is the result of dynamic, complex phenomena where all phases of the system are intimately affected by one another through various processes as discussed earlier. Any system in which biofilm develops can be separated into the following three distinct phases:

a. Bulk Fluid
b. Substratum - Living or Nonliving
c. Biofilm

When investigating some aspect of biofilm development, one cannot ignore the interaction between these three phases. Therefore, rather than listing the many biological (e.g., microbe concentration, type, age), chemical (e.g., nutrients, toxins, pH), and physical (e.g., temperature, fluid velocity) parameters and their effects on overall biofilm development, we will consider how the processes listed in Table 2 may affect the system. It is not our intention to describe the specific dependency of the developmental process on specific parameters.

Bulk Fluid

The bulk fluid with its associated boundary layers serves as a source of nutrients, ions, new or similar organisms, and thermal energy for the biofilm while also removing products of biofilm development. For that reason, the processes of transport and transformation are important (see Characklis, this volume). The bulk fluid phase can also influence biofilm development through parameters affecting detachment processes.

In the case of some living substrata, secretions (i.e., mucus, electrolytes, lectins, antibodies, enzymes, etc.) can be transported within the bulk phase, interact with microbial and host components, and influence the biological and chemical conditions, as well as future events. In addition, motility of the substratum surface and its ordered rate of periodic desquamation can have a significant effect on the nature of the biofilm.

Substratum

A nonliving substratum serves as the supporting surface for the biofilm and can, in irregular-shaped surfaces, provide "shelter" from otherwise adverse bulk phase conditions. The substratum surface texture and/or topography can influence biofilm development (12). These interactions deserve further investigation. The substratum can often supply thermal energy to the biofilm, having direct effects on (metabolic) transformation rates and transport processes. The chemical nature of the inert surface also clearly influences attachment early in development and can affect transformation processes throughout development. Certain substrata can also act as nutrient sources or as primary substrates. How properties of the nonliving substratum affect transport processes within the biofilm, either directly or indirectly, is not clear.

Living substrata differ from nonliving substrata through their active participation with the other two phases. A living surface may actively respond to environmental conditions and can alter those conditions at interfaces and maintain homeostasis.

Biofilm

The biofilm phase responds to environmental parameters communicated between the bulk phase and the substratum. Obviously, the biofilm will be influenced by all processes considered above. However, since research has focused primarily upon transformation processes (see Table 2), there is little information about how the physical, chemical, and biological constitution of the biofilm influences transport, attachment, and detachment processes.

Two situations exist that add complexity when considering environmental influences. First, the existence of three distinct phases in a biofilm system results in gradients within the system. For example, temperature gradients are found in heat exchangers, whereas substrate and oxygen concentration gradients are seen in most biofilm communities. Conditions prevailing in one phase can be entirely different in another part of the same phase or an adjacent phase. This characteristic of "patchiness" is typical of natural systems. Second, the rate and extent of processes influencing biofilm development are time-dependent, even in well controlled experimental systems. This consideration is even more important in natural communities.

PROPERTIES OF BIOFILMS

The two major components of the biofilm are the microbes and their extracellular products. Some of the more significant chemical and physical properties of these two components are given in Table 3A. For both the microbial and extracellular components, there is a three-dimensional, spatial heterogeneity. The quantitative description of the biofilm's microbial content must include the total cellular biomass and the community structure in the overall "anatomy" of the film. Knowledge of the dynamics of the community structure allows the quantitative description of shifts with community development and succession and allows comparisons between films.

Measures of microbial metabolic activities should accompany the description of the biomass and community structure in biofilms. Possibly the most informative measures of microbial metabolic activities are based on the rates of incorporation or turnover of isotopically labeled specific cellular components. The best isotopes may be the mass labels ^{13}C, ^{15}N, ^{18}O, and ^{2}H, although radioactive isotopes have been used (5). The mass labels are nonradioactive, relatively inexpensive, and available carrier free. Thus, concentrations of labeled precursors can be utilized that are just above environmental concentrations. The presence of these isotopes can be readily measured in specific parts of distinctive molecules by utilizing the extraordinary resolution of capillary-gas-liquid chromatography/mass spectrometry.

There is evidence that sampling methods for measuring microbial metabolic activities may introduce artifacts. Consequently, it is useful to measure the nutritional status in situ. Examples are given by White (this volume).

The microbes are clearly important, but their extracellular products may be even more significant in dictating salient features of biofilms.

TABLE 3A - Selected properties of biofilms.[a)]

1) Microbial components of biofilms:
   - biomass
   - community structure (succession)
   - metabolic activity
   - nutritional status

2) Extracellular components of biofilms:

Chemical properties	- primary (monomer compositions)
	- tertiary (specific configurations)

   Physical properties:
   - thickness
   - specific gravity
   - contact interaction
   - bulk modulus
   - shear modulus
   - Donnan potential
   - pore size of network
   - susceptibility to sol-gel transition

[a)] These properties need to be determined in three dimensions on a micro-scale.

In many biofilms, the exopolymer mass may exceed that of the microbes. The chemical structure of these films has been assessed primarily by hydrolysis and chemical analyses. This provides the primary structure. The molecular topology or tertiary structure (using terms for classical protein structure) can be studied using antibodies against the extracellular polysaccharides. Properties of these polymers (such as chelation kinetics and capacity) can also be measured as a function of salts, solvents, and pH. All of these techniques are attempts to describe the physical/chemical properties of these gels. An excellent example of techniques for assessing the properties of gels is the nondestructive laser-light scattering methodology which can provide dynamic and continuous monitoring of the phase properties and transitions of gels (13).

Several important physical properties of an extracellular gel may influence microbial cells: a) the forces the gel exerts on the cell, such as osmotic pressure, contact interactions, and Donnan-type electrostatic interactions, b) the response functions of gel such as osmotic bulk modulus, shear modulus, and susceptibility to sol-gel transition, and finally, c) the pore size of the gel which affects cell activity through its influence on transport phenomena. All of these properties can be derived from knowledge of the physical and chemical structure of a gel using the recently developed theoretical description of phase properties and phase transition of gels (13).

METHODS OF MEASURING THE PROPERTIES OF BIOFILMS

Some techniques to measure the properties of biofilms are listed in Table 3B. To date most of the methods have utilized destructive analysis in which the biofilm must be extracted, hydrolyzed, placed in a high vacuum, and/or bombarded with ions or electromagnetic radiation, probed with microprobes, stained, or shadowed. Some of the newer methods and their validation are reviewed by White (this volume). A review of the more established techniques is presented by Costerton and Colwell (5).

Table 3B - Techniques for measuring composition and properties of biofilms.

A. Destructive Methods
   Chemical analysis (includes isotope incorporation and turnover)
   Secondary Ionization Mass Spectrometry (SIMS)
   MS analysis of large fragments sputtered off
   Auger spectroscopy (Auger electron spectra maps of elemental density) X-ray - scatter; diffraction
   Microprobes and fiberoptic use of laser
   Microscopy
   Autoradiography
   Microcaloremetry
   Immunohistochemistry

B. Nondestructive Methods
   Activity measurements
   Dynamic laser light scattering spectroscopy (Hirokawa et al., this volume)
   Fourier transforming infrared spectrometry (FT/IR) (see White, this volume)
   Nuclear magnetic resonance spectroscopy
   Electron spin resonance (E.S.R.)
   Computer enhanced microscopy

Clearly, the study of important transient phenomena as the biofilms are shifted and subjected to perturbations requires measures that are nondestructive, rapid, and provide specific information. Preferably, these techniques should have sufficient resolution to examine small areas so that spatial heterogeneity can be monitored.

Dynamic laser light scattering spectroscopy will be a powerful tool in studying the physicochemical properties of extracellular gels (14). It is non-invasive and can scan volumes of $2 \mu m^3$ (1, 15). From the amplitude and relaxation rate of the scattered light intensity fluctuations, it is possible to determine the osmotic compressibility and viscous friction between the network and water. These quantities describe the state of a gel within its phase diagram. From the non-fluctuating portion of scattered light intensity, the structural nonhomogeneities permanently embedded in a gel can be determined.

The Fourier transforming infrared spectrometer (FT/IR) can give information about local environments within biofilms. At present the film must be grown on appropriate surfaces such as a germanium crystal and placed in the FT/IR. Because of its extraordinary sensitivity and speed, transient effects can be detected and the IR range expanded to include both the near and far IR. With present equipment the thickness of the films that can be dynamically followed is limited. However, the FT/IR can offer a compromise between the destructive and nondestructive analyses by providing a detailed analysis of biofilms on coupons that can be removed from reactors and examined either by attenuated total reflectance or the photoacoustic spectral unit in which the analysis is independent of the surface topology (8).

Nuclear magnetic resonance (NMR) requires that the biofilm be examined in a carefully controlled magnetic field. If those conditions can be met, the NMR allows a non-invasive mechanism in which to follow specific chemical reactions in microbial consortia as they occur. This should be an increasingly important method of analysis. Electron spin resonance also requires that the sample be placed in a carefully controlled electromagnetic field.

If the biofilm is thin and can be placed in the stage of a light microscope, the technique of computer-enhanced image processing can be adapted to give quantitative measures of growth, structure, etc., particularly if the microscope can be combined with a microspectrophotometer. It can also give information on the physical status of the extracellular gels if the laser light scattering system can be added to the apparatus. In addition, elipsometry can be utilized to measure biofilm thickness in the system. An essential point is that research and resources promoting the development of non-invasive, nondestructive methods of dynamically following the biofilm formation and maturation are necessary if this important area is to progress.

USE OF MODELS TO UNDERSTAND BIOFILM PROCESSES

Previous sections of this report dealt with the relevance of biofilms as well as characteristics and processes of known significance within interfacial communities. The influence of environmental factors was also introduced. As a result, biofilms emerge as complex, multiphasic systems with a variety of internal and environmental interactions. Therefore, the application of models in biofilm research is introduced at this point as a means of increasing our understanding of biofilm communities.

Models may be physical structures, verbal descriptions, or sets of mathematical equations. In some instances models serve as hypotheses, in which case they may be regarded as fundamental to the application of the scientific method. Other types of models are designed to predict the behavior of systems or to help us understand how it operates (heuristic models), or to serve simply as sets of instructions. In the last class, flow charts of computer algorithms and road maps might be included. In no case will a model reflect all aspects of "reality." That is not the function of modeling, nor is its accomplishment a reasonable expectation. A road map details only some small fraction of the geographical area it represents, but this does not detract from its usefulness in showing us how to travel from point A to point B.

Mathematical models can be classified as statistical, lumped-parameter, and phenomenological. Statistical models are "black box" models which are of limited value for extrapolation or prediction. Nevertheless, these models can be quite useful for progressing to more defined experiments. Statistical models can be useful in rather undefined environments for describing overall performance or processes in a restricted range of environments. Phenomenological models attempt to describe a process with physical laws and laws of mass action using intrinsic coefficients to describe process rate and stoichiometry (see Characklis, this volume). Lumped-parameter models are somewhere between statistical and phenomenological models. Both phenomenological and lumped-parameter models are useful for conceptually breaking down a complex process into several relatively fundamental processes described by "intrinsic" coefficients. One very useful application of this type of analysis is the determination of the rate-limiting step in different environments. The behavior of many systems is time-dependent. This is why so many mathematical models take the form of sets of differential equations. In the study of biofilms, the state variables often vary with respect to both time and distance (for example, depth in a trickling filter or in the biofilm itself). In this case, partial differential equations are used.

All models, no matter what their type, should force the investigator to think in a concise manner and consolidate what information is known plus what new information is required. In this way, modeling represents a vehicle to assist in the use of the scientific method. When applied to complex multidimensional systems such as biofilms, models may prove to be of significant value in assessing the interactions between variables and secondary effects.

METHODS FOR BIOFILM CONTROL

Control of biofilm development can be either natural (through one or more of the processes contributing to its formation) or artificial with human intervention. In this context, biofilm control can be defined as a directed response of the biofilm to an imposed stimulus. The intended response may be classified as a) enhancement, b) retardation, or c) destruction. This can encompass either the overall biofilm development, a specific activity of the biofilm, or a chemical/biological component within the biofilm. The study of natural control measures (e.g., sloughing, grazing, etc.) in films on nonliving surfaces or in living systems may lead to more efficient artificial control measures.

Enhancement may refer to increasing the rate of biofilm development or activity. For example, when using slow-growing microorganisms in a film reactor, it may be desirable to decrease the start-up time or the enrichment and maintenance of specific populations within a mixed culture biofilm. It would also be useful to know if chemical transformations, mediated by certain species, can be maximized by controlling the substratum, inoculum, and/or environmental factors. More research is needed in determining how to enhance the rate of biofilm development and activity, as well as the influence of selection pressures within a mixed culture biofilm.

Retardation is the intentional reduction of undesirable biofilm activities. In fixed film reactors excessive thickness is undesirable due to the possibility of substrate transport limitations. Biomass capture particles (Atkinson, this volume) can be used to retard excessive biofilm thickness. Manipulation of the substratum composition, the bulk fluid, and mechanical cleaning are the usual means of control. In some systems with living and nonliving substrata, biofilm development and activity can be retarded physically (abrasive cleaning, i.e., brushing of teeth, or heat exchange surfaces) or chemically (retardant polymer applications to teeth, or continuous oxidant dosage). Chemical control methods are still haphazardly applied to many biofilms on nonliving substrata and represent an important area for research along with the development of safer retardation methods that can be applied to both living and nonliving substrata. The control methods for biofilms from living substrata include enzyme inhibitors, antibodies and lectins, and chelation of crucial nutrients. Cellular control may be mediated by macrophage leukocyte systems or by the maintenance of normal biota.

Destruction is the removal and/or inactivation of the biofilm and may be accomplished by physical or chemical techniques. Physical methods (e.g., brushing of surfaces such as teeth, heat exchanges, and correction of obstructions such as gallstones and urinary calculi) insure relatively complete cleaning but may also abuse the substratum. Chemical destruction methods (e.g., oxidants, biocides, and drugs) are not totally efficient. Biocides (or drugs) may or may not completely terminate all activity within a biofilm. Furthermore, an inactive biofilm may stimulate regrowth of the biofilm. Chemical destructive agents are often applied with little understanding of their relative dosage requirements or resultant effects on the remaining biofilms.

MAINTENANCE OF BIOFILMS

In many living surfaces, the best protection against disease is the maintenance of the normal biota to preclude attachment and invasion by pathogens. This is being increasingly studied in medicine (Savage, this volume) and may become a very useful technique in controlling plant disease (Dazzo, this volume). A better understanding of the mechanism by which biofilms are maintained in living systems may be applied to important nonliving substrata as well.

RESEARCH NEEDS

1. The influence of gradients (mass, momentum, energy) and interactive forces in biofilm systems.
2. Development of theories, unifying concepts, and methods for analysis of biofilm systems including heterogeneous (mixed species) communities.
3. Influence of perturbations on the biofilm environment and processes.
4. Mechanisms and kinetics of particle transport in fluid flow systems.
5. Factors and mechanisms that control sloughing and detachment.
6. Development and application of analytical procedures using mass labeled precursors in study of film metabolism and dynamics.
7. NMR spectrometers for studying biofilms on the scale of microns.

8. Measurements of corrosion on the size scale of microorganisms (microns).
9. Chemical communication within film communities.
10. Genetic exchange between microbes within biofilms.
11. Biological (natural) control strategies that may be used with biofilms.
12. Methods to determine the demand for oxidizable biocides (i.e., Cl^-) in the fluid and substratum phases.
13. The influence of the plane of cell division on subsequent development of biofilms.
14. Variability of polymer gel characteristics with biofilm development.
15. Factors that affect rates of cellular detachment and attachment to biofilms.
16. Definition of biofilm maturation in terms of biological and/or physical changes.
17. The maintenance of steady state within biofilms.
18. The role of the gel phase in biofilm community nutrient dynamics.

REFERENCES

(1) Bazin, M.J.; Cox, D.J.; and Scott, R.I. 1983. Nitrifers in a column reactor: limitations, transient behavior and effect of growth on a solid substrate. Soil Biol. Biochem. 14: 477-487.

(2) Berkeley, R.C.W.; Lynch, J.M.; Melling, J.; Rutter, P.R.; and Vincent, B., eds. 1980. Microbial Adhesion to Surfaces. Chichester: Ellis Horwood.

(3) Bitton, G., and Marshall, K.C., eds. 1980. Adsorption of Microorganisms to Surfaces. New York: John Wiley and Sons.

(4) Characklis, W., and Cooksey, K.E. 1983. Biofilms and microbial fouling. Adv. Appl. Microbiol. 29: 93-138.

(5) Costerton, J.W., and Colwell, R.R., eds. 1979. Native Aquatic Bacteria: Enumeration, Activity and Ecology. Philadelphia, PA: American Society for Testing and Materials.

(6) Costerton, J.W.; Irvin, R.T.; and Chang, K.J. 1981. The bacterial glycocalyx in nature and disease. Ann. Rev. Microbiol. 35: 299-324.

(7) Geesey, C.G. 1982. Microbial exopolymers: Ecological and economic considerations. Am. Soc. Microbiol. News 48: 9-14.

(8) Griffiths, P.R. 1977. Chemical Infrared Fourier Transforming

Spectroscopy. New York: John Wiley and Sons.

(9) Haack, T.K., and McFeters, G.A. 1982. Nutritional relationships among microorganisms in a epilithic biofilm community. Microbial Ecol. 8: 115-126.

(10) Marshall, K.C. 1976. Interfaces in Microbial Ecology. Cambridge, MA: Harvard University Press.

(11) Nickels, J.S., et al. 1981. Effect of manual brush cleaning on biomass and community structure of microfouling film formed on aluminum and titanium surfaces exposed to rapidly flowing seawater. Appl. Envir. Microbiol. 41: 1442-1453.

(12) Nickels, J.S., et al. 1981. Effect of silicate grain shape, structure, and location on the biomass and community structure of colonized marine microbiota. Appl. Envir. Microbiol. 41: 1262-1268.

(13) Tanaka, T. 1981. Gels. Sci. Am. 244: 124-138.

(14) Tanaka, T.; Hocker, L.O.; and Benedek, G.B. 1973. Spectrum of light scattered from a viscoelastic gel. J. Chem. Phys. 59: 5151-5159.

(15) Tanaka, T.; Ishiwata, S.; and Ishimoto, C. 1977. Critical behavior of density fluctuations in gels. Phys. Rev. Lett. 38: 771-774.

Microbial Adhesion and Aggregation, ed. K.C. Marshall, pp. 125-136. Dahlem Konferenzen 1984. Berlin, Heidelberg, New York, Tokyo: Springer-Verlag.

Surface Colonization Parameters from Cell Density and Distribution

D.E. Caldwell
Dept. of Applied Microbiology and Food Sciences
University of Saskatchewan
Saskatoon, Saskatchewan S7N 0W0, Canada

Abstract. Population growth and cell adhesion can occur simultaneously during surface colonization. Consequently, it is difficult to determine whether changes in colonization rate are due primarily to growth, adhesion, or both. To solve this problem, attempts have been made to calculate the rates of growth and attachment from the characteristic cell distribution pattern formed by cells as they colonize surfaces. This approach has been used to study colonization in the chemostat, where the growth rate of planktonic cells is set by the dilution rate and compared to that of cells colonizing immersed surfaces. These studies assume that growth and attachment are the primary processes responsible for the observed distribution pattern. They also assume that the mean rates of cell growth and attachment are initially constant. The accuracy of these assumptions have been tested in a limited number of surface environments. It appears from these studies that colonization might be more accurately described by defining the cell attachment rate as a function of cell surface density and by defining the growth rate as a function of microcolony size. However, a substantial improvement in the accuracy of methods used to gather population data is necessary to define adequately these functions and determine whether they are useful. Culture studies in continuous-flow capillaries, using on-line image processing techniques, may be helpful in this respect.

DERIVATION OF SURFACE COLONIZATION KINETICS

The colonization of surfaces by microorganisms is due to many contributing factors. In the early stages of colonization, the attachment of cells and their subsequent growth are primarily responsible (5). If it is assumed that the mean growth and attachment rates are initially constant, this can be expressed mathematically as:

$$dN/dt = \mu N + A, \quad (1)$$

where A is the attachment rate (cells field^{-1}h^{-1} or colonies field^{-1}h^{-1}), N is the number of cells on the surface (cells field^{-1}), t is the time of surface exposure (h), and μ is the specific growth rate (h^{-1}). Integrating this equation gives the following surface colonization equation:

$$N = (A/\mu)\, e^{\mu t} - (A/\mu). \quad (2)$$

The effect of integration is to sum continuously the number of cells attaching and their progeny. Unfortunately, the surface colonization equation cannot be rearranged to solve for μ and obtain a surface growth rate equation, because μ is present as both an exponent and a term in the equation. However, if A, N, and t are known, a computer program can be used to search for a value of μ which satisfies the equation.

To obtain a surface growth rate equation, the distribution of cells among microcolonies must be considered (7, 8). The rate of change in the number of one-celled microcolonies is equal to the attachment rate minus the rate of formation of two-celled microcolonies. In other words, the growth of a one-celled colony results in the loss of the one-celled colony and the gain of a two-celled colony. The rate of change in the number of one-celled microcolonies is thus equal to the attachment rate (A) minus the growth rate of one-celled colonies (μC_1). This is expressed mathematically as:

$$dC_1/dt = A - \mu C_1, \quad (3)$$

where C_1 is the number of one-celled microcolonies per field.

Initially, all of the microcolonies on the surface will contain one cell. The number of one-celled colonies will increase in number at a rate A. However, as the number of these colonies increases the C_1 term in Eq. 3 will increase due to the increase in the number of one-celled colonies (C_1) as cells attach to the surface. The value of μC_1 will thus continue to increase until a condition develops when the rate of formation of one-celled colonies (A) equals the rate of loss of two-celled colonies (μC_1). Thus:

$$O = A - \mu C_1. \tag{4}$$

Rearranging Eq. 4 gives:

$$\mu = A/C_1. \tag{5}$$

A more general form of Eq. 5 can be written by replacing C_1 with C_i:

$$\mu = A/C_i, \tag{6}$$

where C_i is the number of colonies present in each exponential size class (i.e., size classes of 2, 4, 8, 16, or 32 cells per colony) per field.

This generalization is possible because the number of colonies present in each size class will approach a constant, C_i. For example, the number of two-celled colonies will initially be zero but will gradually increase until the rate of formation of two-celled colonies approaches the rate of formation of four-celled colonies (8). The same is true of eight-celled colonies, sixteen-celled colonies, etc.

Substituting A/C_i for μ in Eq. 2 using Eq. 6, and rearranging, gives the surface growth rate equation:

$$\mu = \frac{\ln (N/C_i + 1)}{t} . \tag{7}$$

Using this equation, the specific growth rate can be calculated from the density of cells on the surface and the density of microcolonies falling in each size class (C_i). By rearranging Eq. 4 to give Eq. 8 the cell attachment rate can also be calculated:

$$A = (\mu)(C_i). \tag{8}$$

The surface growth rate equation thus provides a means of calculating the specific growth rate from the density of cells present on the surface (N) and the density of colonies present in each colony size class (C_i).

The application of the surface growth rate equation is illustrated in Fig. 1. In this series of diagrams the distribution of cells is shown each hour for four hours if three cells attach per hour and each attached cell doubles once each hour. By using the surface growth equation it is possible to

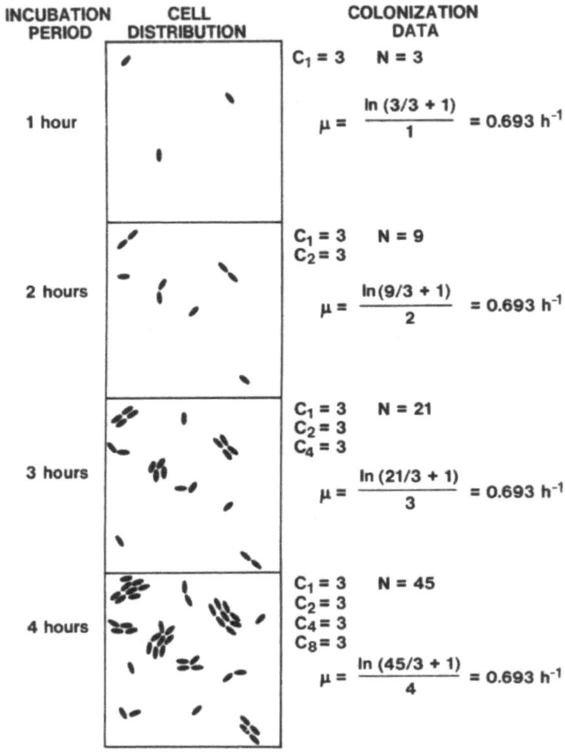

FIG. 1 - Derivation of growth and attachment rate from cell density and distribution. The distribution and density is shown each hour for four hours if three cells attach per hour and each attached cell doubles once each hour as shown; the specific growth rate can be calculated at any point in the incubation. The diagram represents a numerical solution of Eq. 1 and thus cannot be used to predict accurately the attachment rate.

calculate the specific growth rate at any point in the incubation.

APPLICATION OF COLONIZATION KINETICS IN SPECIFIC ENVIRONMENTS

The surface growth rate equation (Eq. 7) and colonization equation (Eq. 2) represent the most elementary mathematical descriptions possible for microbial populations which simultaneously grow and attach during surface colonization. These equations may be adequate in some circumstances but will require additional terms to describe more complex situations. As a result, it is important to determine whether they are sufficient for specific applications. This can be done by evaluating the goodness of fit for the colonization equation (Eq. 2). The linearized plot of Eq. 2 (N versus $e^{\mu t}$) produces a straight line with slope, A/μ, and intercept - (A/μ). Alternatively, both the surface colonization equation and growth rate equation can be tested by determining whether the same number of microcolonies fall in each size class (i.e., if C1 = C2 = C4 = C8, etc.). This number should approach a constant, Ci, if the equations adequately fit the data (7, 8). In this case, the above plot would have a slope equal to Ci ($Ci = A/\mu$) and an intercept of - Ci. Thus, by comparing the slope and intercept of this plot to the actual number of colonies occurring in each size class, it would be possible to determine whether this number does approach Ci as predicted by colonization kinetics (7) and computer simulation (8). However, this analysis has not yet been performed experimentally.

In some cases the presence of reversibly attached, moribund, or dead unicells may interfere with colonization kinetics. Under these circumstances, the number of one-celled microcolonies (attached, growing unicells) can be assumed to equal Ci and this value added to the number of cells in larger colonies to obtain a value N' (N corrected for unicells that are present but not growing or that will subsequently detach), to be used in place of N (total number of cells observed) to calculate a corrected value for μ and A. If the data do not otherwise meet the tests described above, then the surface growth rate equation is not sufficient and additional terms as well as possible substitutions may be required in Eq. 1.

Despite potential difficulties in applying the proposed kinetics in situ, they avoid the pitfall of earlier methods (1-3) which fail to adequately account for the progeny of cells which attach late in the incubation and erroneously inflate growth rates despite UV-irradiated control slides (5).

COLLECTION AND EVALUATION OF DATA

Cells present on surfaces are stained with acridine orange and observed using epifluorescence microscopy (4, 6, 9, 10). If communities are to be studied, then each population having a unique growth and attachment rate must be identified and counted separately using fluorescent conjugated antisera or morphological identification. If the individual populations cannot be distinguished, then an apparent value for A and μ can be obtained for the entire community as a whole. Such values have no exact meaning.

However, they may be useful in comparing the "surface colonization potential" of one habitat to that of another, just as Michaelis-Menten kinetics are used to compare the "heterotrophic potential" of various aquatic environments.

In determining the number of colonies present in each size class, it is assumed that not all colonies will remain synchronous and that colonies outside of these classes occur. Thus instead of observing only colonies with 1, 2, 4, 8, 16, or 32 cells, it is possible to find colonies that fall between these values. As a result, the number of colonies in each size class must also include asynchronous colonies. For example, to determine the number of colonies falling in the four-celled size class (C_4), it is necessary to include half the number of colonies with three cells, all the colonies with four and five cells, as well as half the number of colonies with six cells. Four-celled colonies are considered optimal for determination of C_i, since these are not likely to be mistaken for dead or moribund cells, and they are not so large as to alter the growth rate of cells located within the colony.

Previous studies have shown that variations in attachment rate can occur from one microenvironment to another, either on the same surface or between surfaces. Despite these variations, μ may be relatively constant throughout. Consequently, if data are collected for a large number of fields and averaged, both μ and A will have a high standard deviation. However, it is only the variability in A which is responsible. As a result, both μ and A must be calculated from data collected within each microenvironment (preferably within each microscope field). These values can then be averaged subsequently. Thus if the growth rate of the organism is the same between fields but the attachment rate varies, an accurate mean would still be obtained for both, and the standard deviation would reflect the variability in each. If the data from all fields is averaged before performing the calculations, it is not possible to differentiate accurately growth from attachment, since one of the assumptions mentioned earlier, that A and μ are constant, would not necessarily have been met for data collected between fields, but would be met for data collected within fields. This approach makes it possible also to study differences between differing microenvironment sites. For example, it would be possible to determine the growth and attachment rates for bacteria colonizing the stomata of plant leaves versus other plant cells on the leaf surface.

EVALUATION OF COLONIZATION KINETICS IN CONTINUOUS CULTURE SYSTEMS

Crystals of calcite and pyrite have been studied in thiosulfate-limited continuous cultures (9). The growth rate of the planktonic cells was set at 0.21 per hour by controlling the dilution rate (flow/volume). Thermothrix thiopara, an extremely thermophilic sulfur autotroph isolated from thermal springs, was used as the test organism (4). It was found to colonize pyrite at a much more rapid rate than calcite. The question was whether this net difference was due specifically to attachment, growth, or both. As can be seen in Fig. 2a, the rate of colonization of the pyrite crystal was more than twice the rate for the calcite crystal. However, as shown in Figs. 2b and 2c, the growth rate on both minerals was the same and only the attachment rate differed, the rate for pyrite being more than twice that for calcite. Without the use of colonization kinetics it would not have been possible to differentiate between growth and attachment, nor would it have been possible to study the difference between the growth rate of planktonic and attached cells within a chemostat culture. Similar studies have been conducted using Pseudomonas aeruginosa (10). These suggest that it may be necessary to substitute functions for μ and A in Eq. 1.

Although previous surface colonization studies have relied entirely on traditional chemostat cultures, a new system has been developed which provides additional information and improved control of both the colonization and degradation of particulate substrates. This is a dual-dilution rate continuous culture, and it is presently being used to study microbial degradation of microcrystalline cellulose in suspension. It makes possible the dilution of particulate and aqueous phases of the culture simultaneously, but at different rates. As shown in Fig. 3, this is accomplished using a computer-directed continuous culture. Cellulose is the rate-limiting substrate and is continuously diluted using both supply and outflow pumps. Thus, the mean residence time (volume/flow) can be set to control the mean stage of colonization for the average particle. Unlike the traditional chemostat, the growth rate of the organisms is always at μ_{max} because they are colonization-limited rather than solute-limited. In addition, by using computer control, a reiterative cycle is established in which the cellulose pumps and culture stirrer are turned off at regular intervals allowing the cellulose to settle. An additional pump and filter system then draws off medium while leaving cells and cellulose within the culture. The liquid removed is replaced using a fourth pump. Thus the dilution rate of the solutes can be set at a different rate than that of the particles. This makes it possible to divert cellulase enzyme from cellulose adsorption to the solute outflow. For example, if the solute dilution rate is set at 0.5, then 50% of the free cellulase, normally released into the medium and subsequently adsorbed to cellulose, is diverted to the outflow. This makes it possible to evaluate the importance of soluble enzymes vs. surface-associated enzymes during the colonization process. It is also possible to use a solute outflow filter which passes bacteria but not cellulose particles. In this case, the cell attachment rate is reduced by a known proportion just as in the case of the cellulase adsorption rate. This permits total control and quantitation of the colonization process. The system is useful in evaluating

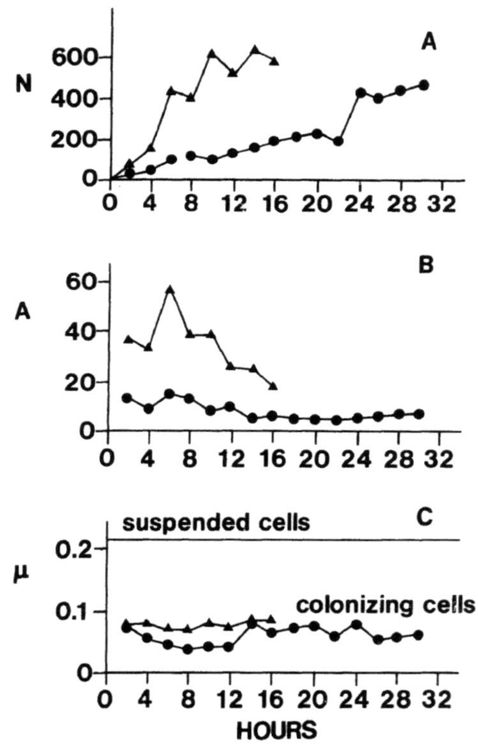

FIG. 2 - Colonization of pyrite (▲) and calcite (●) by Thermothrix thiopara in a thiosulfate-limited chemostat. N = cell density on surface, A = attachment rate to surface, μ = specific growth rate. The rapid colonization of pyrite (A) is due to rapid attachment (B) since the growth rate on both calcite and pyrite is the same (C)(Reproduced with the permission of the Geomicrobiology Journal (9)).

the kinetics of microbial colonization because it provides the independent control of both growth and attachment rates, which is needed to test surface growth kinetics over the full range of potential test values. The attachment rate is controlled by dilution of planktonic cells, the growth rate is controlled by limiting an inorganic nutrient in the culture reservoir.

In this system, the rate of change in the density of planktonic cells (N_p) is equal to the rate of emigration of cells from particles (E), plus the rate of growth of planktonic cells (μNp), minus the rate of immigration to particles (I), minus the rate of dilution (D Np). Mathematically:

$$\frac{dNp}{dt} = E + \mu Np - I - DNp. \tag{9}$$

At steady state:

$$O = E + \mu Np - I - DNp. \tag{10}$$

Rearranging:

$$\frac{E + \mu Np - I}{Np} = D. \tag{11}$$

Thus by setting the dilution rate D, it is possible to set the fraction of planktonic cells lost to washout per unit time. If D is set at 0.2 h^{-1}, then 20% of the planktonic cells will be washed out each hour instead

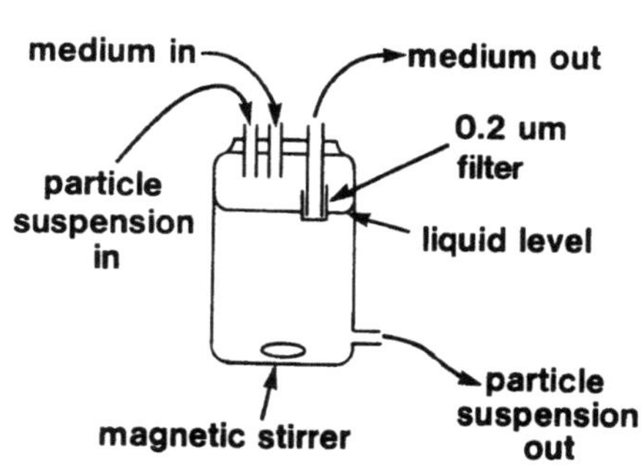

FIG. 3 - Diagram of dual-dilution continuous culture. The particulate fraction is diluted at one rate (flow of particle suspension per culture volume), while the aqueous phase is diluted simultaneously but at a different rate (flow of liquid per culture volume).

of attaching, and the rate of immigration to particles will be reduced proportionately.

In most continuous cultures, surface hydrodynamics are difficult to control and define. Either turbulent or laminar flow can occur at surfaces and is usually determined by the Reynold's Number (<2100 - laminar flow, > 4,000 - turbulent flow). In both turbulent and laminar flow situations, the water near surfaces is relatively quiescent compared to the bulk phase. Within a micrometer of the surface, the flow rate is near zero, and it gradually increases with distance from the surface. While bacterial motility is rarely significant in the bulk phase, it may be important in these various surface boundary layers. It is thus important to control and define this layer in future surface colonization studies if they are

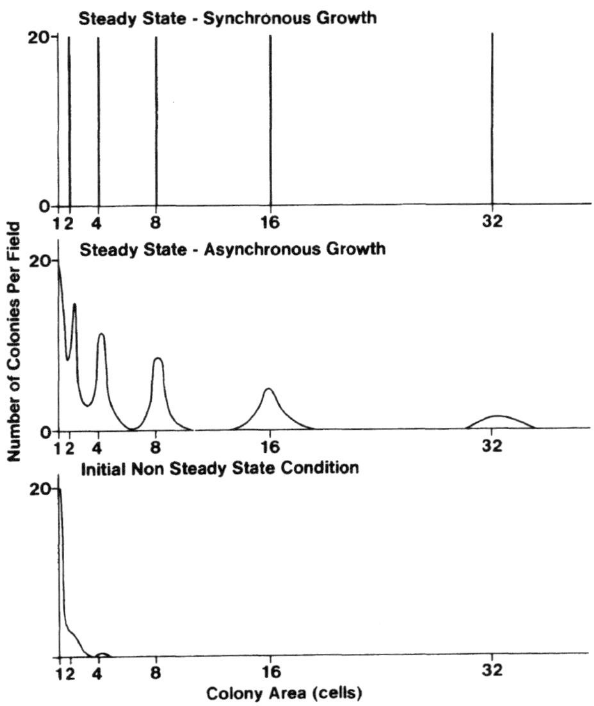

FIG. 4 - Distribution of surface microcolonies in various size classes assuming synchronous growth and asynchronous growth and under initial non-steady state conditions before colonization reaches a steady-state condition.

to be duplicated and correctly interpreted. The continuous flow capillary cultures of Perfilev and Gabe (11) provide an ideal system for such studies. These are very fine rectangular capillaries incubated and observed continuously under the microscope. If this method were combined with computer image analysis, it might be possible to describe more accurately the process of surface colonization under highly defined conditions.

Future studies will rely more heavily on the use of computer image analysis and enhancement in this and other systems. It will also become important to revise data collection procedures to take advantage of the full capacity of image processing. This has resulted in several improvements in the collection of colonization data. Standard image analysis software can dilate and erode the image to maintain the same total colony area. This merges cells less than half a cell length apart into a single unit which it recognizes as a colony. The size distribution of these units is expressed in terms of cell units and plotted as shown in Fig. 4. According to the assumptions of the surface growth rate equation, the size distribution of colonies should not be random but fall into peaks corresponding to 1-, 2-, 4-, 8-, and 16-celled microcolonies (Fig. 4). By analyzing the number and size of entire microcolonies, lower magnifications can be used, resulting in increased depth of field. This type of analysis may be essential in collecting sufficient population data to evaluate adequately existing colonization kinetics as well as to define new functions.

Acknowledgements. Supported in part by an operating grant from the Natural Sciences and Engineering Research Council of Canada. J. Malone is acknowledged for review of the manuscript.

REFERENCES

(1) Bott, T.L. 1975. Bacterial growth rates and temperature optima in a stream with a fluctuating thermal regime. Limn. Ocean. 20: 191-197.

(2) Bott, T.L., and Brock, T.D. 1970. Growth and metabolism of periphytic bacteria: methodology. Limn. Ocean. 15: 333-342.

(3) Bott, T.L., and Brock, T.D. 1970. Growth rate of Sphaerotilus in a thermally polluted environment. Appl. Microbiol. 19: 100-102.

(4) Brannan, D.K., and Caldwell, D.E. 1982. Evaluation of a proposed surface colonization equation using Thermothrix thiopara as a model organism. Microb. Ecol. 8: 15-21.

(5) Caldwell, D.E.; Brannan, D.K.; Morris, M.E.; and Betlach, M.R. 1981. Quantitation of microbial growth on surfaces. Microb. Ecol. 7: 1-12.

(6) Caldwell, D.E.; Kieft, T.L.; and Brannan, D.K. 1984. Colonization of sulfide-oxygen interfaces on hot spring tufa by Thermothrix thiopara. Geomicrobiol. J. 3: 181-200.

(7) Caldwell, D.E.; Malone, J.A.; and Kieft, T.L. 1983. Derivation of a growth rate equation describing microbial surface colonization. Microb. Ecol. 9: 1-6.

(8) Kieft, T.L., and Caldwell, D.E. 1983. A computer simulation of surface microcolony formation during microbial colonization. Microb. Ecol. 9: 7-13.

(9) Kieft, T.L., and Caldwell, D.E. 1983. Chemostat and in situ colonization kinetics of Thermothrix thiopara on calcite and pyrite surfaces. Geomicrobiol. J. 3: 217-229.

(10) Malone, J.A., and Caldwell, D.E. 1983. Evaluation of surface colonization kinetics in continuous culture. Microb. Ecol.

(11) Perfilev, B.V., and Gabe, D.R. 1969. Capillary Methods of Investigating Microorganisms. Transl. J.M. Shewan. Toronto: University of Toronto Press.

Microbial Adhesion and Aggregation, ed. K.C. Marshall, pp. 137-157. Dahlem Konferenzen 1984. Berlin, Heidelberg, New York, Tokyo: Springer-Verlag.

Biofilm Development: A Process Analysis

W.G. Characklis
Institute for Biological and Chemical Process Analysis
Montana State University, Bozeman, MT 59717, USA

Abstract. This paper will provide a framework for understanding the process of biofilm development in the context of stoichiometry and kinetics. Biofilm development is described in terms of selected fundamental rate processes and environmental parameters which influence their rate and extent. The properties of the biofilm and its microenvironment lead to topics of microbial ecology within the biofilm and the physiology of the organisms immobilized within it. These topics will be discussed in terms of unstructured models for the microbial processes.

BIOFILMS

Microbial cells attach firmly to almost any surface submerged in an aquatic environment. The immobilized cells grow, reproduce, and produce exopolymers which frequently extend from the cell forming a tangled matrix of fibers which provide structure to an assemblage which is termed a biofilm. Biofilms sometimes provide a relative, even coverage of the wetted surface and at other times are quite "patchy." Biofilms can consist of a monolayer of cells or can be as much as 30-40 cm thick as observed in algal mats. In the case of thick biofilms such as the algal mats, the biofilm can contain both aerobic and anaerobic environments due to diffusion limitations within the reacting mat. Both aerobes and anaerobes, (e.g., sulfate-reducers) produce significant quantities of exopolymers. Consequently, the term biofilm does not necessarily reflect a surface accumulation which is uniform in time and/or space.

In the simplest case, biofilms are composed of microbial cells and their products such as exopolymers. Such a biofilm generally is a very adsorptive and porous (≥ 95% water) structure. As a result, biofilms observed in many natural environments consist of a large fraction of adsorbed and entrapped materials such as solutes and inorganic particles (e.g., clay, silt). Many such deposits are found to contain less than 20% volatile mass, suggesting that the deposit organic content is a minor component, an observation which may be far from the truth. Organic materials have a specific gravity of approximately 1.05 (cells and carbohydrates), whereas the inorganic materials may have a specific gravity of as much as 2.5. As a result, the organic material is a more significant volume fraction than mass fraction of the biofilm. In any case, the organic matrix may be necessary to bind the inorganic components into a coherent deposit. Consequently, the term biofilm may reflect a surface accumulation which is composed of a significant fraction of inorganic or other abiotic substances held together by a biotic organic matrix.

Biofilms serve beneficial purposes in the natural environment as well as in some modulated or engineered biological systems. For example, biofilms are responsible for removal of soluble and particulate "contaminants" from natural streams and in wastewater treatment plants (fixed film biological systems such as trickling filters, rotating biological contactors, and fluidized beds). Biofilms or mats in natural waters frequently determine water quality by influencing dissolved oxygen levels and serve as a sink for many toxic and/or hazardous materials. These same mats may play a significant role in the cycling of the elements. Biofilm reactors are used in some common fermentation processes (e.g., "quick" vinegar process) and may find considerably more application in the near future (Atkinson, this volume).

Most recent attention focused on biofilms, however, reflects their nuisance roles. Fouling refers to the undesirable formation of inorganic and/or organic deposits on surfaces, which results in unsatisfactory equipment performance or reduces equipment lifetime. Biofilms cause fouling of industrial equipment such as heat exchangers, pipelines, and ship hulls, resulting in reduced heat transfer, increased fluid frictional resistance, and/or increased corrosion. Fouling is of commercial concern in the manufacture of miniaturized electronics, toilet bowl cleaners, rolled steel, and paper. In the medical or dental areas, biofilms are responsible for health problems.

PROCESS ANALYIS

Process analysis, in the context of this paper, refers to the application of thermodynamic and kinetic principles to the recognition and definition of various processes contributing to biofilm accumulation, microbial activity within the biofilm, and biofilm persistence in a given environment. Specifically, the use of mass and energy balances have been most useful in assessing the importance of various fundamental processes contributing to the accumulation and activity of a biofilm. The process analysis generally requires a) mathematical specification of the problem for the given physical situation, b) development of a mathematical model, c) experimental testing of the model, and d) synthesis and systematic presentation of results to ensure full understanding. The process denotes an actual operation(s) or treatment(s) of materials as contrasted with the model which is a mathematical description of the process.

The basis for the process analysis rests in the conservation equations: mass and energy conservation derived from thermodynamics and momentum conservation from Newton's laws of motion. The most important results from a process analysis of a reacting system are expressions which quantitatively describe the rate and stoichiometry of the fundamental reaction processes contributing to the overall process. These expressions, ideally, are useful, regardless of the physical factors imposed on the system (e.g., geometry, flow rates).

In contrast, much of the previous emphasis on biofilm processes has concentrated on biological and chemical aspects of mechanism without resorting to rate and stoichiometric analyses. But the physical, chemical, and biological processes of interest in biofilm development are completed in a certain period of time. For biofilm development, a specified change in fouling of a heat exchanger may signal the shutdown of manufacturing operations and the beginning of cleaning operations. The rate and extent of biofilm development on a tooth surface is reduced by brushing the tooth twice a day. The time required for any specified change is inversely proportional to the rate at which the process occurs. Thus, the rate must play a most important role in the process analysis.

In most, if not all, reported research on biofilms, certain observed or measured quantities are reported: net biofilm accumulation and/or substrate (or oxygen) removal. Substrate removal is frequently determined by removing the biofilm from its environment and conducting a batch test with pulsed substrate (e.g., heterotrophic potential test). Such ex situ kinetic determinations are relatively useless since their accuracy cannot be assessed, nor are they necessarily representative of processes occurring in the sampled environment. Another difficulty with these observed quantities is that they reflect the contribution of several quantities of more fundamental significance. For example, net biofilm accumulation may be the sum of the following processes: a) transport of cells to the substratum, b) attachment of cells to the substratum, c) growth and other metabolic processes within the biofilm, and d) detachment of portions of the biofilm. If, then, the overall process consists of a number of processes in series, the slowest step of the sequence exerts the greatest influence and controls the overall process rate. This step is called the "rate-determining step" or "rate-controlling step."

Which process controls the rate of net biofilm accumulation? Obviously, the rate-limiting step will depend on the environment at the substratum-liquid interface. Parameters such as concentration of nutrients, fluid shear stress, and cell density will determine which fundamental process will control the rate of net biofilm accumulation. For example, the rate-limiting process for biofilm accumulation may be different in a tube enclosing a turbulent flow as compared to a glass slide immersed in a quiescent fluid, even if the rates of accumulation are the same.

FUNDAMENTAL PROCESSES CONTRIBUTING TO BIOFILM DEVELOPMENT

In this discussion, biofilm development will be considered to be the net result of the following physical, chemical, and biological processes:

1. Transport of organic molecules and microbial cells to the wetted surface.
2. Adsorption of organic molecules to the wetted surface resulting in a "conditioned" surface.
3. Attachment of microbial cells to the "conditioned" surface.
4. Metabolism by the attached microbial cells resulting in more attached cells and associated material.
5. Detachment of portions of the biofilm.

These processes can be further classified into transport and transformation processes. Transport processes describe the transfer of energy or material between the environment and the system in question or the transfer of energy or material from the system boundary through the system. For example, if the system is defined as the biofilm, 1 and 2 above are transport processes describing transfer of materials between the environment and the system. Transformation processes are those processes resulting in molecular transformations of matter, i.e., chemical or biochemical reactions. Transformation processes are described by rate equations or, more specifically, constitutive or kinetic equations of the following form:

$$r = r(c_1, \ldots c_n), \tag{1}$$

in which $c_1 \ldots c_n$ are concentrations of the various reacting components.

Of the fundamental processes listed above, all can be classified as transport processes except for metabolism by the attached cells and the production of more attached cells and associated material which refers to transformation processes. These transformation processes can be broken down further into the following processes: a) microbial growth and reproduction, b) product formation (e.g., exopolymers), c) maintenance, and d) decay. Note that these are unstructured processes in that specific intracellular transformations need not be considered. Otherwise, our models would be considerably more complex. In a system experiencing rapid environmental fluctuations, homeostasis (balanced growth) is not possible and "unstructured" models will not suffice to describe the system in phenomenological terms. In fact, batch reactors sometimes suffer from the disadvantage of creating these transient conditions which make modelling more difficult.

The categorization of these processes as fundamental is somewhat relative. However, at this time, I have not been able to further subdivide these processes while maintaining my dependence on observable quantities for the analysis. Another criterion for a fundamental process, in this paper, is that accepted mathematical descriptions for the process exist.

CONSERVATION OF MASS

Generally, in systems such as biofilm "reactors," transport and transformation processes are combined in a general type of model based on the conservation of mass. The equations in such a model have the following form:

$$\text{In} - \text{Out} + \text{Conversion} = \text{Accumulation} \tag{2}$$

or

$$\text{transport} + \text{transformation} = \text{Accumulation} \tag{3}$$

In batch "reactors," the transport term is zero since there is no exchange with the environment.

The conservation equation is especially useful since it permits the in situ determination of the rates, especially the transformation rates, in a natural or experimental system. The conservation equation also incorporates the thermodynamic (e.g., biofilm mass and specific gravity) and constitutive (e.g., reaction and transport coefficients of the biofilm) properties of the system.

TRANSPORT PROCESSES

Transport processes, as indicated in the previous section, include transport of organic molecules and microbial cells to the substratum, cell attachment to the substratum, and detachment of cells from the substratum. These processes are closely related, especially during early events in the accumulation of a biofilm. Transport processes discussed in this section will be limited to those relevant to early events in the establishment of a biofilm. Other transport processes, such as diffusion of reactants within the biofilm, are significant at a later stage in biofilm development and are discussed by Atkinson (this volume).

Significant effort has been invested in studying the mechanisms by which cells accumulate at the substratum. For example, substratum surface and bulk properties, microbial cell surface properties, exopolymers, and type or extent of the "conditioning" film are variables that have been considered as influencing attachment of cells to the substratum.

However, less effort has been used to determine the rate and extent of attachment and the influence of physical factors in the environment. In order to explore such matters, consider the following events which represent a composite of the observations of many investigators for the past forty to fifty years:

1. Organic molecules are transported from the bulk fluid and adsorbed to the substratum.
2. A fraction of the suspended microbial cells are transported to the conditioned substratum. In a quiescent environment, the predominant mechanism for transport may be sedimentation or motility of the organism. In laminar flow, the primary transport mechanism may be diffusion (Brownian motion), whereas in turbulent flow convective (that related to fluid motion) mechanisms may dominate.
3. A fraction of the cells that strike the substratum attach to the substratum for some finite time, "the critical residence time," and then detach. This process will be termed reversible attachment. Detachment may result from fluid shear forces, but other detachment mechanisms are certainly plausible.
4. A fraction of the reversibly attached cells remain attached beyond

the "critical residence time." These cells are irreversibly attached.

Transport and Adsorption of Organic Molecules

Transport and adsorption of organic molecules to the surface is very rapid as compared to transport of microbial cells. As a consequence, microbial cell attachment is essentially always to a "conditioned" surface. Therefore, the rate of "conditioning" is instantaneous as compared to other rate processes. Hence, the term substratum in this paper will specifically refer to a "conditioned" substratum. Debates continue regarding the necessity of the conditioning film for attachment of microbial cells.

Transport of Microbial Cells

As far as this author knows, no one has measured the rate of transport of suspended microbial cells to a surface. In fact, little is known about transport rates of any particles in aqueous systems. The quantity which is inevitably measured is net rate of accumulation, i.e., the number of cells that are transported to the surface and which irreversibly adhere. The number of cells that strike the surface is unknown. The ratio of the number of adhesions to the number of strikes can be considered as the adhesion efficiency and could serve as an assay for the adhesiveness of a microbial cell under a variety of environmental conditions.

Few reports exist regarding mechanisms of cell transport in various liquid flow regimes. Certainly, the dominant mechanism for cell transport is different in turbulent flow than it is in a quiescent system. In fact, flow may significantly influence adhesion efficiency. For example, experimental data for net accumulation rate as compared to calculated transport rates, assuming the mechanism for transport was known, are presented in Table 1. Fletcher (3) measured net accumulation rate under quiescent conditions in which sedimentation could have been the dominant transport mechanism. Under these conditions, detachment is negligible. Powell and Slater (6) conducted their experiments in laminar flow where cell transport is presumably controlled by diffusion. Using these assumptions, transport rates of cells to the substratum were calculated (Table 1). Several observations are striking:

1. Calculated rate of transport is significantly greater in the quiescent system. Microscopic observations in our laboratory confirm the dramatic difference in the striking rate between these two flow regimes (Nelson, unpublished results). In laminar flow, the transport rates increase with increasing shear stress. The rate of transport is greater in the high shear stress laminar experiment than in the low shear

TABLE 1 - A comparison between measured net cell accumulation rate at a surface and the calculated cell transport rate to the surface.

Reference	Fluid Shear Stress ($N\ m^{-2}$)	Rate of Transport	Rate of Accumulation	Cell Concentration cells m^{-3} X 10^{-13}	Presumed Transport Mechanism
		(cells $m^{-2}s^{-1}$ X 10^{-4})			
Fletcher (3)	0	5000	4170	10.0	sedimentation
Powell and	0.80*	472	3	1.4	diffusion
Slater (5)	0.09**	167	31	1.4	diffusion

* Reynolds number = 11
**Reynolds number = 1.2

stress experiment. The apparent difference in transport rates between quiescent and laminar flow may be attributed to a change in the dominant transport mechanism.

2. Net rate of accumulation is significantly greater in the quiescent system. Net rate of accumulation is greater in the laminar flow with lower shear stress than in the higher shear stress experiment. One apparent explanation for these results is that detachment rate increases with increasing shear stress (6).
3. The calculated adhesion efficiency in the various experiments is proportional to shear stress (Fig. 1).

Transport is also influenced by the macroscopic geometry of the experimental system. Figure 2 is a composite of several environments that are encountered in a power plant condenser system. The fluid forces in the various environments are quite different and will influence transport rates significantly. The resulting microenvironments, e.g., crevices, will also determine the type (e.g., aerobic or anaerobic) of microbial activity in the biofilm.

Attachment of Cells to the Substratum

Reversible attachment occurs after the cell is transported to the substratum. The cell may be held at the surface by relatively weak bonds such as electrostatic interactions. If the cell does not detach after a certain time period, the cell is irreversibly attached. Does the cell require a finite or critical residence in the reversible state to irreversibly adhere

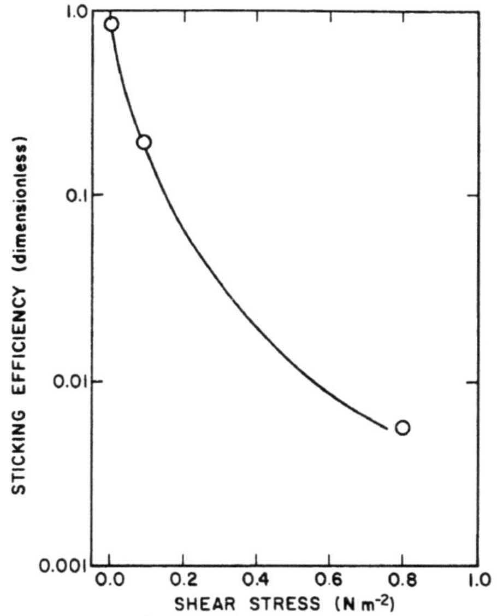

FIG. 1 - Influence of fluid shear stress on adhesion efficiency of bacterial cells (data from (3, 6)).

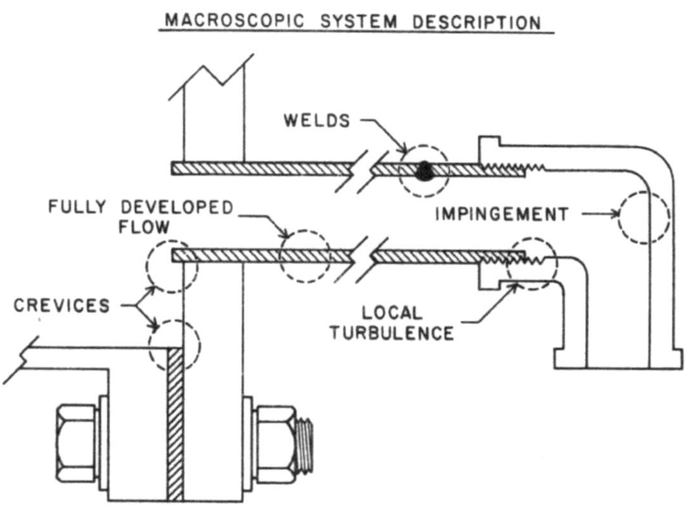

FIG. 2 - Schematic composite diagram of various geometries existing within a condenser/heat exchanger.

or is irreversible attachment spontaneous? Is exopolymer production a necessary prerequisite for irreversible attachment? If so, is it a specific polymeric substance? If exopolymer production is necessary, then a finite time will be required for its synthesis and critical residence time appears to be a viable concept. Other explanations have been offered (see Robb, this volume).

The extent of attachment provides other tantalizing questions. Fletcher (3) observed approximately 45% coverage of the substratum surface at the saturation point in a quiescent system. Powell and Slater (6) observed 1-5% coverage at saturation in laminar flow. Experiments in our laboratory (Nelson, unpublished results) in turbulent flow resulted in approximately 0.01% coverage at saturation and uniform distribution of cells across the surface. The uniform distribution and low saturation coverage suggest that there exists a "zone of influence" around the attached cells possibly defined by exopolymers or other chemosensory mechanisms. In any case, saturation coverage appears to be strongly influenced by flow regime. Kjelleberg (personal communication) reports less than 5% saturation coverage in all samples he has observed.

The physiological state of the organism influences the rate and, possibly, the extent of attachment (3). Bryers and Characklis (2) observed that attachment rate was directly proportional to growth rate in a mixed culture system when feeding the biofilm reactor from a chemostat. Nelson (unpublished results) has observed a decrease in attachment rate with increasing specific growth rate for a Pseudomonas species (pure culture) in a similar experimental system.

The surface roughness of the substratum may be a significant factor which has been overlooked in experimental studies purporting to simulate a specific environment. Microroughness of various roll finishes in stainless steel tubes is described schematically in Fig. 3. The figure clearly emphasizes the probable role of microroughness in both transport and attachment of microbial cells. Convective transport rates near the microrough surface will be greater than at a smooth surface. Once the cell arrives at the surface, attachment rates will probably be higher for several reasons. Two reasons which support the observations that net cell accumulation rate is greater on rougher surfaces are: a) detachment due to shear forces will be reduced since the cells are shielded from the bulk fluid flow, and b) more substratum surface area may be available for contact with the cell.

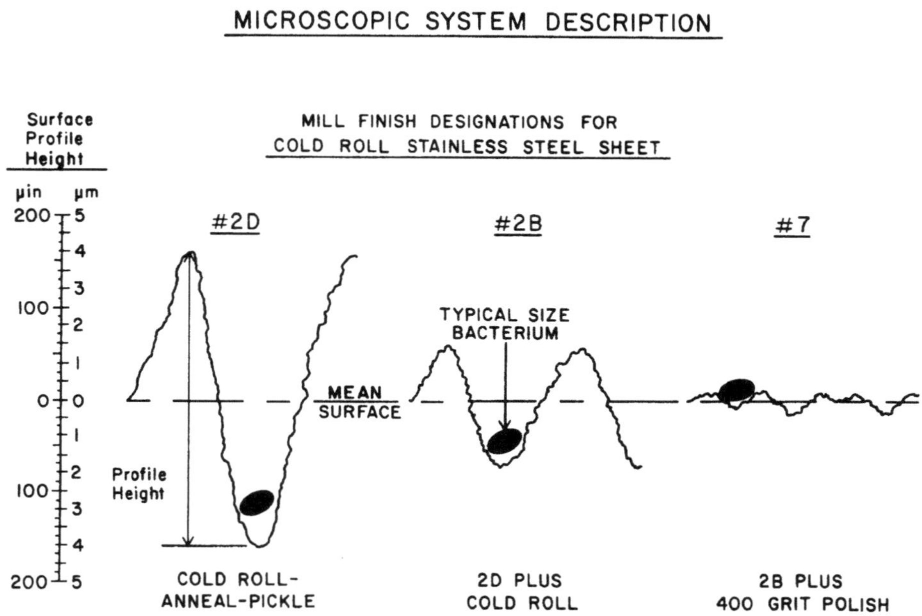

FIG. 3 - Schematic diagram of microroughness in stainless steel tubing as compared to a microbial cell.

Detachment from the Substratum

During early events in biofilm formation, hydrodynamic influences control detachment (5, 7). Detachment increases with increasing fluid shear stress at the substratum surface. Macro- and microroughness may significantly influence detachment since cells attached in cavities will be sheltered from severe hydrodynamic shear.

Shear forces along a conduit are exerted parallel to the surface. Hence, a cell detached from the surface due to shear may be transported close to the surface (in the viscous or laminar sublayer) for a significant distance, i.e., it may bounce or roll. This behavior will result in numerous collisions with the surface and, hence, more opportunities for reattachment. In turbulent flow, a significant lift force is exerted which may transport the detached cell quickly into the bulk liquid. The lift forces, however, are several orders of magnitude smaller than the shear forces.

Detachment may also result from a chemical treatment (e.g., chelants, oxidants). Little fundamental information is available on kinetics of detachment, despite their potential usefulness as comparative criteria in testing chemical control treatments.

TRANSFORMATION PROCESSES

Transformation processes refer to those processes in which molecular rearrangements occur, i.e., reactions. After a significant number of cells attach to a substratum, microbial transformation or metabolic processes will become significant if nutrients are available and "colonization" will ensue (1).

Studies of transformation processes within biofilms have generally relied on a relatively unstructured approach to analysis of the biomass component. Unstructured models characterize the biotic component only in terms of biomass with little attention given to the various biofilm components, physiological state of the organisms, microbial species, or exopolymer content. Processes in unstructured models are limited to those that can be evaluated without considering activities within the biofilm, i.e., a more macroscopic approach.

Fundamental and Observed Processes

Recent studies (Robinson et al., submitted; Bakke et al., submitted) have provided more information on biofilm structure by using pure cultures (Pseudomonas aeruginosa), measuring exopolymer content, and culturing cells in a chemostat. To obtain these results, four fundamental rate processes were identified: a) growth, b) product formation, c) maintenance and/or endogenous decay, and d) death and/or lysis. Any or all of these processes may be occurring in a biofilm at any time. Growth refers to cell growth and multiplication. The cells also form products, some of which are retained in the biofilm (e.g., exopolymers) and some of which diffuse out into the bulk fluid. The cells also have to maintain their internal structure, another energy-consuming process. If nutrients are depleted or toxic substances are present, death and/or lysis ensues.

The rates of the fundamental microbial processes are difficult to measure directly and are generally inferred from more easily observed rate processes. The more familiar observed rate processes include a) substrate consumption, b) electron acceptor consumption, c) biomass production, and d) product formation. The relationship between fundamental and observed process rates is presented in Table 2. The stoichiometry of the process is qualitatively represented by each row in the matrix

TABLE 2 - A matrix representation for the fundamental microbial rate processes.

FUNDAMENTAL PROCESS		STOICHIOMETRY						
		REACTANTS			=	PRODUCTS		
Process	Rate Coefficient	Substrate s	Nutrient z	Electron Acceptor e	Biomass x	Product Pe	Product Pi	Metabolite a
Growth	μ	-	-	-	+	+	(+)	+
Maintenance								
exogenous	m	-	-	-		+		+
endogenous	k_e		+	-	-	+	-	+
Product Formation	k_p	-	-	-		+	+	+
Death	k_d	(+)	(+)		-	+		
OBSERVED RATE		q_s	q_z	q_e	μ_n	q_p		q_a

q = specific production or removal rate (t^{-1})
μ = specific growth rate (t^{-1})
x = biomass concentration (ML^{-3})
Pe = extracellular microbial product concentration (ML^{-3})
Pi = intracellular microbial product concentration (ML^{-3})
s = substrate concentration (ML^{-3})
z = nutrient concentration (ML^{-3})
e = electron acceptor concentration (ML^{-3})
μ_n = net specific biomass production rate (t^{-1})

(- refers to reactants and + refers to products). The columns of the matrix indicate the fundamental rate processes that may contribute to the observed rates (last row in the matrix). For example, substrate removal (column 1) is the net result of growth, maintenance, and product formation.

Several investigators ((2, 8), Robinson et al., submitted; Bakke et al., submitted) have used similar experimental techniques in a laboratory biofilm reactor to quantify the rate and stoichiometry of the fundamental processes within a biofilm. The reactor is a continuous flow stirred tank reactor which provides a relatively uniform surface environment for microbial attachment and biofilm activity. Their results suggest the following:

1. The growth rate of cells (Ps. aeruginosa) in the biofilm can be estimated from their growth rate in chemostats when substrate concentration in the microenvironment of the cell is equal in both conditions. Results

of experiments in a chemostat and in a biofilm reactor are compared in Fig. 4. Results indicate that specific substrate removal rate is the same for Ps. aeruginosa at the same growth rate in a chemostat and in the biofilm. The linear slope suggests constant energy metabolism in both microbial environments. Maximum specific growth rate, saturation coefficient, and yield are the same in a chemostat and in the biofilm. One important restriction requires that no significant diffusional resistances exist in the biofilm, which was the case in this study. Another concern requires that the substratum surface not release any components which influence metabolism. In these studies, acrylic plastic was the substratum.

2. Exopolymer formation rate and yield by biofilm cells (Ps. aeruginosa) was essentially the same for dispersed cells. However, exopolymer accumulation rate in the biofilm was quite high. Therefore, exopolymer may be the dominant component in the biofilm (up to 90% of the biofilm organic carbon in this study).
3. Maintenance requirements are essentially negligible until the biofilm becomes very thick. Even then, the results of exopolymer formation or anaerobic metabolism deep within the biofilm may be mistaken for maintenance energy requirements if sufficient measurements are lacking.

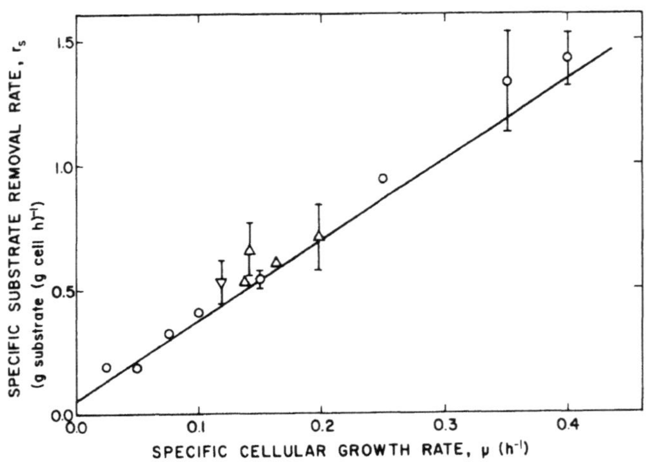

FIG. 4 - Relationship between steady state specific substrate removal rate and specific growth rate of Ps. aeruginosa in a chemostat (○) and a biofilm (Δ, D = $6h^{-1}$; ∇, D = $3h^{-1}$). Definition of specific growth rate in a biofilm is expressed by Eq. 11 (see Table 3).

One of the important conclusions from these results is that chemostat-derived rate and stoichiometric coefficients may be used for modelling biofilm processes in some cases. However, important questions still remain. Accumulation rate also depends on detachment rate and few useful rate expressions are available for detachment. The results do indicate that substrate removal rate alone is not a sufficient criterion for comparing the activity of attached and dispersed cells since substrate removal is the net result of several fundamental processes. The measurements described above were accomplished in situ. Removing the cells from the surface or removing the surface from its environment obviates any relevance in subsequent measurements which purportedly describe the activity of attached cells.

Mass Transfer and Diffusion

Analyses of biofilm process rates and stoichiometries are frequently complicated by significant mass transfer resistances in the liquid or diffusional resistances within the biofilm. Trulear and Characklis (9) have observed that substrate removal rate increases in proportion to biofilm thickness up to a critical thickness beyond which removal rate remained constant. The critical or "active" thickness was observed to increase with increasing substrate concentration. This behavior has been observed by others and attributed to diffusional resistance within the biofilm. Once the biofilm thickness exceeds the depth of substrate penetration into the biofilm, the removal rate is unaffected by further biofilm accumulation.

The biofilm rate processes may also be controlled by mass transfer limitations in the bulk fluid phase (10). For example, substrate removal rate is dependent on fluid velocity past the biofilm. At low fluid velocities, a relatively thick mass transfer boundary layer can cause a fluid phase mass transfer resistance which decreases substrate concentration at the fluid-biofilm interface and, thereby, decreases substrate removal rate. Two factors may result in low mass transfer rates from the bulk fluid to the biofilm: a) low fluid velocities, and b) dilute liquid phase concentrations of the material being transported. Much biofilm research has been conducted at relatively low flows or in quiescent conditions. Mass transfer may be the rate-controlling step in the overall process in these studies and, without further analysis, may be confused with the rates of more fundamental processes such as growth rates, adsorption rates, etc. In highly turbulent systems, mass transfer in the liquid phase is rarely a significant factor.

Summary of Biofilm Transformation Processes

The microbial processes occurring in a biofilm are more complex than suggested by the four fundamental processes defined above. However, this classification has been useful in determining, to some extent, the flow of substrate energy through the biofilm. Mathematical description of the kinetic expressions has also been accomplished (Table 3). Further structuring of biofilm processes may await more sophisticated methods for observing the processes within the biofilm (as opposed to the influence of the processes on the overlying liquid phase) and more specific identification of the products being formed.

Bakke and Characklis (submitted) have observed a remarkable biofilm phenomenon which demands more attention. The supply of growth substrate was increased stepwise to a biofilm with the following results: a) biofilm material was immediately detached, b) biofilm cell numbers remained constant, and c) specific substrate removal rate and product formation rate increased instantaneously. These observations cannot be described with unstructured models but suggest that the biofilm organisms may slough their exopolymers in response to the "shock." In addition, the attached cells seem to possess a "reaction potential" which is expressed in response to instantaneous increase in substrate loading. The experiments clearly indicate the need to observe microbial physiology while the organisms are attached in their growth environment (in situ) and lead to questions regarding the use of other techniques, such as heterotrophic potential, for determining biofilm activity. More effort must be expended to understand these transient biofilm phenomena because of their relevance to natural and technological phenomena and their potential for providing keys to the understanding of mechanisms.

DETACHMENT OF BIOFILM

Detachment of microbial cells and related biofilm material occurs from the moment of initial attachment (see above). However, the macroscopic observation of biofilm detachment is easier as the biofilm becomes thicker.

Detachment phenomena can be arbitrarily categorized as shearing or sloughing. Shearing refers to continuous removal of small portions of the biofilm which is highly dependent on fluid dynamic conditions. Under these circumstances, rate of detachment increases with increasing biofilm thickness and fluid shear stress at the biofilm-fluid interface (9). Sloughing refers to a random, massive removal of biofilm generally attributed to nutrient or oxygen depletion deep within the biofilm (4) or some dramatic change in the immediate environment of the biofilm (see previous

TABLE 3 - System of equations describing biofilm processes in a continuous flow stirred tank reactor with a sterile substrate feed.

Compound	Net Rate of Accumulation	Net Rate of Transport Out of Reactor	Net Rate of Transformation	Units	Equation No.
Bulk Liquid Substrate Concentration, s	$\frac{ds}{dt}$ =	$D(s_i - s)$	$-(\frac{\mu}{Y_{x/s}} + \frac{r_p}{Y_{p/s}})x_b\frac{A}{V}$	$ML^{-3}t^{-1}$	(4)
Suspended Cellular Mass, x	$\frac{dx}{dt}$ =	$-Dx$	$+r_{dx}x_b\frac{A}{V}$	$ML^{-3}t^{-1}$	(5)
Biofilm Cellular Areal Density, x_b	$\frac{dx_b}{dt}$ =		$\mu x_b - r_{dx}x_b$	$ML^{-2}t^{-1}$	(6)
Suspended Exopolymer Mass, p	$\frac{dp}{dt}$ =	$-Dp$	$+ r_{dp} P_b \frac{A}{V}$	$ML^{-3}t^{-1}$	(7)
Biofilm Exopolymer Areal Density, p_b	$\frac{dp_b}{dt}$ =		$r_p x_b - r_{dp}p_b$	$ML^{-2}t^{-1}$	(8)

At steady state, Eq. 4 becomes

$$r_s^* = \frac{VD(s_i - s)}{x_b A} = (\frac{\mu}{Y_{x/s}} + \frac{r_p}{Y_{p/s}}) = \text{specific substrate removal rate} \qquad (9)$$

Eq. 5 becomes

$$r_{dx} = \frac{D x V}{A x_b} \qquad (10)$$

Combining Eqs. 10 and 6 at steady state

$$\mu = \frac{D x V}{A x_b} = \text{specific growth rate in the biofilm} \qquad (11)$$

s = substrate concentration (ML^{-3})
x = cellular concentration in liquid (ML^{-3})
p = exopolymer concentration in liquid(ML^{-3})
x_b = cellular concentration in biofilm (ML^{-2})
p_b = cellular concentration in biofilm (ML^{-2})
V = liquid volume (L^3)
A = wetted surface area (L^2)
D = dilution rate (t^{-1})
μ = specific growth rate (t^{-1})
r_{dx} = cellular detachment rate (t^{-1})
r_{dp} = exopolymer detachment rate (t^{-1})
r_p = exopolymer formation rate (t^{-1})
$Y_{x/s}$ = cellular yield
$Y_{p/s}$ = product yield

section). Sloughing is more frequently witnessed with thicker biofilms developed in nutrient-rich environments. Shearing is probably occurring under the same conditions under which sloughing is occurring but no such direct measurements have been attempted.

Hydrodynamic Influences

Trulear and Characklis (9) have observed that biofilm detachment increases with both fluid shear stress and biofilm mass. Both Powell and Slater (5) and Timperley (7) conducted studies to determine the influence of fluid dynamics on detachment. Both investigators observed an increase in detachment with an increase in Reynolds number. Timperley also considered different tube sizes and, within that context, concluded that mean fluid velocity was more significant in determining cleaning effectiveness than Reynolds number.

As fluid velocity increases, the viscous sublayer thickness decreases. Consequently, the region near the tube wall subject to relatively low shear forces (i.e., the viscous sublayer) is reduced. As a result, there may be some upper limit to the effectiveness of any cleaning operation based on fluid shear stress. The viscous sublayer may provide a valuable a priori criterion for predicting the maximum effectiveness (the minimum thickness attainable) of any cleaning technique dependent on fluid dynamic forces.

Detachment processes must also be significant in the processes of cell turnover (individual cell residence time) in the biofilm. As a biofilm develops, succession in species is observed. Trulear (8) developed a biofilm of Pseudomonas aeruginosa in conditions of relatively high shear stress and then challenged it with Sphaerotilus natans. The Sphaerotilus quickly became the dominant species within the biofilm. Detachment, influenced strongly by fluid shear stress, may serve to "wash out" organisms from the biofilm. Growth processes may also dilute out the slower growing organisms.

Chemical Treatment

Detachment may occur for reasons other than hydrodynamic forces. Bakke and Characklis (submitted) have observed massive detachment when substrate loading to the biofilm was instantaneously doubled. One hypothesis suggests that cell membrane potential plays a key role in the phenomena. Turakhia et al. (10) have observed dramatically increased detachment upon the addition of a calcium-specific chelant (EGTA), suggesting the importance of calcium to the cohesiveness of the biofilm. Many other chemical treatments have been used to detach biofilm material with varying success, including: a) oxidizing biocides (e.g., chlorine), b) uv radiation, c) surfactants, and d) non-oxidizing biocides.

Summary of Detachment Processes

Detachment processes must play a major role in the ecology of the biofilm. Detachment from and absorption into the biofilm of microorganisms provides the means for interaction between dispersed (planktonic) organisms and the biofilm. Detachment of biofilm is the major objective of many antifouling additives used in manufacturing processes.

Very little is known regarding the kinetics of detachment and the factors affecting the removal. Such kinetic expressions would be useful for modelling purposes and for serving as comparative criteria in testing of antifouling treatments.

SUMMARY

Biofilm processes have been discussed in terms of the more fundamental physical, chemical, and biological processes which contribute to biomass accumulation at a surface. The purpose was twofold: a) to present a framework for analysis of the rate of biofilm development and extent of biofilm development, and b) to stimulate fundamental investigations on topics related to biofilm processes.

The purpose of the analysis and the investigations is a predictive "model" for biofilm accumulation and activity.

Biofilms are emerging as a most critical factor affecting natural aquatic systems, water distribution systems, wastewater treatment, heat exchangers, fuel consumption by ships, and even human diseases. More attention must be directed to their behavior. Some topics that require more attention include the following:

1. Physical, chemical, and biological properties and structure of biofilms as a function of water quality and hydrodynamic regime.
2. Mathematical models relating process rates to bulk water concentrations, surface characteristics, and microbial species.
3. Population dynamics within the biofilm and its relationship to microbial populations in the bulk water.
4. Methods to enhance biofilm accumulation and activity in terms of substrate removal (wastewater treatment) or product formation (fermentation).
5. Structure, properties, and function of exopolymers in biofilms. What are the properties of special relevance to adhesion?
6. Is the surface "sensed" by the cell? If so, how? Sensing may lead to the synthesis of exopolymers. Is sensing the initial step in synthesis of adhesive?
7. Does cross-feeding take place in the biofilm? If it does, which organisms are producers and which are consumers?
8. Does the close proximity of the cells and their extended intimate

contact facilitate genetic exchange?

9. Little is known about biofilm activity toward particulate substrates found in natural and technological applications. Most studies are accomplished with soluble substrates.
10. Do biofilms serve as a sink for heavy metals and xenobiotic compounds in natural waters (surface or groundwaters)?

Biofilms play important roles, beneficial and detrimental, in many natural and technological processes. Development of methods to effectively inhibit or exploit biofilm processes will present a considerable challenge but satisfying rewards in the ensuing years.

Acknowledgments. Supported in part by U.S. National Science Foundation (CPE 80-17439), Office of Naval Research (N00014-80-C-0475), the Calgon Corporation, and the Montana State University Engineering Experiment Station. R.O. Lewis prepared Figures 2 and 3.

REFERENCES

(1) Brannan, D.K., and Caldwell, D.E. 1982. Evaluation of a proposed surface colonization equation using Thermothrix thiopara as a model organism. Microb. Ecol. 8: 15-21.

(2) Bryers, J.D., and Characklis, W.G. 1982. Processes governing primary biofilm formation. Biotech. Bioeng. 24: 2451-2476.

(3) Fletcher, M. 1977. The effects of culture concentration and age, time, and temperature on bacterial attachment to polystyrene. Can. J. Microbiol. 23: 1-6.

(4) Howell, J.A., and Atkinson, B. 1976. Sloughing of microbial film in trickling filters. Water Res. 10: 307-315.

(5) Powell, M.S., and Slater, N.K.H. 1982. Removal rates of bacterial cells from glass surfaces by fluid shear. Biotech. Bioeng. 24: 2527-2537.

(6) Powell, M.S., and Slater, N.K.H. 1983. The deposition of bacterial cells to solid surfaces. Biotech. Bioeng. 25: 891-900.

(7) Timperley, D. 1981. Effect of Reynolds number and mean velocity of flow on the cleaning-in-place of pipelines. In Fundamentals and Applications of Surface Phenomena Associated with Fouling and Cleaning in Food Processing, eds. B. Hallstrom, D.B. Lund, and C. Tragardh, pp. 402-412. Tylosand, Sweden: Lund University.

(8) Trulear, M.G. 1983. Cellular reproduction and extracellular polymer formation in the development of biofilms. Ph.D. Thesis, Montana State University, Bozeman, Montana, USA.

(9) Trulear, M.G., and Characklis, W.G. 1982. Dynamics of biofilm processes. J. Water P. C. 54: 1288-1301.

(10) Turakhia, M.H.; Cooksey, K.E.; and Characklis, W.G. 1983. Influence of a calcium-specific chelant on biofilm removal. Appl. Envir. Microbiol. 46: 1236-1238.

Microbial Adhesion and Aggregation, ed. K.C. Marshall, pp. 159-176. Dahlem Konferenzen 1984. Berlin, Heidelberg, New York, Tokyo: Springer-Verlag.

Chemical Characterization of Films

D.C. White
Center for Biomedical and Toxicological Research
Florida State University, Tallahassee, FL 32306, USA

Abstract. Biofilms represent a complex assembly of different groups of attached microbes with their excretory products. Sensitive measures of the biomass, community structure, nutritional status, and metabolic activities, as well as the chemical characterization of their extracellular polymers, have given insight into the ecology of this system. Exposure of surfaces to flowing waters produces a succession of microbes whose community structure and metabolic activity is affected by the chemistry, biodegradability, and microtopography of the surface, as well as the shear forces and nutrient content of the flowing waters. The grazing of the biofilm by predators or the mechanical or chemical disruption of the biofilm greatly affect the metabolic activity and the potential for secondary reaccumulation and the secretion of extracellular polymers. The extracellular polymers formed by the biofilm microbes are particularly important as they greatly increase the resistance of the microbes to biocides and the efficiency of heat transfer. The metabolic activity of thin biofilm can create microanaerobic sites that facilitate the growth of fermenters and hydrogen utilizers whose acidic fermentation products can greatly facilitate corrosion. The microanalytical methods currently available require the destruction of the biofilm in its assay. With progress in instrumentation the destructive sampling can be supplanted by nondestructive continuous monitoring, possibly utilizing a Fourier transforming infrared system which may provide insights into the chemical basis of adhesion and interactions between the components of the biofilm microbial assembly.

INTRODUCTION

Clean surfaces exposed in this microbial world rapidly become coated with films of microbes and their products. Microbes generally initiate the biofilm formation and this review will be limited to their activity. Biofilms become particularly significant in aqueous environments where the attachment to a surface may often greatly benefit the microbes. These films are of great importance when bound to epithelial surfaces. They may perform essential functions in processing nutrients, as in nitrogen metabolism at the surface of the rumen, or form a hostile environment that precludes the attachment and subsequent infection by pathogenic microbes. Biofilms are often a great economic consideration, as the products of their biomass and metabolic activity may inhibit heat transfer in condensers, induce drag on ships, and facilitate the corrosion of metals. Consequently, it is important to begin to understand the biological process of attraction, adhesion, secretion of extracellular materials, and metabolic activities of the microbial components of the films, as well as the succession and stability of the films. To begin to understand these biological processes, the biofilm must be defined quantitatively. The quantitative definition of biofilm biomass, community structure, and metabolic activity will be the subject of this essay.

THE PROBLEM

The ideal analytical system would provide accurate estimates of the microbial content of the biofilm, measures of the metabolic activity, and of the extracellular polymers that are the products of the microbes that make up the biofilm. The methods should be quantitative and sufficiently sensitive that small areas associated with corrosion or specific activities could be delimited within the films. The sensitivity should be adequate to describe the initial microfouling community and to detect transient responses to changes in various parameters. The ideal system would be one in which the integrity of the system could be maintained so that the rates of various processes could be correlated with shifts in the microbial community structure. The methods should preserve the interactions between the various physiological types of microbes that make up the microcolonies in the films if understanding of the ecology of the whole association is to be realized. Although this appears to be a tall order, the advances in analytical instrumentation are moving the analysis of biofilms ever closer to this goal.

CLASSICAL METHODS

The classical methods to assess the numbers of microbes in a sample were to culture them on agar plates and count the colonies after a period of incubation. This was shown to be an inadequate method which underestimated the numbers of organisms by direct microscopic count of between 10 and 10^5 (reviewed in (27, 28)). This method also destroys the interactions between the species in microcolonies and can give great distortions in understanding the community structure because of the selectivity in the culturing methods. The necessity of removing the organisms from the biofilm and of dispersing them without selection also could induce errors. Recently, the technique of staining the nucleic acids with acridine orange or 4,6'-diamidino-2-phenylindole and then counting the fluorescence created by epifluorescent illumination (42, 61) has been widely utilized in microbial ecology. With some sediments, particularly those with significant contents of clay, the recovery of microbes after various blending or dispersion techniques has been shown to be neither quantitative nor reproducible (33). Removal of the organisms from the substrate and subsequent dilution and filtration onto membranes greatly facilitates the direct counting. The removal must be tested for the release of cytoplasmic enzymes to be sure that cell rupture has not occurred with the vigorous disruptive techniques. These disruptive techniques also destroy the microstructure of mixed microcolonies that appear to be present in biofilms. Microbes come in a limited number of shapes - rods, spheres, and spirals, for example - and their morphology gives no clue to their physiological function or metabolic status. The presence of cells in the process of cell division can be utilized to give an indication of the growth rates in natural systems (23). Examination of biofilms microscopically can give useful information as to the distribution of organisms, and various "factors" can be utilized to correct for the organisms not visualized in attempts to quantitate their numbers (44). Tagged antibodies or lectins can be utilized for specific portions of the microbiota. Radioautography is especially useful in determining the proportions of cells in biofilms that are active. Scanning electron microscopy, with its depth of focus and magnification over five orders of magnitude, has given a new perspective to the structure of biofilms. There are two problems with scanning electron microscopy in the analysis of biofilms. The first problem is that not all microbes are quantitatively preserved in the fixation process (this is particularly striking in the case of protozoa). The second problem results from the fixation in which there is irreproducible shrinkage in the dehydration process that distorts important parameters such as film thickness.

THE PHILOSOPHY OF THE CHEMICAL ANALYSIS OF BIOFILMS

The general philosophy of the chemical analysis of biofilms is to devise sensitive and reproducible analytical methods for various components of the microbial cells and their products. Cellular components that are found in all cells can be utilized to measure biomass. Components restricted to a subset of the microbial community can be used as "signatures" for the respective groups of organisms. Often these "signatures" can be further subdivided into "fingerprints" of strains within the specific groups (52). These components, limited to subsets of the microbial community, can be used to define quantitatively the community structure.

The biomass and community structure give measures of the total microbial component. The extracellular materials can often be defined by unique components such as the uronic acids or sialic acids of the exopolysaccharides. Many microbes in nature exist in a state of potential activity with only a small proportion showing metabolic activity at any one time. Consequently, measures of the metabolic activity are important as a second parameter in the examination of the biofilms. Possibly the best method is to follow the incorporation or turnover of labeled precursor molecules into the specific "signature" components of the association. A particularly exciting and as yet little exploited method involves the use of mass labeled components containing ^{13}C, ^{18}O, or ^{15}N. Precursor molecules with specific activities approaching 100% can be generated and used in environments where radioactivity would be proscribed. The high specific activities mean that smaller concentrations of precursor molecules can be utilized with less possibilities of distorting normal metabolic pathways. The use of mass labeled precursors and their detection in "signatures" allows the extraordinary resolving power of capillary gas-liquid chromatography (CGLC) to be utilized with detection by mass spectrometry. This methodology has been utilized to document the growth of bacteria in the microbial films on detritus (11).

THE CHEMICAL COMPONENTS OF BIOFILMS

Components with universal or near universal distribution in cells can be used to estimate the biomass of the microbes in the biofilms. All membranes contain phospholipids which can be measured as the extractable lipid phosphate (57). This has proved a relatively simple method to assess the microbial content of biofilms, as bacteria generally have about 50 μmoles lipid phosphate per gram dry weight (55). Some pseudomonads, when subjected to the extreme stress of phosphate limitation in a chemostat, can increase the acylated ornithine lipids or uronic acid-containing glycolipids with reductions of phospholipid (32, 59), but these extreme conditions are unlikely to be found in biofilms encountered in nature. The phospholipids have a relatively rapid turnover in living organisms in biofilms (29), so the phospholipid content estimates the "viable" biomass. The usual method of measurement involved the colorimetric analysis of perchloric digests of the lipid extracts. This is sensitive to 0.05 µmoles (57). A new methodology utilizing the analysis of glycerol phosphate derived from the phospholipid after mild alkaline methanolysis increases the sensitivity by three orders of magnitude and allows estimation of the phosphatidyl glycerol content of the films (17). The lipid extract utilized for the analysis of glycerol and glycerol phosphate yields the ester-linked phospholipid fatty acids. Palmitic acid is a ubiquitous fatty acid (26) and can be utilized for estimating the biomass of the biofilm organisms with a sensitivity of 0.1 pmoles (36). The lipid extraction provides a lipid fraction that can be utilized for the analysis of the phospholipids, a residue and the water-soluble components that are concentrated in the water portion of the lipid extraction. The water soluble fraction of the lipid extract contains the components of the cytoplasm of the cells of the biofilm. It proved possible to derivatize the adenosine nucleotides of these cells for detection by their fluorescence after separation by high pressure liquid chromatography (7). The ATP content can be utilized to provide a measure of the biomass, which correlates with the phospholipid, the muramic acid, and various enzyme activities of the biofilm (54, 55). The chemical derivatization and fluorimetric determination allow measures of the other adenine-containing components and could be utilized directly to measure the adenosine energy charge of the biofilm organisms and to estimate one of the homeostatic mechanisms by which the energy charge is maintained (as reflected in the adenosine to ATP ratio) (7). The ratio of adenosine to ATP proved a much more sensitive measure of stress to the biofilm organisms than the energy charge. Measurements of the nucleotides must be utilized with care as microorganisms excrete nucleotides into the surrounding medium as part of the homeostatic compensation for low rates of ATP formation (7). This is a serious problem in sediments where the energy charge seems to indicate a moribund biota, but it can be minimized by using enzymes to destroy extracellular nucleotides prior to their assay (LaRock, unpublished data). Certain enzyme activities give excellent correlations with other measures of biomass (54, 55).

The third component of the lipid extraction is the lipid-extracted residue. If the residue is treated with mild acid and re-extracted and the hydroxy fatty acids recovered, they may be utilized to estimate the lipopolysaccharide lipid A of the Gram-negative bacterial component of the biofilm (48). Not only the biomass but some indication of the community structure of the Gram-negative organisms can be derived from the fingerprint of the fatty acids present (40). A second portion of the lipid-extracted residue can be hydrolyzed by cold concentrated hydrofluoric acid which is specific for phosphate esters and then by mild acid for the recovery and gas-chromatographic analysis (GLC) of glycerol and ribitol of the teichoic acids of the Gram-positive bacterial components of the films at a sensitivity of 10 pmoles (18). Finally, the residue may be acid hydrolzyed and the amines recovered for GLC analysis for muramic acid, glucosamine, and other amines (11). Muramic acid is a unique component of the prokaryotic cell wall (27), where it occurs in a 1:1 ratio with glucosamine. Higher ratios of glucosamine to muramic acid indicate the presence of microeukaryotes with chitin walls (fungi, etc.).

COMMUNITY STRUCTURE

The ester-linked phospholipid fatty acid composition after analysis by CGLC with verification by mass spectrometry can give insight into the community structure of the biofilms. Certain specific fatty acids seem to be concentrated in specific groups of microbes. The definition of these specific groups rests initially on studies of monocultures of the bacteria. Care must be exercised to utilize lipid-free media and to determine the effect of shifting the carbon sources on the phospholipid fatty acid composition. These suggested "signatures" for groups of microbes are then validated after analysis of films or sediments that have been manipulated by shifting the concentration of terminal electron acceptors, the nutrients and minerals added to the culture, the pH, the Eh, and various specific inhibitors to select for various ecologically important groups of microbes. Such groups could include the photoautotrophs, the sulfate-reducers, the methanogenic bacteria, the facultative and obligate aerobes, the denitrifiers, etc. This has proved successful (4, 36, 41, 52, 56). From these studies some useful associations between phospholipid fatty acid composition and ecological functional groups have emerged. The aerobic bacteria contain cyclopropane fatty acids in their phospholipids. These components accumulate when the aerobic films age or are released from grazing pressure (52). The short 12 and 14 carbon saturated fatty acids seem to accumulate in facultative aerobes grown anaerobically (41). Sulfate-reducing bacteria in sediments accumulate iso branched 15 carbon saturated fatty acids (41), and hydroxy fatty acids (15), with 10-methyl palmitic acid signifying the acetate-utilizing Desulfobacter and a 17 carbon iso branched monoenoic fatty acid signifying lactate-utilizing Desulfovibrio desulfuricans (4, 49). Anaerobic growth conditions stimulate the formation of the ω-7-monounsaturated fatty acids in the phospholipids (41, 52). Long chain and polyenoic fatty acids are associated with the microeukaryotes

(4, 52, 58), with the diatoms being particularly enriched in the 16 and 20 carbon fatty acids with ω-3-unsaturation (3). Another useful "signature" is the 16 carbon ω-13-trans fatty acid found exclusively in phosphatidyl glycerol of photosystem 1. These biomarkers for the ester-linked phospholipid fatty acids have not only been shown to associate with the various functional groups of microbes in experiments in which the biofilms were manipulated, but have been detected in caging experiments on mud-flat sediment microbes which were significantly altered by excluding the epibenthic predators that are at the top of the estuarine food chain (10).

Certain other lipid components can be utilized to define physiologically active components of the microbial food chain. In the microbial world the presence of vinyl ether bonds in phospholipids is associated with the anaerobic fermenters (48, 55). These lipids yield fatty aldehydes on mild acid hydrolysis and their concentration is greatly increased in microbial films subjected to anaerobic selection (22). The finding of significant levels of plasmalogen phospholipids in microbial films after short exposure to seawater supports the notion that microzones of anaerobic growth can exist in thin films exposed to rapidly flowing seawater. This finding has significant implications to the microbial facilitation of corrosion. Increases in the plasmalogen content of the biofilms associated with mucus have proved to be a sensitive indicator for potentially reversible stress of pollution exposure (39). The sequence of events in this disease seems to involve the irritation of the coral which induces shifts to greater mucus volumes and density. This is followed by the metabolic creation of anaerobic niches by stimulation of the aerobic heterotrophic bacteria with the subsequent growth of anaerobes. The application of high pressure liquid chromatographic analysis (HPLC) for the detection of the unique phytanylglycerol ether phospholipids of the methanogens can be used to show that these obligate anaerobes are found at the surface of putatively aerobic sediments (31). This finding of anaerobic microniches in aerated films is further reinforced by the detection of respiratory naphthoquinones in the lipids from these films. The utilization of potentiometric detectors capable of detecting femtomolar concentrations of respiratory quinone isoprenologues after HPLC has made possible this study (Hedrick, unpublished data). The presence of naphthoquinones in the respiratory chain indicates low potential terminal electron acceptors, whereas the benzoquinone respiratory pigments signify the high potential nitrate or oxygen as the terminal oxygen acceptors (24, 30).

These combinations of techniques have made possible the demonstration of successional changes in community structure in the microfouling films that attach to surfaces exposed to rapidly flowing seawater (35). They also made possible the quantitative finding that different surfaces accumulate different microfouling films when exposed to the same seawater source (2). It should be emphasized that only a few of the multitude of components found on these chromatographs have been assigned as "signatures." Many have yet to be identified. The means by which "signatures" have been validated is reviewed in (52).

NUTRITIONAL STATUS

Chemical methods can be utilized not only to describe the community structure of biofilms but to gain some indication of the nutritional status of the organisms. The quantitative estimation of the nutritional status can give some indication of the recent history. Endogenous storage components of the cells accumulate under specific environmental conditions. Triglycerides, waxes, and glycogen accumulate when food is plentiful in the microeukaryotes (45). Estimations of the triglyceride glycerol to phospholipid ratio has been utilized to determine that the amphipods grazing microbial films in nature are starving relative to organisms in microcosms in the laboratory with abundant food resources (16). The detection of triglycerides is a positive indicator for the presence of microeukaryotes in biofilms, as triglycerides have not been reported in bacteria (16). In bacteria with adequate nutrients the cells divide. In some specific groups of aerobic bacteria, if a critical nutrient becomes limiting in the presence of adequate carbon and oxygen, the cells accumulate poly-β-hydroxy butyrate polymers. The polymer should more accurately be called poly-β-hydroxy alkanoate (PHA), as it contains several short chain hydroxy fatty acids (12). Biofilms accumulate PHA relative to their total phospholipid under conditions of unbalanced growth, for example, in the presence of humic acids that chelate trace metals (29). A specific chemical marker for the dwarfed and adhesive copiotrophic bacteria has yet to be found.

A second polymer accumulates under conditions of metabolic stress. This is the extracellular polymer secreted by microbial films. Providentially these extracellular films contain a relatively unique component "signature," the uronic acids (9). With the development of a quantitative CGLC-mass spectrophotometric assay for the uronic acids in these polymers (9), it was possible to show that at least one microbe of the marine microfouling biofilm secretes the largest amounts of uronic acid-containing exopolymers during conditions of nutritional stress (50).

The conditions that stimulate secretion of exopolymers are particularly important as brushing or otherwise disturbing the microfouling biofilm forming on surfaces exposed to rapidly moving seawater greatly stimulates the formation of these polymers (36, 51). This polymer has a much higher heat flow resistance per gram than the microbial film that generates it (53). These types of polymers can greatly modify the local environment surrounding bacterial microcolonies (8), causing great increases in their resistance to biocide treatments (6) and possibly changing the corrosion potential when adhering to metallic surfaces (37, 38).

Although chemical methods can quantitatively describe the biomass, community structure, and nutritional status of the organisms in a biofilm, it seems that most of the organisms in these films are metabolically inactive much of the time. The measurement of the metabolic activity can induce a disturbance artifact that is easily detectable in muddy anaerobic sediments. In these sediments, the microbes consume reduced carbon food sources and inorganic nutrients, but ATP generation and rapid growth are limited by the diffusion of high potential terminal electron acceptors. Findlay could not detect the effects of prior disturbance of the mud flat on the rate of phospholipid synthesis either by removing the sediment and mixing it in flasks (the usual heterotrophic potential measurement technique) or by allowing isotope-containing anaerobic pore water to percolate through cores (13). By gently injecting isotope-labeled precursors into the core through small holes in the sides, he was able to detect a significant increase in phospholipid synthesis in the sediments that had been disturbed prior to the experiment. Furthermore, he was able to quantitatively define the disturbance by showing the ratio of PHA to phospholipid synthesis from ^{14}C-acetate increases as the degree of disturbance is increased (14). Using this measure of disturbance, it is far easier to measure various metabolic activities by the incorporation or turnover of specific "signatures" in thin biofilms than in sediments.

Growth rates of microbes in biofilms can be measured with great sensitivity by the incorporation of ^{32}P or ^{14}C-acetate into the phospholipids (27). In aerobic free-living bacteria in the water column, phospholipid synthesis parallels the incorporation of thymidine into DNA or the increase in intracellular ATP. It has also been shown (Moriarty, unpublished data) that the incorporation of ^{3}H-methyl thymidine into DNA with short exposures in sediments is not sensitive to the disturbance artifact. However, comparing the rate of labeling of DNA to that of the phospholipid indicates that not all microbes, particularly the anaerobes, incorporate thymidine into DNA. The turnover of the phospholipids in the biofilm in "pulse-chase" type experiments gives measures of the minimum rates of replacement by nonradioactive precursors (29). Of the phospholipids, the phosphatidyl glycerol backbone glycerol-phosphorylglycerol has a turnover rate that closely parallels the growth rate in a number of bacterial monocultures (46, 47, 58), and it shows a more rapid turnover in biofilms than other lipids (29). The biofilms formed on plastic strips exposed in estuaries contain phosphatidyl choline, whose glycerolphosphorylcholine backbone shows no turnover with "pulse-chase" experiments (29, 34). This means that once formed this lipid is metabolically stable in the cellular membrane. The rate of loss of glycerolphosphorylcholine derived from the phosphatidyl choline can be utilized to measure quantitatively the rate of grazing of the detrital microbial film (34).

The ability to detect all the adenosine-containing components in the cytoplasm of microbial films by HPLC of the fluorescent derivatives (7) has been utilized to show the remarkable resilience of the microfouling biofilm to anoxia. The film recovered its adenosine energy charge after a period of anoxia at least as fast as the measurements could be made (7). Use of the combination of assays of rates of synthesis and turnover of lipids, nucleotides, and PHA, together with the adenosine nucleotides, can give insight into the metabolic status of the biofilm community.

NONDESTRUCTIVE MICROSAMPLING

The newest and potentially most exciting development in the chemical analysis of biofilms is the possibility that a nondestructive assay focused on small patches of the biofilm could provide monitoring capability for metabolic activity and community structure.

Microelectrodes with diameters near 10 µm have recently been utilized to determine oxygen utilization and pH changes in biofilms (43). These microelectrodes both interact directly with the film (use oxygen) and are larger than some microbial niches.

Metabolic activity in films can be measured using nuclear magnetic resonance spectroscopy. Rates of formation or turnover of specific carbon, phosphorous, or hydrogen atoms in particular molecules can be monitored with minimal disturbance to the biofilm organisms. The measurements, however, emphasize the entire biofilm and so represent an average of the multiple possible interactions and must be performed inside the magnetic field of the instrument. Aside from requiring that the reactions occur within the magnetic cavity of the device, NMR is non-invasive and the thickness or mass disturbance of biofilms can be measured nondestructively using ellipsometry. Changes in refraction coefficients and relative phase differences of polarized light upon reflection from the biofilm can be used to calculate the refractive index of the surface (25). This requires a smooth surface (roughness < 50 to 5000 A) and can be utilized on glass surfaces. This technique can possibly be combined with measurements of changes in capacitance after rinsing to give indications of strongly bound materials in the biofilms.

Dynamic laser light scattering spectroscopy can be utilized to provide information on the physical properties of extracellular polymer gels (Hirokawa et al., this volume). In combination with fiber optic light pipes, this procedure could give in situ information on gel strength, binding, pore size, etc., with biofilm formation and maturation.

The potential of infrared spectrometry studies of adhesion have been pioneered by Baier (1). The development of Fourier transforming infrared spectrometers (FT/IR), with their extraordinary increase in sensitivity and spectral resolution, have made possible the utilization of a much wider spectral range that includes the far infrared (2000-10 cm^{-1}) in which the weak vibrations of heavy metal carbon or oxygen bonds can be readily detected. With these instruments the beam size can approach the theoretical limits near 25 μm^2 for studies of spatial heterogeneity and the vibration-rotation activity of specific chemical bonds can be examined in smooth biofilms by total attentuated reflectance or with photoacoustic analysis that is independent of the surface roughness (21). The formation of biofilms on germanium surfaces can be monitored nondestructively in situ with FT/IR, and the sequence of protein adsorption to this surface has been measured in the blood of a living sheep (12, 19, 20, 60). The extraordinary spectral resolution of this instrument permits detailed analysis of changes in the local environment surrounding specific chemical bonds.

CONCLUSIONS

This essay has dealt predominantly with the techniques for quantitatively measuring the chemical structure of biofilms. Quite simply, our insight into the formation and function of biofilms has been limited by the ability to measure essential parameters that can define these complex associations of microbes and their extracellular products. The tools are being developed and the prospects are truly exciting. The study of microbial films is in reality the study of what is important in microbial ecology, as most of the microbial business of this planet occurs by its attached biofilms.

Acknowledgements. The research reported here was supported by contracts N00014-82-C-0404 and N00014-83-K-0056 from the Department of the Navy, Office of Naval Research, grants CR810292-01 and CR-80994-02-0 from the Gulf Breeze Environmental Research Laboratory and the Robert S. Kerr Environmental Research Laboratory of the Environmental Protection Agency, grant NAG2-149 from the Advanced Life Support Office, National Aeronautics and Space Administration, and grant OCE-80-19757 from the Biological Oceanography section of the National Science Foundation.

REFERENCES

(1) Baier, R.E. 1980. Substrata influences on adhesion of microorganisms and their resultant new surface properties. In Adsorption of Microorganisms to Surfaces, eds. G. Bitton and K.C. Marshall, pp. 59-104. New York: Wiley-Interscience.

(2) Berk, S.G.; Mitchell, R.; Bobbie, R.J.; Nickels, J.S.; and White, D.C. 1981. Microfouling on metal surfaces exposed to seawater. Int. Biod. Bull. 17: 29-37.

(3) Bobbie, R.J.; Nickels, J.S.; Smith, G.A.; Fazio, S.A.; Findlay, R.H.; Davis, W.M.; and White, D.C. 1981. Effect of light on the biomass and community structure of the estuarine detrital microbiota. Appl. Envir. Microbiol. 42: 150-158.

(4) Bobbie, R.J., and White, D.C. 1980. Characterization of benthic microbial community structure by high-resolution gas chromatography of fatty acid methyl esters. Appl. Envir. Microbiol. 39: 1212-1222.

(5) Boon, J.J.; DeLeeuw, J.W.; v. d. Hoek, G.T.; and Losjan, J.H. 1977. Significance and taxonomic value of iso and anteiso monoenoic fatty acids and branched beta hydroxy acids in Desulfovibrio desulfuricans. J. Bacteriol. 129: 1183-1191.

(6) Costerton, J.W.; Irwin, R.T.; and Cheng, K.-J. 1981. The bacterial glycocalyx in nature and disease. Ann. Rev. Microbiol. 35: 299-324.

(7) Davis, W.M., and White, D.C. 1980. Fluorometric determination of adenosine nucleotide derivatives as measures of the microfouling, detrital and sedimentary microbial biomass and physiological status. Appl. Envir. Microbiol. 40: 539-548.

(8) Dudman, W.F. 1977. The role of surface polysaccharides in natural

environments. In Surface Carbohydrates of the Prokaryotic Cell, ed. I.W. Sutherland, pp. 357-414. New York: Academic Press.

(9) Fazio, S.A.; Uhlinger, D.J.; Parker, J.H.; and White, D.C. 1982. Estimations of uronic acids as quantitative measures of extracellular polysaccharide and cell wall polymers from environmental samples. Appl. Envir. Microbiol. 43: 1151-1159.

(10) Federle, T.W.; Livingston, R.J.; Meeter, D.A.; and White, D.C. 1983. Modifications of estuarine sedimentary microbiota by exclusion of epibenthic predators. J. Exp. Mar. Biol. Ecol. 73: 81-94.

(11) Findlay, R.H.; Moriarty, D.J.W.; and White, D.C. 1983. Improved method of determining muramic acid from environmental samples. Geomicrob. J. 3: 133-150.

(12) Findlay, R.H., and White, D.C. 1983. Detection of changes in metabolic activity of sedimentary microbiota induced by small scale disturbance. Abs. Am. Soc. Microbiol. 1983: 168.

(13) Findlay, R.H., and White, D.C. 1983. Polymeric betahydroxy-alkanoates from environmental samples and Bacillus megaterium. Appl. Envir. Microbiol. 45: 71-78.

(14) Findlay, R.H., and White, D.C. 1983. The disturbance artifact in the measurement of microbial activity in sediments. Abstracts Third International Symposium on Microbial Ecology, Michigan State University, East Lansing, Michigan, 1983: 49.

(15) Fredrickson, H.L. 1981. Lipid characterization of sedimentary sulfate reducing communities. Abs. Am. Soc. Microbiol. 1981: 205.

(16) Gehron, M.J., and White, D.C. 1982. Quantitative determination of the nutritional status of detrital microbiota and the grazing fauna by triglyceride glycerol analysis. J. Exp. Mar. Biol. Ecol. 64: 145-158.

(17) Gehron, M.J., and White, D.C. 1983. Sensitive measurements of phospholipid glycerol in environmental samples. J. Microbiol. Methods 1: 23-32.

(18) Gehron, M.J.; Davis, J.D.; Smith, G.A.; and White, D.C. 1984. Determination of the Gram-positive content of soils and sediments by analysis of teichoic acid components. J. Microbiol. Methods 2: 165-176.

(19) Gendreau, R.M.; Leininger, R.I.; and Jakobsen, R.J. 1980. Molecular

level studies of blood protein-materials interactions. In Biomaterials 1980, eds. G.D. Winter, D.F. Gibbons, and H. Plenk, Jr. New York: John Wiley and Sons.

(20) Gendreau, R.M.; Winters, S.; Leininger, R.I.; Fink, D.; Hassler, C.R.; and Jakobsen, R.J. 1981. Fourier transforming infrared spectroscopy of protein adsorption from whole blood: Ex vivo dog studies. Appl. Spectr. 35: 353-357.

(21) Griffiths, P.R. 1977. Chemical Infrared Fourier Transform Spectroscopy. New York: John Wiley and Sons.

(22) Guckert, J.B.; Mancuso, C.A.; Martz, R.F.; and Nickels, J.S. 1983. Factors affecting the anaerobic-aerobic community structure in sediments. Abstracts Third International Symposium on Microbial Ecology, Michigan State University, East Lansing, Michigan, 1983: 49.

(23) Hagström, A.; Larsson, U.; Hörstedt, P.; and Normark, S. 1979. Frequency of dividing cells, a new approach to the determination of bacterial growth rates in aquatic environments. Appl. Envir. Microbiol. 37: 805-812.

(24) Holländer, R.; Wolf, G.; and Mannheim, W. 1977. Lipoquinones of some bacteria and mycoplasmas with considerations on their functional significance. Antonie van Leeuwenhoek 43: 177-185.

(25) Jonsson, U.; Ivarsson, B.; Lundstrom, I.; and Berghem, L. 1982. Adsorption behavior. J. Coll. Interface Sci. 90: 148-163.

(26) Kates, M. 1964. Bacterial lipids. Adv. Lipid Res. 2: 17-90.

(27) King, J.D., and White, D.C. 1977. Muramic acid as a measure of microbial biomass in estuarine and marine samples. Appl. Envir. Microbiol. 33: 777-783.

(28) King, J.D., and White, D.C. 1978. Muramic acid as a measure of microbial biomass in Black Sea sediments. Initial Rep. Deep Sea Drill. Proj. 42B: 765-770.

(29) King, J.D.; White, D.C.; and Taylor, C.W. 1977. Use of lipid composition and metabolism to examine structure and activity of estuarine detrital microflora. Appl. Envir. Microbiol. 33: 1177-1183.

(30) Mannheim, W. 1981. Taxonomically useful test procedures pertaining to bacterial lipoquinones and associated functions, with special

reference to Flavobacteria and Cytophaga. In Gesellschaft für Biotechnologische Forschung mbH Braunschweig-Stöckheim, eds. H. Reichenbach and O.B. Weeks, pp. 115-124. Deerfield Beach, FL: Verlag Chemie.

(31) Martz, R.F.; Sebacher, D.I.; and White, D.C. 1983. Biomass measurement of methane-forming bacteria in environmental samples. J. Microbiol. Methods 1: 53-61.

(32) Minnikin, D.W.; Abdolrahimzadeh, H.; and Baddiley, J. 1974. Replacement of acidic phospholipids by acidic glycolipids in Pseudomonas diminuta. Nature 249: 268-269.

(33) Moriarty, D.J.W. 1980. Measurement of bacterial biomass in sandy sediments. In Biogeochemistry of Ancient and Modern Environments, eds. P.A. Trudinger, M.R. Walter, and B.J. Ralph, pp. 131-138. Canberra: Australian Academy of Sciences.

(34) Morrison, S.J., and White, D.C. 1980. Effects of grazing by estuarine gammaridean amphipods on the microbiota of allochthonous detritus. Appl. Envir. Microbiol. 40: 659-671.

(35) Nickels, J.S.; Bobbie, R.J.; Lott, D.F.; Martz, R.F.; Benson, P.H.; and White, D.C. 1981. Effect of manual brush cleaning on the biomass and community structure of the microfouling film formed on aluminum and titanium surfaces exposed to rapidly flowing seawater. Appl. Envir. Microbiol. 41: 1442-1453.

(36) Nickels, J.S.; Parker, J.H.; Bobbie, R.J.; Martz, R.F.; Lott, D.F.; Benson, P.H.; and White, D.C. 1981. Effect of cleaning with flow-driven brushes on the biomass and community composition of the marine microfouling film on aluminum and titanium surfaces. Int. Biod. Bull. 17: 87-94.

(37) Obuekwe, C.O.; Westlake, D.W.S.; Plambeck, J.A.; and Cook, F.D. 1981a. Corrosion of mild steel in cultures of ferric iron reducing bacterium isolated from crude oil. I. Polarization characteristics. Corrosion 37: 461-467.

(38) Obuekwe, C.O.; Westlake, D.W.S.; Plambeck, J.A.; and Cook, F.D. 1981b. Corrosion of mild steel in cultures of ferric iron reducing bacterium isolated from crude oil. II. Mechanism of anodic depolarization. Corrosion 37: 632-637.

(39) Parker, J.H.; Nickels, J.S.; Martz, R.F.; Gehron, M.J.; Richards, N.L.; and White, D.C. 1984. Effect of oil and well-drilling fluids on the physiological status and microbial infection of the reef building

coral Montastrea annularis. Arch. Env. Contam. Toxicol. 13: 113-118.

(40) Parker, J.H.; Smith, G.A.; Fredrickson, H.L.; Vestal, J.R.; and White, D.C. 1982. Sensitive assay, based on hydroxy-fatty acids from lipopolysaccharide lipid A for gram negative bacteria in sediments. Appl. Envir. Microbiol. 44: 1170-1177.

(41) Parkes, R.J., and Taylor, J. 1983. The relationship between fatty acid distributions and bacterial respiratory types in contemporary marine sediments. Est. Coas. Shelf Sci. 16: 173-189.

(42) Porter, K.G., and Feig, Y.S. 1980. The use of DAPI for identifying and counting aquatic microflora. Limn. Ocean. 25: 943-948.

(43) Revsbech, N.P., and Ward, D.M. 1983. Oxygen microelectrode that is insensitive to medium chemical composition: Use in an acid microbial mat dominated by Cyanidium caldarium. Appl. Envir. Microbiol. 45: 755-759.

(44) Rublee, P., and Dornseif, B.E. 1978. Direct counts of bacteria in the sediments of a North Carolina salt marsh. Estuaries 1: 188-191.

(45) Sargent, J.R.; Lee, R.F.; and Nevenzel, J.C. 1976. Marine waxes. In Chemistry and Biochemistry of Natural Waxes, ed. P.E. Kolattukuey, pp. 49-91. New York: Elsevier.

(46) Short, S.A., and White, D.C. 1970. Metabolism of the glucosyl diglycerides and phosphatidyl glucose of Staphylococcus aureus. J. Bacteriol. 104: 126-132.

(47) Short, S.A., and White, D.C. 1971. Metabolism of phosphatidylglycerol, lysylphosphatidylglycerol and cardiolipid of Staphylococcus aureus. J. Bacteriol. 108: 219-226.

(48) Short, S.A.; White, D.C.; and Kaback, H.R. 1972. Mechanisms of active transport in isolated bacterial membrane vesicles. V. The transport of amino acids by membrane vesicles prepared from Staphylococcus aureus. J. Biol. Chem. 247: 298-304.

(49) Taylor, J., and Parkes, R.J. 1983. The cellular fatty acids of the sulphate-reducing bacteria, Desulfobacter sp., Desulfobulbus sp. and Desulfovibrio desulfuricans. J. Gen. Microbiol. 129: 3303-3309.

(50) Uhlinger, D.J., and White, D.C. 1983. Relationship between the physiological status and the formation of extracellular polysaccharide

glycocalyx in Pseudomonas atlantica. Appl. Envir. Microbiol. 45: 64-70.

(51) White, D.C. 1981. The effects of brush cleaning upon microfouling in an OTEC simulation system. Final report, ANL/OTEC-BCM-025, pp. 1-121. Argonne, IL: Argonne National Laboratory.

(52) White, D.C. 1983. Analysis of microorganisms in terms of quantity and activity in natural environments. In Microbes in Their Natural Environments, eds. J.H. Slater, R. Whittenbury, and J.W.T. Wimpenny, pp. 37-66. Cambridge: Cambridge University Press.

(53) White, D.C., and Benson, P.H. 1982. Determination of the biomass, physiological status, community structure and extracellular plaque of the microfouling film. Washington, DC: U.S. Naval Institute Press.

(54) White, D.C.; Bobbie, R.J.; Herron, J.S.; King, J.D.; and Morrison, S.J. 1979. Biochemical measurements of microbial mass and activity from environmental samples. In Native Aquatic Bacteria: Enumeration, Activity and Ecology, eds. J.W. Costerton and R.R. Colwell, ASTM STP 695, pp. 69-81. Philadelphia: American Society for Testing and Materials.

(55) White, D.C.; Bobbie, R.J.; King, J.D.; Nickels, J.S.; and Amoe, P. 1979. Lipid analysis of sediments for microbial biomass and community structure. In Methodology for Biomass Determinations and Microbial Activities in Sediments, eds. C.D. Litchfield and P.L. Seyfried, ASTM STP 673, pp. 87-103. Philadelphia: American Society for Testing and Materials.

(56) White, D.C.; Bobbie, R.J.; Nickels, J.S.; Fazio, S.A.; and Davis, W.M. 1980. Nonselective biochemical methods for the determination of fungal mass and community structure in estuarine detrital microflora. Botan. Marin. 23: 239-250.

(57) White, D.C.; Davis, W.M.; Nickels, J.S.; King, J.D.; and Bobbie, R.J. 1979. Determination of the sedimentary microbial biomass by extractible lipid phosphate. Oecologia 40: 51-62.

(58) White, D.C., and Tucker, A.N. 1969. Phospholipid metabolism during bacterial growth. J. Lipid Res. 10: 220-233.

(59) Wilkinson, S.G. 1972. Composition and structure of the ornithine-containing lipid from Pseudomonas rubescens. Biochim. Biophys. Acta 270: 1-17.

(60) Winters, S.; Gendreau, R.M.; Leininger, R.I.; and Jakobsen, R.J. 1982. Fourier transform infrared spectroscopy of protein adsorption from whole blood: II. Ex vivo sheep studies. Appl. Spectr. 36: 404-409.

(61) Zimmerman, R., and Meyer-Reil, L.-A. 1974. A new method for fluorescence staining of bacterial populations on membrane filters. Kiel. Meeresforsch. 30: 24-27.

Microbial Adhesion and Aggregation, ed. K.C. Marshall, pp. 177-188. Dahlem Konferenzen 1984. Berlin, Heidelberg, New York, Tokyo: Springer-Verlag.

Effects of Network Structure on the Phase Transition of Acrylamide-Sodium Acrylate Copolymer Gels

Y. Hirokawa, T. Tanaka, and S. Katayama
Dept. of Physics and Center for Materials Science and Engineering
Massachusetts Institute of Technology, Cambridge, MA 02139, USA

Abstract. We report preliminary studies on the effect the relative amount of ionizable groups incorporated in the network, main polymer-constituent molecules, and cross-linking agent has on the volume phase transition of acrylamide-sodium acrylate copolymer gels. The experimental results agree with the theoretical prediction that the phase transition is characterized mainly by the number of ionized groups per effective polymer chain.

INTRODUCTION

Several polymer gels having ionizable groups are known to undergo a discontinuous volume change upon changes in temperature (5, 8, 10, 11), solvent composition (4, 6, 7), pH (11), or ionic composition (7). The volume transition is also induced by applying a small electric field across an ionized gel (13). These phenomena can be interpreted as a phase transition in the system consisting of cross-linked polymer network and solvent (1, 8, 11). This view is also supported by observation of critical phenomena in the concentration fluctuations of polymer network measured by the technique of laser light scattering spectroscopy (3, 9, 12).

In this paper we report our studies on the influence of the polymer network structure, represented by the relative amounts of ionizable groups, polymer constituent monomers, and cross-linking monomers, on the phase transition. We also studied the effect of ionization equilibrium on the phase transition by examining the pH dependence of the gel volume.

MEAN-FIELD THEORETICAL CONSIDERATIONS

The equilibrium condition for a gel immersed in a large volume of solvent is that the osmotic pressure difference between inside and outside of the gel should be zero. Zero osmotic pressure difference is also necessary for the free energy of the gel, F, to be minimized since $\Pi = \partial F/\partial V$, where V is the volume of the gel. In the mean field theory the osmotic pressure of a gel is described by the Flory-Huggins formula (2, 11).

$$\Pi = -\frac{NkT}{v}\left[\phi + \ln(1-\phi) + \frac{\Delta F}{2kT}\phi^2\right] + \upsilon kT\left[\frac{\phi}{2\phi_o} - \left(\frac{\phi}{\phi_o}\right)^{\frac{1}{3}}\right] + \upsilon f kT\left(\frac{\phi}{\phi_o}\right), \qquad (1)$$

where N is Avogadro's number, k is the Boltzman constant, T is the temperature, v is the molar volume of the solvent, ϕ is the volume fraction of the network, ΔF is the free-energy decrease associated with the formation of the contact between polymer segments, ϕ_o is the volume fraction of the network at the condition the constituent polymer chains have random-walk configurations, υ is the number of constituent chains per unit volume at $\phi = \phi_o$, and f is the number of dissociated counter-ions per effective polymer chain. From Eq. 1, the equilibrium condition is expressed as

$$\tau \equiv 1 - \frac{\Delta F}{kT} = -\frac{\upsilon v}{N\phi^2}\left[(2f+1)\left(\frac{\phi}{\phi_o}\right) - 2\left(\frac{\phi}{\phi_o}\right)^{\frac{1}{3}}\right] + 1 + \frac{2}{\phi} + \frac{2\ln(1-\phi)}{\phi^2} . \qquad (2)$$

The left side of Eq. 2 corresponds to the reduced temperature, which changes with temperature and solvent composition. The equation then determines the equilibrium-reduced temperature as a function of the network concentration. For certain values of the reduced temperature, however, Eq. 2 is satisfied by three values of ϕ, corresponding to two minima and one maximum of the free energy. The value of ϕ corresponding to the lower minimum represents the stable equilibrium value. A discrete volume transition occurs when the two free-energy minima have the same value. The equilibrium volume of the gel is inversely proportional to its equilibrium concentration

$$\frac{V}{V_o} = \frac{\phi_o}{\phi} , \qquad (3)$$

where V_o is the volume of the gel at $\phi = \phi_o$.

In Fig. 1 the reduced temperature, τ , is calculated and plotted as a function of the equilibrium gel volume for different values of f which

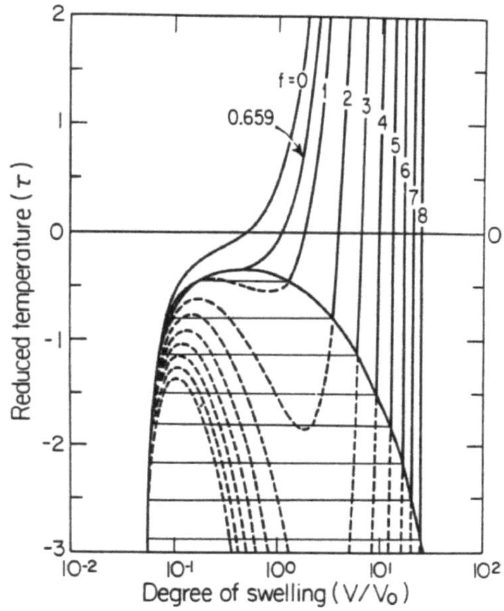

FIG. 1 - Equilibrium degree of swelling V/V_0 is plotted as a function of reduced temperature for gels having various numbers of ionized groups per chain. Equations 1-4 in the text were used for the calculation (taken from (11)).

represents the amount of ionized groups incorporated in the network. Figure 1 shows that with a larger number of ionized groups per chain, the size of volume transition becomes larger and the reduced temperature at the transition becomes lower.

The criterion which determines whether a volume change of a gel is discontinuous or continuous can be seen by expanding the logarithmic term $\ln(1-\phi)$ in Eq. 2, retaining terms up to order ϕ^3. This expansion yields

$$t = S\,(\rho^{-\frac{5}{3}} - \frac{\rho^{-1}}{2}) - \frac{\rho}{3} \quad , \tag{4}$$

where

$$t \equiv \frac{(1-\Delta F/kT)(2f+1)^{\frac{3}{2}}}{2\phi_o} \quad , \tag{5}$$

$$S \equiv \frac{\nu \upsilon}{N \phi_o^{\,3}} (2f + 1)^4 \equiv S_o (2f + 1)^4, \tag{6}$$

and

$$\rho \equiv (\phi/\phi_o)(2f + 1)^{\frac{3}{2}} . \tag{7}$$

Equation 4 is the equation of state of a gel, that is, the relationship between the renormalized reduced temperature, t, and the renormalized concentration, ρ. In this approximation, the equation of state is determined by a single parameter S. The parameter S solely determines whether the volume change is continuous or discontinuous. The ratio of the two volumes at either side of the discontinuous transition is also a function of S. A microscopic interpretation of the parameter S can be seen by the following simple argument. Let us consider an effective polymer chain to consist of n freely jointed segments of radius a and persistent length b. The RMS end-to-end distance of the single polymer chain is $R \sim n^{1/2} b$. Since R is also the RMS distance between neighboring cross-links, we see that $\upsilon \sim ba^2/R^3$, and $\phi_o \sim nba^2/R^3$. From these relations we have

$$S = \left(\frac{b}{a}\right)^4 (2f + 1)^4. \tag{8}$$

Equation 8 shows that the S-value depends on two elements, the number of ionized groups per chain, f, and the ratio of the persistent length, b, to the effective radius of the polymer chain, a. The ratio b/a reflects the stiffness of the chain. Equation 8 reveals that the S-value changes very rapidly with f and b/a. In order for a gel to have a large S-value and thus to undergo a discontinuous phase transition, the constituent polymer chains must have sufficient stiffness and/or a sufficient amount of ionized groups. Both the counter-ions to the ionized groups and the stiffness of the polymer chains increase the osmotic pressure acting to expand the polymer network, resulting in a discontinuous volume change. This situation is similar to a gas-liquid phase transition which can be either continuous or discontinuous depending on the external pressure exerted on the system.

Let us now consider how the S-value of a gel depends on the network structure. For the first approximation we may assume that the chain stiffness is independent of the network structure. Then the S-value depends only on f, which is the number of ionized groups per effective polymer chain (or per half the number of effective cross-linkings). The f-value may be estimated using the following simple argument. The total number of ionized groups in one liter is Ac · N, where Ac is the molar concentration of ionized groups. The maximum possible number of cross-linkings in one liter is B · N, where B denotes the molar concentration of cross-linking molecules. The probability for a cross-linking molecule to be incorporated simultaneously in two polymer chains can be estimated by considering a "reaction volume," v_{reac}, which defines a volume around a cross-linking molecule. If the polymer end having a free radical passes through the volume, the reaction takes place between the polymer and the cross-linking molecule. This probability is given by $[(Am + Ac)v_{reac}]^2$, where Am is the molar concentration of main polymer-constituent molecules. Thus, the f-value is given by

$$f = \frac{2Ac}{B[(Am + Ac)v_{reac}]^2} \quad . \tag{9}$$

This equation shows that the volume transition is larger for gels containing a higher concentration of ionized groups, Ac, a lower concentration of cross-linking molecules, B, and/or a lower concentration of polymer constituent, Am.

SAMPLE PREPARATION

Monomers of acrylamide (main constituent molecule), N, N'-methylenebisacrylamide (cross-linking molecule), and sodium acrylate (ionizable molecule) were purchased (Biorad, electrophoresis grade) and used without any further purification. These monomers were dissolved in pure water to a final volume of 100ml, to which were added 40mg of ammonium persulfate (initiator). The gels were made in micropipettes (inner diameter 1.35mm) at a temperature of 70°C. The gels were then removed from the micropipettes and immersed in a large amount of pure water for five days to wash away unreacted substances from the polymer network. The gels were then cut into pieces 1cm long and immersed in acetone-water mixtures having various compositions. The equilibration time for swelling took from several hours to several days, depending on the solvent composition. The diameter of the gels was measured under a microscope, from which we calculated the degree of swelling:

$$\frac{V}{V_o} = \left(\frac{d}{d_o}\right)^3 , \tag{10}$$

where d_0 is the inner diameter of the micropipette in which the gelation took place, and d is the equilibrium diameter of the gel.

EXPERIMENTAL RESULTS

Effects of Network Structure

Effect of ionizable monomers

Figure 2 shows the degree of swelling of the gels made using constant concentrations of acrylamide, Am (700mM), bisacrylamide, B (8.6mM), and various concentrations of sodium acrylate, Ac (0-320mM). The gels having Ac smaller than 4.3mM undergo a continuous volume change, whereas gels having larger concentrations of Ac show discontinuous volume transitions. The volume change and the acetone composition at the transition monotonically increases with Ac.

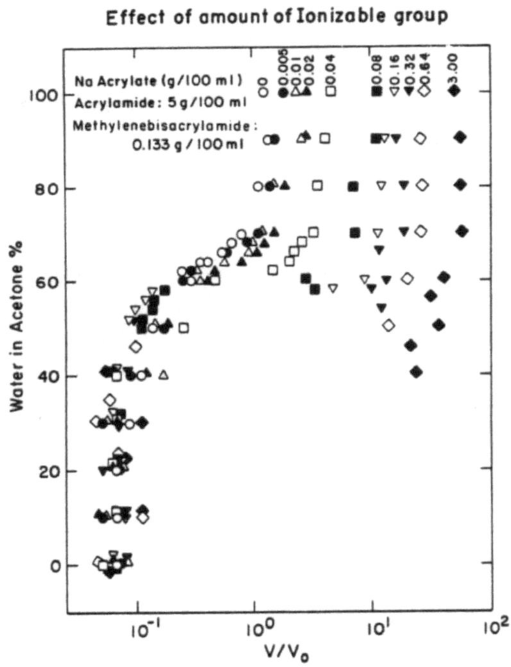

FIG. 2 - Equilibrium degree of swelling of acrylamide-sodium acrylate copolymer gels having fixed concentrations of acrylamide, Am (700mM), bisacrylamide, B (8.6mM), and various concentrations of sodium acrylate, Ac (0-320mM).

Effect of constituent monomers

Figure 3 shows the degree of swelling of the gels having fixed concentrations of Ac (17mM) and B (8.6mM), and various amounts of Am (280-2800mM). The volume change of the gels having Am amount larger than 1400mM was continuous. Discontinuous volume transitions were observed for gels having Am smaller than 990mM. The size of the volume change and the acetone composition at the transition monotonically decrease with Am.

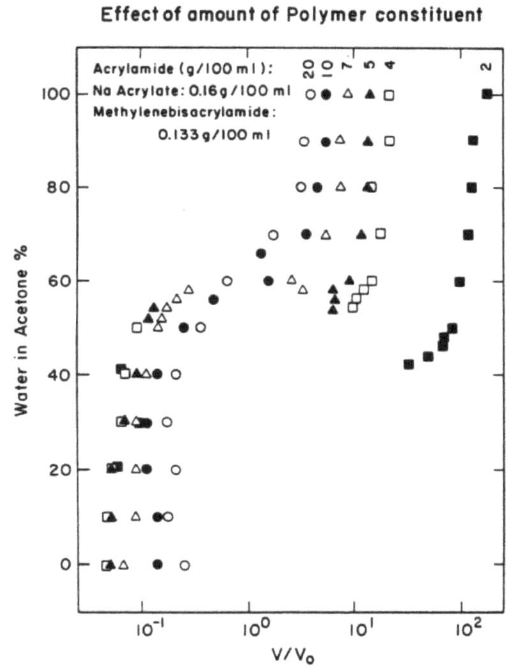

FIG. 3 - Equilibrium degree of swelling of acrylamide-sodium acrylate copolymer gels having fixed concentrations of sodium acrylate, Ac (17mM), and bisacrylamide, B (8.6mM), and various concentrations of acrylamide, Am (280-2800mM).

Effect of cross-linking monomers

Figure 4 shows the effect of the concentration of cross-linking molecules, B (2.2-22mM), on the phase transition of the gels. The concentrations of Am (700mM) and Ac (34mM) were fixed. The volume transitions were all discontinuous. The volume change and the acetone concentration at the transition increase monotonically with B.

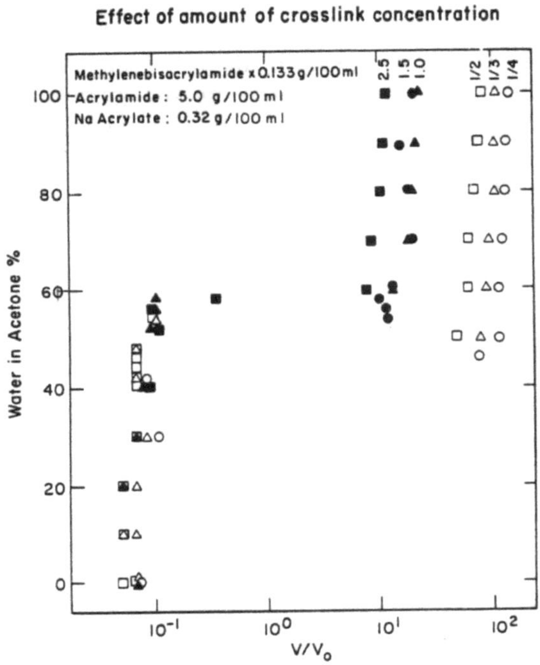

FIG. 4 - Equilibrium degree of swelling of acrylamide-sodium acrylate copolymer gels having fixed concentrations of acrylamide, Am (700mM), sodium acrylate, Ac (34mM), and various amounts of bisacrylamide, B (2.2-22mM).

In order to check the quantitative validity of Eq. 9, we plotted in Fig. 5 the f-value observed in the swelling experiments as a function of $Ac/[B(Am + Ac)^2]$, which is determined by the preparation conditions. If the reaction volume, v_{reac}, is constant, the points will fall on a straight line. All the points are indeed found on a straight line except for three points for the gels having Ac/Am larger than 0.03. In these gels the high concentration of ionizable groups might reduce the ionization due to charge repulsion, resulting in the smaller effective f-value.

EFFECT OF IONIZATION EQUILIBRIUM (pH EFFECT)

In order to further check the effect of ionization on the phase transition of the gels, we also studied the pH dependence. In the experiments shown above, we assumed that all the ionizable groups were ionized. This assumption is valid for gels immersed in solvents having a pH higher

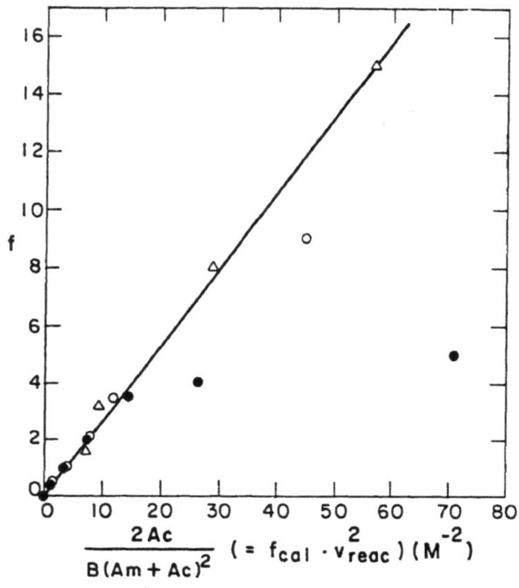

FIG. 5 - The f-value (number of ionized groups per effective polymer chain) of acrylamide-sodium acrylate copolymer gels determined from swelling experiments is plotted as a function of $Ac/B(Am + Ac)^2$, which is determined by the preparation conditions. The solid circles (●) represent the data for gels prepared for constant B (8.6mM), Am (700mM), and various values of Ac (0-320mM). The open circles (o) are for constant Ac (17mM), B (8.6mM), and various values of Am (280-2800mM). The triangles (Δ) are for gels prepared with constant Ac (34mM), Am (700mM), and various values of B (2.1-41mM).

than pK_a of acrylic acid (pK_a = 4.5), which was the case in the experiments described in the previous section. Degree of dissociation, α, of ionizable groups depends on pH. It is given by

$$\alpha = \frac{1}{10^{(pK_a - pH)} + 1} \quad . \tag{11}$$

The f-value is then expressed by the following formula:

$$f = \frac{f_o}{10^{(pK_a - pH)} + 1} \quad , \tag{12}$$

where f_o is the f-value at 100% dissociation. According to Fig. 1, when the f-value changes, a gel undergoes a continuous volume change at reduced temperatures higher than -0.36. However, a discontinuous transition will be observed for reduced temperatures lower than -0.36.

In these swelling experiments, we used gels prepared with constant concentrations of Am (700mM), Ac (34mM), and B (8.6mM). The pH of the mixtures of acetone-water used as solvents were controlled using 1mM phosphate buffers. The results are shown in Fig. 6. As we predicted above, the gels in the mixtures of lower acetone composition show continuous volume changes, whereas the gels in the solvents having high acetone compositions undergo discontinuous transitions. In Fig. 6 we also plotted the theoretical curves for the degree of swelling calculated using Eqs. 2 and 12. We have a good quantitative agreement between the experimental data and the theoretical prediction.

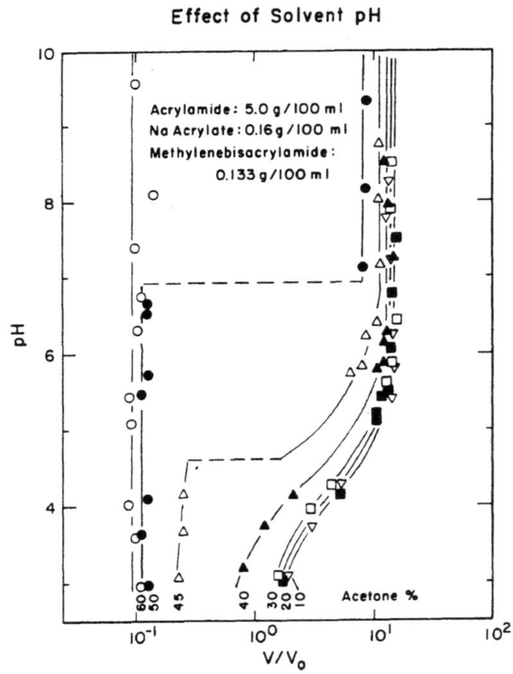

FIG. 6 - The pH dependence of the equilibrium degree of swelling of acrylamide-sodium acrylate gels in acetone-water mixtures. A gel having 700mM acrylamide, 34mM sodium acrylate, and 8.6mM bisacrylamide in 100 ml of volume was used. The lines are theoretically calculated using Eqs. 2 and 12.

CONCLUSION

The experimental observations presented in this paper are quantitatively consistent with our theoretical picture that the discontinuity of the phase transition of acrylamide-sodium acrylate copolymer gels is uniquely determined by the number of ionized groups per effective polymer chain. The studies presented here have clearly shown that through careful examination of the phase transition of polymer gels it is possible to obtain important information on their network structure.

Acknowledgement. This work has been supported by Nippon Zeon Co. Ltd.

REFERENCES

(1) Dusek, K., and Patterson, D. 1968. Transition in swollen polymer networks induced by intramolecular condensation. J. Polym. Sci. A-2,6: 1209.

(2) Flory, P.J. 1953. Principle of Polymer Chemistry. Ithaca, NY: Cornell University Press.

(3) Hochberg, A.; Tanaka, T.; and Nicoli, D. 1979. Spinodal line and critical-point of an acrylamide-gel. Phys. Rev. Lett. 43: 217-219.

(4) Hrouz, J.; Ilavsky, M.; Ulbrich, K.; and Kopecek, J. 1981. The photoelastic behavior of dry and swollen networks of poly (N, N-Diethylacrylamide) and of its co-polymer with N-tert butylacrylamide. Eur. Polym. J. 17: 361-366.

(5) Ilavsky, M. 1982. Phase-transition in swollen gels. 2. Effect of charge concentration on the collapse and mechanical-behavior of polyacrylamide networks. Macromolec. 15: 782-788.

(6) Ilavsky, M., and Hrouz, J. 1982. Phase-transition in swollen gels. 4. Effect of concentration of the crosslinking agent at network formation on the collapse and mechanical-behavior of polyacrylamide gels. Polym. Bull. 8: 387-394.

(7) Ohmine, I., and Tanaka, T. 1982. Salt effects on the phase-transition of ionic gels. J. Chem. Phys. 11: 5725-5729.

(8) Tanaka, T. 1978. Collapse of gels and critical endpoint. Phys. Rev. Lett. 40: 820-823.

(9) Tanaka, T. 1978. Dynamics of critical concentration fluctuations in gels. Phys. Rev. A-17: 763-766.

(10) Tanaka, T. 1981. Gels. Sci. Am. 244(1): 124.

(11) Tanaka, T.; Fillmore, D.J.; Sun, S.-T.; Nishio, I.; Swislow, G.; and Shah, A. 1980. Phase-transitions in ionic gels. Phys. Rev. Lett. 45: 1636-1639.

(12) Tanaka, T.; Ishiwata, S.; and Ishimoto, C. 1977. Critical behavior of density fluctuations in gels. Phys. Rev. Lett. 38,14: 771-774.

(13) Tanaka, T.; Nishio, I.; Sun, S.-T.; and Ueno-Nishio, S. 1982. Collapse of gels in an electric-field. Science 218(4571): 467-469.

Microbial Adhesion and Aggregation, ed. K.C. Marshall, pp. 189-200. Dahlem Konferenzen 1984. Berlin, Heidelberg, New York, Tokyo: Springer-Verlag.

Colonization by Higher Organisms

R. Mitchell
Laboratory of Microbial Ecology
Division of Applied Sciences, Harvard University
Cambridge, MA 02138, USA

Abstract. In natural waters pelagic invertebrate larvae and motile algal spores preferentially colonize substrata coated with microorganisms. Chemotactic responses to bacterial metabolites are involved in the attraction of invertebrates and some algae to surfaces. Settlement and possibly metamorphosis of spirobid larvae is triggered by lectins reacting with specific bacterial exopolymers. A similar mechanism appears to apply to algal settlement. Neither chemotaxis nor lectins control settlement and metamorphosis of red abalone larvae. Induction requires contact. A neurotransmitter, γ-aminobutyric acid, is involved. Nothing is known about genetic control of the cues for development of invertebrates or algae produced by periphytic bacteria.

INTRODUCTION

In aquatic habitats the motile forms of sessile organisms colonize surfaces coated with populations of microorganisms. A specific population of periphytic bacteria apparently controls development of a significant portion of the aquatic population living on surfaces, including many invertebrates and algae. The biochemical processes controlling the interactions are not understood. An explanation of the role of the surface bacteria in the colonization and metamorphosis of eukaryotic organisms would yield important insights into both their developmental biology and ecology. In addition, it would provide new tools for both mariculture and for control of marine fouling.

MICROBIAL CUES FOR INVERTEBRATE SETTLEMENT

Bacterial films appear to be important in the early development of invertebrate communities. Settlement and metamorphosis of the pelagic larvae of many marine invertebrates is dependent upon the presence of a bacterial biofilm on the surface (8). The spirorbid polychaetes seem to be particularly good models for the study of development of invertebrate communities. Larvae of individual species of Spirobis settle preferentially on different surfaces (5). One species develops on stones covered with microbial films, another on the alga, Fucus, while a third species requires the alga, Corallinia. The spirorbid polychaete, Janua, develops on mussel shells, on the green alga, Ulva lobata, and on eel grass, Zostera marina (8).

Janua brasiliensis is ideal for studies of larval settlement since it is small, between two and four millimeters in length, hermaphroditic, and abundant throughout the year. Larvae are obtained readily from external brood sacs located in the operculum. A high percentage settle and undergo metamorphosis within two to four hours. The larvae develop on multispecies microbial films. The films are prepared by immersing surfaces in seawater for seven days at ambient temperatures. The polychaetes do not grow on surfaces devoid of bacterial films or on pennate diatom films. Invertebrate development is typically stimulated by bacterial populations on the order of 10^8 bacteria cm^{-2} (8).

Not all epibenthic bacteria stimulate invertebrate larval development equally. Whereas some bacteria actively stimulate settlement, others are totally inactive. In our studies we found that only about twenty percent of the bacteria growing on surfaces actively stimulated development (8).

In the case of Janua, the microbial film does not have to be actively metabolizing in order to stimulate larval settlement and metamorphosis. Treatment of the film with a combination of chloramphenicol and ampicillin, in order to inhibit prokaryote protein and cell wall synthesis, did not prevent larval settlement. Nor was settlement significantly inhibited by formalin treatment of biofilms (8). Apparently, a viable bacterial film is not necessary for induction of larval development. In contrast to Janua, metamorphosis of the red abalone, Haliotis rebuescens, is induced by metabolic products of crustose coralline algae (15). Active metabolic products of Vibrio also induce morphogenesis of Cassiopea andromeda (17).

The stimulation of settlement by bacterial products may be complicated by the wettability of the surface. Experiments with bryozoan larvae indicate that they prefer to settle on non-wettable substrata (12). However, wettable substrata become attractive to the larvae after formation of bacterial films. Two sensory stimulants may be involved in settlement, wettability and bacterial films (12). When bacterial films form on surfaces, stimulation by the bacterial products appears to override wettability.

Chemotaxis

Most organisms living in aquatic habitats have chemoreceptors, enabling them to respond to low concentrations of specific chemicals in the environment. Larvae of the oyster, Crassostrea virginica, are attracted to multispecies bacterial films (27). Bacterial metabolites, including glucose and amino acids, also attract oyster larvae. It must be emphasized that, at least in the case of oysters, the chemotactic attractant does not provide the necessary metamorphic cue. Metamorphosis only occurs in the presence of bacterial films.

Attraction of invertebrate larvae to microbial products may be complicated by the phenomenon of negative chemotaxis. Oyster larvae are strongly repelled by very low concentrations of alginic acid (27). This chemical is commonly excreted by algae. No information is available about the repulsion of other invertebrate larvae by either bacterial or algal metabolites.

The change from searching behavior to metamorphosis is controlled by a different set of signals. In the case of bottom-living marine invertebrates, where metamorphosis occurs rapidly as a direct consequence of finding an appropriate surface, the sensory cue involves a neurosecretory process (7). This process may involve an external hormone produced by the periphytic microbiota.

Lectins

Lectins are proteins that bind to specific carbohydrate groups. Typically, binding is reversible and lectins have more than one specific carbohydrate binding site. Because lectins are multivalent, they can couple complimentary saccharide-containing compounds within cells or on cell surfaces, thereby influencing cellular or tissue organization (1). Lectins appear to play an important role in developmental regulation. For example, they induce mitogenesis in resting lymphocytes (2). They have also been implicated in differentiation of cellular slime molds (1). Three distinct lectins have been detected during development of chicken embryos (1). Cellular recognition between legume roots and the nitrogen-fixing symbiont, Rhizobium, also involves specific complimentary lectin-polysaccharide interactions (4). Attachment of rhizobia to legume host root hairs occurs by means of hapten-reversible interactions between specific lectin-binding polysaccharides associated with the bacterial cells and lectins on the plant root hairs (26).

Larvae of spirobid polychaetes attach to specific bacterial exopolymers on surface films. Lectins are involved in the attachment process (9). Settlement and metamorphosis of Janua larvae is blocked either by treatment of the larvae with low concentrations of sugars or by immersion of the bacterial film in solution of lectins. Oxidation of the bacterial glycoprotein by periodate also prevents attachment. These data are summarized in Table 1.

Janua larvae apparently produce lectins on their surfaces which bind them to specific glycoproteins produced by the bacterial film (9). Recent experiments using fluorescent antibodies have demonstrated the presence

TABLE 1 - Inhibition of settlement of Janua larvae by blockage of either larval lectins or the bacterial lectin receptors.

Larval Treatment	Inhibition of Settlement
Glucose	+++
2-Deoxyglucose	+
Galactose	--
N-Acetylglucosamine	--
Bacterial Film Treatment	
Periodate	+++
Concanavalin A (Glucose-binding lectin)	+++
AHA (Galactose-binding lectin)	--

+++ More than 70% inhibition
\+ More than 20% inhibition

of lectins at specific sites on the larval surface. The question of the relationship between lectin-triggered settlement and metamorphosis has not been resolved. In the Janua studies the bacterial polymer cues settlement and metamorphosis. However, it is possible that the polymer contains more than one cue. Alternatively, a single cue may trigger complete development of the invertebrate.

The specificity of lectin binding of invertebrate larvae requires more extensive study. Sugar-blocking experiments have shown that bryozoan lectins are species-specific. Chemical studies are needed to determine both the structure of the invertebrate lectins and their bacterial glycoprotein receptors. Physiological control of glycoprotein production by the periphytic bacteria also requires elucidation.

Neurotransmitters

Not all invertebrate development is controlled by chemotaxis to the bacterial film or by lectin binding to bacterial glycoproteins. An entirely different model has been used to explain settlement and metamorphosis of red abalone larvae. Coralline red algae serve as nursery grounds for juveniles of abalone. Chemotaxis to the algae is not observed. The alga does not release soluble or freely diffusible inducers for larval settlement. The larvae are trapped by direct contact with the recruiting alga (15). Extracts of the crustose coralline red alga contain inducers of settlement for red abalone larvae. The active component is γ-amino butyric acid (GABA), a neurotransmitter in higher animals (15). Settlement occurs within seven minutes after the addition of one mM of GABA. Fifty percent settlement is observed with concentrations as low as 10^{-6} molar. The neurotransmitter induces developmental metamorphosis in addition to settlement, including rapid growth of new shell. The accessory photosynthetic pigments responsible for coloration of the coralline red algae, phycoerythrobilin, and its specific protein, phycoerythrin, contain two analogs of GABA. Both pigments are strong inducers of abalone settlement. Other linear and cyclic tetrapyrroles and proteins are inactive (15). Morse and his colleagues have shown an absolute requirement for the carboxyl and amino groups of GABA, and for specific substitution of the amino group at the gamma position (16). Increasing or decreasing the aliphatic chain length of GABA homologs progressively decreased their activity in the induction of metamorphosis. Table 2 shows some of these data.

Stimulation of settlement and complete metamorphosis of abalone by a simple neurotransmitter produced by algae suggests that molecules

TABLE 2 - Specificity of induction of settlement and metamorphosis of abalone larvae (from (15)).

Inducer	Percentage Settled
None	0
Lithothamnium and Lithophyllum species	82
γ- Aminobutyric acid (GABA)	≥99
α- Aminobutyric acid	0
γ- Hydroxybutyric acid	58
δ- Amino-n-valeric acid	89
ε - Amino-n-caproic acid	74

specialized for signal transduction and regulation will be found to mediate interspecific control of larval dispersion, settlement, and metamorphosis in other marine and freshwater species (15). Planktonic larvae of the black chiton, Katharina tunicata, are also induced to settle and undergo metamorphosis by GABA (20). In a study of the stereochemical specificity for GABA for Haliotis metamorphosis, its sensitivity to inhibition by noninducing analogs, and the effects of neuropharmacological probes of GABA function, Morse and his colleagues have shown the involvement in metamorphosis of GABA receptors and/or early processing sites similar to those in mammalian brain (16).

Biochemical, histological, and electronmicrographic analyses suggest that cyclic AMP, calcium, and a glycopeptide secretion from the cephalic sensory complex may mediate transduction of the GABA signal in the control of the behavioral and morphogenic changes induced by this environmentally deployed transmitter substance (14). It would be useful to determine if GABA and its analogs are produced by periphytic bacteria or attached unicellular algae.

SETTLEMENT OF ALGAE

Very little is known about the development of algal communities on surfaces. The adhesion of Chlorella to solid surfaces is enhanced by bacterial polymers. The evidence suggests that the macromolecules responsible are bacterial glycoproteins (25). Periphytic marine bacteria stimulate colonization of surfaces by motile spores of the filamentous marine alga, Enteromorpha (24). As in the case of the polychaete, Janua, there is a high degree of specificity in stimulation of settlement. Some bacteria actively discourage settlement.

There is evidence that chemosensory systems are used to attract algae to surfaces (3). Algae are frequently attracted to metabolites produced by bacteria. For example, the marine dinoflagellate, Crypthecondinium, is attracted to sucrose, glycine, alanine, and fructose (6). The green alga, Dunaliella, is strongly attracted to specific amino acids. The responses occur at concentrations as low as 10^{-6} molar (23). The narrow chemotactic responses of this alga suggest that attachment may be controlled by the presence of specific bacterial metabolites on the surface. Adhesion of algae to bacterial films may be complicated by negative chemotaxis. Bacterial metabolites, including benzoic acid, frequently repel algal cells and spores and prevent adhesion to biofilms. Field studies using sublethal concentrations of repellents showed that algal colonization of surfaces in the sea is prevented by the presence of repellents (13).

The green alga, Dunaliella tertiolecta, attaches more rapidly and in larger quantities to surfaces coated with bacterial films. It is probable that protein-binding sites, or lectins, are involved in the permanent attachment of Dunaliella to bacterial films. The bacteria probably produce the necessary polysaccharides to specifically bind the lectins produced by the algae. We have observed that minimal settlement occurs when the lectin concanavalin A is added to substrata coated with bacteria. Recently Klutt et al. have shown that Limulus lectins agglutinate cells of Dunaliella. Their data indicate the presence of glycoprotein in the surface coat of this alga (10). It is possible that, during adhesion of the algae, glycoproteins on both the bacterial and algal surfaces may combine with their complementary lectins.

Lectins appear to be important in algal settlement. An anti-sialic acid agglutinin has been isolated from the marine red algae Solieris chordalis (19). Glycoprotein binding sites for lectins are found in a wide variety of algae. The lectin-binding patterns of the glycoprotein are species- but not group-specific. Often closely related species are characterized by different lectin-binding properties (22).

The importance of lectins in algal adhesion to surfaces is further emphasized by data obtained in the study of lichens (11). Phycobionts freshly isolated from a lichen show a strong affinity for specific lectins. The mycobionts similarly have lectin binding sites. It would appear that lectin receptors on either or both symbionts are responsible for recognition and attachment of the alga and fungus, yielding the lichen symbiosis.

NEMATODE-TRAPPING FUNGI

The nematode-trapping fungus Arthrobotrys oligosphora captures nematodes, using adhesive capture organs. Capture starts with the firm adhesive of a nematode touching the capture organ and continues with the penetration of the nematode cuticle, followed by invasion and digestion of the nematode by the fungus (13). Studies of these interactions provide useful insights into the adhesion of higher organisms to microbial surfaces. In the presence of the nematode, the fungus produces an increased secretion of a mucilaginous substance (18). This coat is confined to the trap and is not present on hyphae.

Nematophagous fungi are chemotactic for their prey nematodes. Chemotaxis is abolished when the fungi are killed by ultraviolet radiation. Non-nematophagous fungi do not attract nematodes. It seems that chemotaxis is an important phenomenon in the attraction of nematode-trapping fungi.

There is evidence for the presence of a developmentally regulated lectin on the fungal traps (18). Capture of nematodes is specifically inhibited by N-acetyl-galactosamine. The lectin receptors can be demonstrated on the nematode surface. In addition, a model prey containing lectin receptors on the surface, fresh red blood cells, adheres preferentially to the fungal traps.

The evidence clearly indicates that a developmentally regulated lectin is present on the traps of nematophagous fungi and binds to carbohydrates on the nematode surface. This binding initiates an enzymatic alteration of the nematode surface leading to penetration of the nematode cuticle by the fungus (18).

GENETICS

The abalone studies suggest the involvement of a hormonal component in attachment and metamorphosis of sessile organisms. The chemical cue in abalone is GABA, whereas in polychaetes the trigger is a lectin. In both cases, an external chemical cue signals the animal to attach to the surface and undergo metamorphosis. In the case of abalone the genetic control may be either in the larva or in the corralline alga.

Regulation of gene expression in planktonic larvae by GABA is possible and has been suggested by Morse and his co-workers (15).

Genetic control by the bacterial film in lectin-mediated attachment and metamorphosis is also a distinct possibility. Attachment of specific Rhizobia to legumes appears to be encoded on large transmissable plasmids of Rhizobium that control surface polysaccharide production. Conjugal transfer of the nodulation plasmid from Rhizobium to agrobacteria results in a hybrid bacterium which nodulates clover roots (4). Nothing is known about genetic control of exopolymer production in bacterial films. Studies by Silverman and his co-workers (this volume) are aimed at elucidation of genetic control of adhesion of bacteria to surfaces and their production of specific metabolic products.

CONCLUSIONS

Microbial films provide important cues for settlement and development of sessile organisms in aquatic habitats. Metabolic products of periphytic bacteria provide chemotactic signals for attraction of algal spores and invertebrate larvae. Bacterial glycoproteins are involved in settlement of some algae and invertebrate larvae by linkage with the eukaryotic lectins. It is not clear if the bacterial exopolymers are involved in metamorphosis. Another model involving neurotransmitters explains the settlement and metamorphosis of abalone larvae. A very small number of algae and invertebrates have been studied. It remains to be determined if the lectins and neurotransmitters are common cues for settlement and development in planktonic organisms and if they are produced by microorganisms attached to substrata.

Coral development is of particular interest because of the sensitivity and steady decline of coral reefs. Despite the production of huge numbers of coral larvae, colonization rates are very slow. Recently, settlement and metamorphosis of larvae of a temperate soft coral, Alcyonium, was shown to be induced by contact with crustose algae (21). The relationship between development of coral larvae and exogenous cues requires further study.

Invertebrate larvae frequently harbor pure cultures intracellularly. Larvae of the bryozoan, Watersipora, possess large numbers of mycoplasma-like organisms in the visceral coeloms. These microorganisms are internalized during metamorphosis, inoculating the incipient colony (28). Other bryozoans harbor pure cultures of Moraxella in their sinus cavity. These intracellular microbial populations may be functional in development of the invertebrates.

I have confined my discussion to growth of higher organisms on microbial films in aquatic habitats. However, it appears likely that similar phenomena occur in the soil. Very little is known about the contribution of microbial films on soil particles to the metamorphosis of the invertebrate population. This interaction merits study as a means of both providing better insight into the role of periphytic bacteria in developmental biology and obtaining a better understanding of soil biology.

Acknowledgment. Work reported from my laboratory was supported by the U.S. Office of Naval Research, Oceanic Biology Program, and by NOAA Sea Grant.

REFERENCES

(1) Barondes, S. 1983. Developmentally regulated lectins. In Cell Interactions and Development Molecular Mechanisms, ed. K.M. Yamada, pp. 185-202. New York: John Wiley & Sons.

(2) Brown, J.C., and Hunt, R.C. 1978. Lectins. Int. Rev. Cyt. 52: 277-349.

(3) Chet, I., and Mitchell, R. 1976. Ecological aspects of microbial chemotactic behavior. Ann. Rev. Microbiol. 30: 221-239.

(4) Dazzo, F.B., and Truchet, G.L. 1983. Interactions of lectins and their saccaride receptors in the rhizobium-legume symbiosis. Membr. Biol. 73: 1-16.

(5) deSilva, P. 1962. Experiments on choice of substrate by Spirorbis larvae (Serpulidae). J. Exp. Biol. 39: 483-490.

(6) Hauser, D.C.R.; Levandowsky, M.; Hutner, S.H.; Chunosoff, L.; and Hollwitz, J. 1975. Chemosensory responses by the heterotrophic marine dinoflagellate Crypthecodinium cohnii. Microb. Ecol. 1: 246-256.

(7) Highnam, K.D. 1981. A survey of invertebrate metamorphosis. In Metamorphosis a Problem in Developmental Biology, eds. L.I. Gilbert and E. Frieden, pp. 43-72. New York: Plenum Press.

(8) Kirchman, D.; Graham, S.; Reish, D.; and Mitchell, R. 1982. Bacteria induce settlement and metamorphosis of Janua (Dexiospira) brasiliensis Grube (Polychaeta: Spirorbidae). J. Exp. Mar. Biol. Ecol. 56: 153-163.

(9) Kirchman, D.; Graham, S.; Reish, D.; and Mitchell, R. 1982. Lectins may mediate in the settlement and metamorphosis of Janua (Dexiospira) brasiliensis Grube (Polychaeta: Spirorbidae). Mar. Biol. Lett. 3: 131-142.

(10) Klut, M.E.; Bisalputra, T.; and Antia, N.J. 1983. Agglutination of the chlorophycean flagellate Dunaliella tertiolecta by treatment with lectins or divalent cations at alkaline pH. J. Phycology 19: 112-115.

(11) Marx, M., and Peveling, E. 1983. Surface receptors in lichen symbionts visualized by fluorescence microscopy after use of lectins. Protoplasma 114: 52-61.

(12) Mihm, J.W.; Banta, W.C.; and Loeb, G.I. 1981. Effects of adsorbed organic and primary fouling films on bryozoan settlement. J. Exp. Mar. Biol. Ecol. 54: 167-179.

(13) Mitchell, R. 1975. Negative chemotaxis: a new approach to marine fouling control. Office of Naval Research Technical Report. Cambridge, MA: Harvard University Press.

(14) Morse, D.E.; Duncan, H.; Hooker, N.; Baloun, A.; and Young, G. 1980. GABA induces behavioral and developmental metamorphosis in planktonic molluscan larvae. Fed. Proc. 39: 3237-3241.

(15) Morse, D.E.; Hooker, N.; Duncan, H.; and Jensen, L. 1979. γ-Aminobutyric acid, a neurotransmitter, induces planktonic abalone larvae to settle and begin metamorphosis. Science 204: 407-410.

(16) Morse, D.E.; Hooker, N.; and Duncan, H. 1980. GABA induces metamorphosis in Haliotis, V: Steriochemical specificity. Brain Res. Bull. 5: 381-387.

(17) Neumann, R. 1979. Bacterial induction of settlement and metamorphosis in the planula larvae of Cassiopea andromeda (Cnidaria: Scyphozoa, Rhizostomeae). Mar. Ecol. Prog. Ser. 1: 21-28.

(18) Nordbring-Hertz, B., and Mattiasson, B. 1979. Action of a nematode-trapping fungus shows lectin-mediated host-microorganism interaction. Nature 281: 477-479.

(19) Rogers, D.J., and Topliss, J.A. 1983. Purification and characterization of an anti-sialic acid agglutinin from the red alga Soliera chordalis (C. Ag.) J. Ag. Botan. Mar. 26: 301-305.

(20) Rumrill, S.S., and Cameron, R.A. 1983. Effects of gamma-aminobutyric acid on the settlement and metamorphosis of the black chiton, Katharina tunicata. Mar. Biol. 72: 243-247.

(21) Sebens, K.P. 1983. Settlement and metamorphosis of a temperature soft-coral larva (Alcyonium siderium verrill) induction by crustose algae. Biol. Bull. 165: 286-305.

(22) Sengbusch, P.V., and Miller, U. 1983. Distribution of glycoconjugate at algal cell surfaces as monitored by FITC-conjugated lectins. Studies on selected species from Cyanophyta, Pyrrhophyta, Raphidophyta, Euglenophyta, Chromophyta, and Chlorophyta. Protoplasma 114: 103-113.

(23) Sjoblad, R.D.; Chet, I.; and Mitchell, R. 1978. Chemoreception in the green alga Dunaliella teriolecta. Curr. Microbiol. 1: 305-307.

(24) Thomas, R.W.S.P., and Allsopp, D. 1983. The effects of certain periphytic marine bacteria upon the settlement and growth of Enteromorpha, a fouling alga. In Biodeterioration 5, eds. T.A. Oxley and S. Barry, pp. 348-361. New York: John Wiley & Sons.

(25) Tosteston, T.R., and Corpe, W.A. 1975. Enhancement of adhesion of the marine Chlorella valgaris to glass. Can. J. Microbiol. 21: 1025-1031.

(26) Tsien, H.C., and Schmidt, E.L. 1980. Accumulation of Soybean lectin-binding polysaccharide during growth of Rhizobium japonicum as determined by hemagglutination inhibition assay. Appl. Envir. Microbiol. 39: 1100-1104.

(27) Young, L.Y., and Mitchell, R. 1973. The role of microorganisms in marine fouling. Int. Biod. Bull. 9: 105-109.

(28) Zimmer, R.L., and Woollacott, R.M. 1983. Mycoplasma-like organisms: occurrence with the larvae and adults of a marine bryozoan. Science 220: 208-210.

Standing, left to right:
Tsutomu Hattori, Keith Cooksey, Ronald Gibbons, Dwayne Savage, Allan Hamilton, Hans Güde, Friedrich Eckhardt, Zdenek Filip

Seated, left to right:
Hans-Georg Hoppe, Madilyn Fletcher, Moshe Shilo, Ann Matthysse, John Breznak

Microbial Adhesion and Aggregation, ed. K.C. Marshall, pp. 203-221. Dahlem Konferenzen 1984. Berlin, Heidelberg, New York, Tokyo: Springer-Verlag.

Activity on Surfaces
Group Report

J.A. Breznak, Rapporteur
K.E. Cooksey
F.E.W. Eckhardt
Z. Filip
M. Fletcher
R.J. Gibbons
H. Güde
W.A. Hamilton
T. Hattori
H.-G. Hoppe
A.G. Matthysse
D.C. Savage
M. Shilo

INTRODUCTION

Natural environments contain a bewildering array of surfaces potentially available for attachment and colonization by microorganisms. Such surfaces range from inorganic to organic, animate to inanimate, to artificial (i.e., man-made) forms. In many cases the substratum to which microbes attach can also constitute a metabolizable substrate.

In an effort to explore the mechanisms and consequences of microbial adhesion and aggregation, this discussion group addressed its attention to the activities of microorganisms on surfaces. Topics discussed ranged from the influence of surface attachment on microbial activities to unique adaptations of attached organisms. Various surface types were discussed, ranging from inorganic particles to living tissue surfaces of plants and animals. An attempt was made to point out deficiencies in our knowledge, as well as important areas for future research. The results of these discussions constitute the substance of the present report.

DO ATTACHED BACTERIA EXHIBIT PREDICTABLE PATTERNS OF ACTIVITY?

The effect of attachment on bacterial physiology has been investigated by comparing the activities of attached cells with those of homologous cells in the bulk liquid phase. Activities measured have included substrate uptake and utilization, respiration (as measured by O_2 uptake or CO_2 release), heat generation, viability, and growth. However, no consistent pattern of activity has been discernible, inasmuch as attachment can result in enhancement or depression of activity or no effect at all, depending on the specific microorganisms being studied, the surface to which they are attached, or the phenomenon being measured. Indeed, the existing literature is fraught with inconsistencies at the present time. A major source of these inconsistencies may be the lack of methods for critically distinguishing between attached versus unattached cells (Fletcher, this volume).

To better understand the effects of attachment on microbial activity, it would help greatly to know more about the microenvironment in which attached cells reside. However, our knowledge about microenvironments is meager, because methods to analyze microenvironments are not well developed. Consequently, if progress is to be made in this area, more consideration must be given to the experimental model system and the specific kind of activity one wishes to measure. This entails a) limitation and better characterization of variables including the type of microorganism being studied, the substratum, and the type of activity being measured; b) more complete characterization of the experimental system with respect to mass, heat, momentum, and interactive force-transfer boundary layers; and c) more attention to the kinetic characteristics of the activity under consideration. Even with these strategies, one still must be careful that procedures used for attachment of microbial cells to substrata have not selected for a subpopulation that is inherently more or less active. This has been found to occur for some marine pseudomonads (3). By using these guidelines, recognizable and reproducible patterns of activity might be elucidated and afford predictive value for more complex, natural environments typified by soil, where properties such as mass, surface tension, shape, and surface area of the substratum are either poorly characterized or difficult to estimate.

Many activities critical to growth or survival of attached bacteria are associated with the cell envelope (e.g., nutrient uptake, respiration, DNA replication), and it is possible that attachment somehow causes transient disorder of cell membranes (Fletcher, this volume). Nevertheless, it was thought that it would be premature to embark on sophisticated approaches to test this hypothesis (e.g., use of spin-labeled probes) until more is known about attachment effects on activity by using better defined systems as mentioned above.

ADSORBED MACROMOLECULES AND EXTRACELLULAR ENZYME ACTIVITY

Although surfaces commonly provide a site for adhesion by aquatic bacteria, careful microscopic examinations have revealed that not all particles in some aquatic biotopes are colonized. Indeed, in many environments inorganic particles are devoid of attached bacteria, whereas those particles seen to contain an attached biota are usually organic in nature (Hoppe, this volume). Since solutes adsorbed to inanimate surfaces in aquatic biotopes are more likely to be macromolecules (proteins, polysaccharides, and nucleic acids) than simple organic compounds, more studies are needed on the precise nature of such adsorbed macromolecules, because organisms that can metabolize them are more likely to be at such surfaces than in the bulk phase. In this regard, not only the chemical nature of the macromolecules should be identified, but the manner in which they adsorb as well, since this may affect their degradability by an attached biota. On the other hand, both the ion species and concentration in the bulk phase, as well as ionic saturation of exchange sites (and consequently the electrokinetic charge) should be considered factors regulating microbial adhesion on inorganic surfaces. In addition, the role of bacteria in loose contact with adsorbed macromolecules should not be overlooked. For example, spirochetes of the genus Leptospira have been observed to "crawl" over surfaces and utilize fatty acids in the process (12).

If macromolecules serve as important substrates for attached bacteria, it seems likely that extracellular enzymes should be of major importance for the nutrition of the cells. By "extracellular enzymes," it is meant those enzymes that are located external to the cell membrane and which may be either periplasmic, cell surface-bound, or diffusible. The above hypothesis has received some experimental support. In one system examined by Hoppe (this volume), bacteria attached to particles in marine habitats were found to possess increased extracellular hydrolytic activity compared to unattached bacteria, as assessed by using methylumbelliferon-linked sugars and amino acids as substrates. Similar observations were made for the extracellular enzymatic decomposition of dissolved radiolabeled proteins (10). Hydrolysis of adsorbed polymers by such enzymes could provide nutrients not only for the attached cells, but for organisms in the bulk phase as well, because a considerable fraction of the hydrolytic products is not immediately incorporated by the attached cells (10, 11). More work is needed to evaluate extracellular enzymes that hydrolyze native macromolecules, including assessments of whether such enzymes catalyze exo- or endocleavage of the substrates. It should be borne in mind, however, that extracellular enzymes may become inactivated on some substrata. For example, fungal decay of polyurethane, based on enzymatic hydrolysis, is inhibited by clays (Filip and Hattori, this volume). Indeed, extensive studies should be done to elucidate the role of particulate substrata in the microbial formation and environmental activity of extracellular enzymes - particularly those enzymes (e.g., phenoloxidases) which affect the fate of xenobiotics in soils and sediments.

NUTRIENT AVAILABILITY AT SURFACES IN OLIGOTROPHIC ENVIRONMENTS

Bacteria in the bulk phase of nutrient-poor habitats may have adaptations that allow them to survive and grow there. In this regard, two general groups of bacteria can be defined: copiotrophs and oligotrophs ((16), and Kjelleberg, this volume). Copiotrophs are those organisms that do not grow (and are probably starving) in dilute nutrient environments, but which have the ability to rapidly oxidize substrates if the concentration of nutrients increases. Oligotrophs are those that appear to grow in dilute environments, albeit slowly, but which do not respond as rapidly to changes in nutrient flux as do copiotrophs. Among the most significant adaptations of oligotrophs is their possession of high affinity (i.e., low k_S) systems for nutrient uptake. In addition, the attachment of cells (either oligo- or copiotrophs) to a solid substratum may confer an additional advantage in terms of nutrient availability. This was suggested by early experiments with E. coli, whereby cells in very dilute media could not grow unless glass beads were added (9).

Bacteria in aquatic environments are predominantly Gram-negative, and such cells may increase their hydrophobicity in oligotrophic conditions (Kjelleberg, this volume). This would enhance their ability to stick to hydrophobic surfaces after random collisions. If the surfaces contain adsorbed macromolecules, these are probably exploited by attached cells via extracellular hydrolytic enzymes. However, the degree of availability to attached cells of low molecular weight nutrients will depend, in part, on whether such nutrients and the substratum are charged. For example, it can be estimated that a doubling of the electrostatic potential at the substratum surface will lead to an approximate 7-fold increase in the concentration of counter-ions at the surface for a fixed bulk counter-ion concentration. This phenomenon may be quite beneficial to bacteria attached to charged surfaces, because nutrients most often limiting in oligotrophic habitats are charged species (i.e., NH_4^+, NO_3^-, or PO_4^{-3}). In this regard, it is pertinent to note that stimulation of nutrient assimilation by attached bacteria in vitro has been observed with certain charged compounds (21).

By contrast, with respect to low molecular weight uncharged nutrients, the chemical activity and hence availability is not likely to be greater at the uncolonized surface than in the bulk phase at equilibrium. Nevertheless, despite the fact that uncharged low molecular weight compounds may not accumulate at surfaces, bacteria which attach may still enhance their ability to use such nutrients. For example, if soluble, freely diffusible substrates are utilized by organisms in the bulk phase, the concentration of such substrates will be reduced in the immediate vicinity of the organism (Filip and Hattori, this volume). Replenishment of this zone of lower concentration will take place by diffusion. If the rate of diffusion is less than the intrinsic rate of utilization by the organism, a layer known as the mass transfer boundary layer will be formed. Mixing of the bulk phase will have little effect on the size of this layer because bacteria are of similar density to the bulk phase and, therefore, inertial forces will be minimal (i.e., the bacteria move with the fluid). If the bacterium attaches to a particle of increased density, the effective inertia of the cell increases and mixing will alter its microenvironment. When the particle is infinitely large with respect to the bacterium (i.e., a surface) and when the flow of liquid is highly turbulent relative to the surface, the mass transfer resistance will be greatly diminished and the concentration of nutrient adjacent to the cell will approach that of the bulk phase.

Yet another potential advantage to an attached microbiota is the ability to exert better regulation or control of its microenvironment. This could have significant effects on a substratum consisting of living cells or tissue.

In view of these considerations, enhanced availability of nutrients at the surface may result from adhesion to charged surfaces, the development of nutrient concentration gradients created by the cells, and liquid flow. These factors could be augmented by intrinsic high affinity uptake systems present in the cells themselves.

ACTIVITIES THAT MAY BE UNIQUE TO ATTACHED MICROORGANISMS

Microorganisms have evolved an intriguing variety of strategies that permit them to exploit as well as escape from the attached state. This is not surprising, inasmuch as attachment to surfaces may impose a variety of constraints as well as advantages to microorganisms, and adaptation to such situations would constitute a seemingly strong selective pressure. This should be true not only for free-floating, nonmotile organisms, but those with well adapted positioning mechanisms as well (e.g., gas vacuoles, chemo-, photo-, magnetotaxis, etc.). Primary adaptations are therefore seen as those that allow the cells to control the formation and maintenance of adhesion mechanisms, to exploit the potential benefits of adhesion, and to provide for means of detachment and dispersion.

The ability of bacterial cells to modify their surface chemistry under adverse nutrient conditions has been observed. For example, certain copiotrophic marine bacteria have been found to develop an increasingly hydrophobic surface when starved (Kjelleberg, this volume). This would increase their affinity for hydrophic surfaces and should enhance colonization, possibly affording better nutritive conditions at the surface. Indeed, it has been generally observed that survival of bacteria is improved if they are attached to surfaces. Another adaptation to a surface is seen with the marine bacterium Vibrio parahaemolyticus. These cells induce the synthesis of lateral flagella when they contact a solid surface. This enables them to swarm over surfaces in marine environments and presumably is of survival advantage (Silverman et al., this volume).

Bdellovibrio is a spectacular example of a bacterium that has attachment as a necessary phase of its life cycle in nature. In addition, this predatory bacterium has a remarkable strategy for exploitation of its substratum, which also constitutes its substrate (i.e., a bacterial host cell). Upon attachment of Bdellovibrio to a substrate cell, a cascade of enzymes is synthesized including LPSase, glycanase, deacetylase and O-acylases (18). These enzymes are induced only after attachment to its substrate cell and allow Bdellovibrio to penetrate its host, seal itself within the substrate cell, exclude potential competitors, stabilize the substrate cell, and grow within its periplasmic space.

In contrast to the situation for Bdellovibrio, we know very little about adaptations in bacteria that do not have attachment as a necessary phase in their life cycle. However, it seems almost certain that mechanisms for detachment and dispersal from substrata would be paramount among these. This type of adaptation is readily seen with organisms that produce motile daughter cells (e.g., Caulobacter and Hyphomicrobium). Many cyanobacteria form hormogonia that possess a hydrophilic surface layer as opposed to the hydrophobic layer of the mother filament. This allows the hormogonium to escape from a hydrophobic surface to which the mother filament may be attached (4). Mechanisms for detachment of cyanobacteria under unfavorable conditions have also been observed. For example, hydrophobic cyanobacterial cells can be induced to change their surface to a hydrophilic nature in the presence of chloramphenicol; they then detach from a substratum. Still other intriguing strategies have been seen with bacteria that have no special life cycles. This includes the production of emulsifiers which are amphipathic and which coat the otherwise hydrophobic surface, permitting detachment. Emulsan produced by bacteria (19) attached to oil appears to function in this way. Phase variation of fimbriae present on host-associated organisms (e.g., Neisseria) provides not only a means for detachment and dispersal, but also evasion of surface antibodies secreted from host tissues.

Other adaptations relevant to the attached state of bacteria involve metabolic flexibility. Indeed, the very attachment of cells to particles imposes the requirement that cells be metabolically flexible, if they are to survive (20). This is essential for cells that are moved into unfavorable environments as a result of movement of the particles to which they are attached. Again, good examples are seen with benthic cyanobacteria. These adaptations involve high tolerance to H_2S, facile shifts from oxygenic to anoxygenic photosynthesis, and ability to readily respire or ferment polyglucose reserves or to carry out reduction of sulfur compounds to H_2S in the dark. In addition, some cyanobacteria produce bioflocculants which are thought to clear the path of downwelling light by removal of shadowing organisms or particles in the overlying water column (1).

In view of the remarkable variety of strategies utilized by microbes to cope with or exploit the attached state, a better understanding of the physiological, biochemical, and genetic control of these activities is sorely needed.

MICROORGANISMS ATTACHED TO LIVING SURFACES

Associations between attached bacteria and their substratum become increasingly complex when the substratum is a living cell or animal or plant host tissue. Intrinsic to such situations is the dynamic state of the association. These dynamics range from host factors designed to limit or avoid colonization by microbes, to cooperative effects between the host and its associated microbiota.

Living surfaces can be roughly divided into three general categories: cuticular, squamous, and absorptive/secretory. Each has rather unique characteristics. Cuticular surfaces are those that are reasonably "inert" and are typified by tooth surfaces. Such surfaces are nonsecretory, although they may be bathed by secretions arising elsewhere. Squamous surfaces mimic cuticular surfaces in that they approximate inert surfaces, but are dynamic in that they continually slough and are replaced. An example is the squamous epithelium of animal intestinal tracts. As with cuticular surfaces, these also are bathed with secretions usually arising elsewhere in the ecosystem. The cuticular and squamous surfaces are probably the closest living tissue analogue of inert surfaces encountered in aquatic and soil habitats. Absorptive-secretory surfaces are by far the most complex. These surfaces in animals should actually be thought of as dynamic, three-dimensional compartments consisting of a glycocalyx layer of glycoproteins and glycolipids, and an overlying "cloud layer" of mucus (glycoproteins) which may be several μm thick. The "cloud layer," in which associated microorganisms exist, is an extremely complex microenvironment. It may be a viscous polymer gel or zone created by the secretion-absorption functions of the tissue cells. Such gels may have ion exchange and molecular sieve properties. The microenvironment of the cloud layer undoubtedly contains pH, ionic and nutrient gradients created by the tissue itself, the attached microbiota, and the flow of secretions through it. A marked difference between absorptive-secretory surfaces and the quasi-inert surfaces mentioned above is the reversed nutrient gradients that occur due to uptake of potential microbial nutrients by the tissue itself. Indeed, the colonized surfaces of animal intestinal tracts, although sometimes thought of as "nutrient-rich," are probably nutrient-poor for the associated organisms. In these sites, attached microbes undoubtedly find themselves in fierce competition with the host for nutrients. Microbes in the oral cavity or gut may only divide a few times daily.

Given these properties of living surfaces, adhesion of microbes to such surfaces may involve dynamic processes where bonds continuously form and break unless they are somehow stabilized by architectural changes in the cell membranes and microbial wall. It is not surprising, then, that a number of oral and gut organisms, as well as plant-associated bacteria, exhibit rather diverse mechanisms for adhesion to their living substratum, including the possession by the microbes of adhesive proteins, lipoteichoic acids, and insoluble polysaccharides (Jones and Dazzo, both this volume). In many cases this binding is facilitated and exhibits specificity by the involvement of lectins produced and secreted by the substratum or by the adhering microbes.

Microorganisms attached to living surfaces must contend with constant "cleansing forces" operative on the substratum and which reflect the host's effort to limit or avoid colonization. These forces are both mechanical and biochemical. Mechanical forces include desquamation or sloughing of animal cells mentioned above. Some microbes cope with this by rapidly colonizing newly exposed surfaces before the epithelial cells completely detach. A striking example of sloughing is seen with invertebrates such as insects, where periodic molting involves sloughing of the entire chitinous lining of the proctodeum and most of its attached biota. Presumably, recolonization occurs before the entire lining is voided, although this has not been carefully studied. Behavioral idiosyncracies of some social insects (e.g., termites), however, insures recolonization as newly molted larvae solicit hindgut contents from intermolt colony mates. This constitutes a survival advantage for the insect, too, however, inasmuch as the hindgut microbiota is essential for wood digestion in the insect.

A major mechanical factor of animals is fluid movement over epithelial surfaces. Such flow may be facilitated by tongue and cheek movement (mouth), ciliary movement (upper respiratory tract), and peristalsis (alimentary tract). Flow is particularly rapid in the vertebrate small intestine. However, in rodents, morphological adaptations of filamentous bacteria in this region allow them to insert end-on into the tissue surface by a plug-and-socket mechanism and presumably resist washout in this way. Morphogenetic factors, including endospore formation by these filaments, is undoubtedly another adaptation to this environment (Savage, this volume).

Chemical "cleansing" factors of animal surfaces include the secretion of acids and bile salts, as well as enzymes of digestion and lysozyme. These may modify the surfaces of bacteria that might otherwise bind to the tissues. Shedding of tissue receptors, as well as production by tissues of adhesin-masking antibodies, also inhibit bacterial attachment. The production of chelators by host tissue (lactoferrin, statherin, Ca^{2+}-binding proteins) is a further competitive factor imposed on the bacteria.

Plant root surfaces display phenomena analogous to desquamation as the root cap cells abrade when the root grows through the soil. In addition, fluid flow from mucigels, as well as environmental abrasion (rain, wind, and freezing) are factors with which plant surface-associated bacteria must contend.

Chemical factors of plant surfaces that limit colonization include cuticular waxes, fatty acids, and different toxic compounds such as low molecular weight compounds induced by injury (e.g., phytoalexins).

Despite the array of colonization-limiting factors imposed by host tissues and surfaces, there are certain cooperative interactions that suggest that certain of these associations are highly co-evolved. Noteworthy is the fact that germfree animals are unusual in almost every physiological and biochemical respect from their normally colonized counterparts. Indeed, in termites this co-evolution is underscored by the digestive symbiosis that has become established and which allows the insects to exploit a relatively refractile food resource.

In plant systems, specific attachment of rhizobia to root surfaces is critical to the complex changes that occur during the development of the N_2-fixing symbiosis. Indeed, leghemoglobin synthesis in this symbiosis is an activity shared by both the rhizobia (heme synthesis) and plant (globin synthesis). Globin mRNA synthesis by the plant only takes place 3-4 days after infection of root hairs by the rhizobia. A somewhat similar cooperative scenario is seen during Agrobacterium infection of plants, wherein transcription of tumorigenic genes carried by the bacteria occurs only in the plant cells.

Such cooperative interactions between animals or plants and their associated bacteria add a significantly greater and exceedingly complex dimension to studies of "bacterial attachment and aggregation." While such systems are obviously highly co-evolved, available information is not yet sufficient to draw firm conclusions regarding mechanisms of co-evolution. This is an important area of future investigation that should be emphasized.

IMPORTANCE OF MOTILITY AND CHEMOTAXIS TO ATTACHMENT AND ACTIVITY

At the present time it is impossible to make a general conclusion regarding this issue, because data are scarce. Certainly tactic phenomena must be important to survival of free-living bacteria in nature, and it is not difficult to imagine how magnetotaxis, for example, might position bacteria close to aquatic sediments, whereas photo- or aerotaxes might favorably position bacteria at or close to air-water interfaces. However, the exact conditions under which such taxes play a role in attachment is not known. Indeed, most bacteria at air-water interfaces appear to be transported there by rising air bubbles, whereas larger particulate materials in aquatic systems settle too fast to allow motile bacteria to swim towards them and firmly attach.

Regarding attachment and activity of bacteria, it may be better to think of advantages from motility/chemotaxis in terms of distance of the bacteria from a prospective surface. Considering that a boundary layer present at a surface may be on the order of 100 or more µm thick (23), motile bacteria present in this region might make effective use of cues emanating from the surface (possibly from chemicals liberated by attached microbes) and either swim chemotactically to the surface and attach there, or continue to migrate in the boundary layer in response to favorable gradients present in the region. By contrast, motile chemotactic bacteria further than, say, 100 µm from a surface would probably be able to reach the surface through directed movement only if the surface were stationary with respect to the cell. In the bulk phase of flowing aquatic systems, it seems unlikely that motility or chemotaxis would have any significant effect. Moreover, a general observation is that most bacteria seen attached to soil particles are coccoid and nonmotile.

In contrast to the situation for free-living bacteria, motility and chemotaxis appear to play an important role in tissue colonization by host-associated intestinal bacteria. This seems to be true for pathogens as well as nonpathogens, wherein it appears that many nonmotile or non-chemotactic mutants have a reduced ability to locate near and adhere to host tissues compared to the wild type strain ((22), and references therein). However, the exact mechanism by which motility/chemotaxis aids colonization is unknown. In the rhizosphere, motility and chemotaxis could also be significant and might be strongly influenced by plant root secretions.

EFFECT OF ATTACHMENT ON MICROBIAL CONSORTIA

In nature, decomposition of organic compounds, as well as cycling of inorganic elements, often involve the combined activities of two or more bacterial species. This is particularly apparent for decomposition of biopolymers by anaerobic microbial communities. Such natural communities or "consortia" are often strikingly complex (Hirsch, this volume), showing temporal and structural heterogeneity, but exhibiting functional homogeneity based on intercellular fluxes of organic carbon compounds, inorganic electron acceptors, and reducing equivalents. Nature has apparently selected for physiological groups of organisms designed to carry out particular phases of the decomposition. Although many of the species can be grown in pure culture, some cannot because the conversions they carry out are thermodynamically unfavorable. Examples of such organisms are the fatty acid-oxidizing, obligate proton reducers. These bacteria oxidize fatty acids of three carbon chain length or longer to acetate and H_2 (propionate is also a product with odd-number, long-chain fatty acids). The overall reaction is thermodynamically unfavorable under standard conditions, and these bacteria can only be grown in the presence of avid H_2-consumers (e.g., methanogens or sulfate reducers) which shift the thermodynamics to a favorable free energy change (25).

It might be expected that such organisms would occur together as attached consortia - i.e., either attached together as microbial aggregates, or attached together on surfaces (biofilms). In fact, studies on the anaerobic decomposition of sugar waste from beet-processing plants have shown that "pellets," consisting of aggregated bacteria, can carry out a complete fermentation of such waste to CO_2 and CH_4. Little is known about the morphological and functional structure of such pellets, but it seems certain that the aggregated state is critical to effective bioconversion.

Consortia in the form of biofilms are particularly important in biodeterioration processes, such as anaerobic microbial corrosion of steel. When corrosion coupons are exposed in a range of aquatic environments, they develop over a period of weeks (the exact time being influenced by the organic loading, pollution, etc., of the particular environment) a complex microbial surface community. These biofilms are found to contain aerobic and facultative heterotrophs, anaerobes, and, most particularly, sulfate-reducing bacteria. Even with relatively thin films, the oxygen uptake by the aerobic species cannot be compensated for by the oxygen diffusion and, consequently, anaerobic conditions are established within the biofilm. As one direct consequence of this, the heterotrophic catabolism of the primary nutrients present in the bulk phase will be incomplete with the formation of a range of fermentation products; evidence from sediments suggest that H_2 and acetate may be especially important. These conditions of anaerobiosis and the products of partial breakdown of primary nutrients are ideal for the growth and activity of methanogens, or, in sulfate-rich environments (e.g., marine), the sulfate-reducing bacteria. Sulfate-reducing bacteria are directly implicated in the anaerobic corrosion of metals, but clearly it is the formation and activity of the biofilm as a complete microbial consortium that is of fundamental importance. It is interesting to note that the substratum itself might play a direct role both in the corrosion process and in the biological activity of the biofilm. The classical theory of the mechanism of anaerobic corrosion (currently subject to considerable questioning) proposes that H_2, produced at the cathode of the electrochemical corrosion cell, is used by the sulfate-reducing bacteria and oxidized as a source of energy (8).

Our understanding of natural microbial consortia is in its infancy, but this should be recognized as an important topic for further research. New and unique experimental methods will clearly be needed for such analyses, particularly methods that exert little or no perturbation of the natural community. Light and electron microscopy will continue to be useful as a means to enumerate and, for cells with distinctive morphology, identify organisms in consortia, but will reveal limited information about the activity of the microbes. Conversely, the classical methods of bacterial isolation and quantitation alone may indicate which microbes are present in consortia, but will tell little about their three-dimensional distribution in the consortia and their dynamic physiological relationships to other organisms. Similar drawbacks exist with radiotracers, although these techniques as well as NMR analyses will give clues as to the dynamics of biological activities. Fourier transforming infrared spectroscopy (FT/IR) is one of the newest developments for nondestructive micro-sampling and holds promise as an important and sensitive tool for studying microbial consortia (White, this volume). Such a technique may be far more important than microprobes to future research on consortia and biofilms, because of minimal perturbation effects.

DO MICROENVIRONMENTS RESULTING FROM ATTACHMENT EXERT PROTECTIVE OR INHIBITORY EFFECTS ON THE ATTACHED MICROBIOTA?

Microenvironments can possess or create a tremendous array of potential situations for an attached biota, and it is important to consider whether protection from or susceptibility to environmental stresses will accrue to such populations. However, it is presently impossible to predict for every case whether or not microbes will be protected by attachment. The potential abiotic and biotic factors involved are too complex. While it is known that attachment of a microbe in a biofilm or floc will protect such cells from attack by viruses and Bdellovibrio, the response to other factors are poorly understood. Accordingly, Table 1 is intended to provide only a general overview of probable consequences of attachment, by environmental parameter: a "best guess," so to speak, that is for some cases supported by experimental data. It should be emphasized that many different environmental situations exist in nature, and the effects will undoubtedly vary from habitat to habitat. Therefore, Table 1 should be thought of in relative terms, e.g., bacteria within the biofilm of a soil particle will probably be grazed upon by a predator, but perhaps less efficiently than would be a loosely attached or free bacterium. Moreover, it should be borne in mind that certain advantages of attachment (protection from predation) might involve trade-offs (exposure to oxygen deficiencies). By and large, however, it was the consensus of our group that the protective aspects derived from attachment may well be one of the driving forces selecting for the attached/aggregated state in aquatic environments and perhaps in soil.

PHYSICAL MODELS FOR STUDYING THE ACTIVITY OF ATTACHED ORGANISMS

Abiotic surfaces may be composed of organic or inorganic substances, or both. The substratum may also constitute a substrate. Extensively used model substrata have included glass beads, clay minerals, and ion exchange resins. Materials like polystyrene or latex may also be used, and the composition and properties of these can be defined reasonably accurately. Moreover, polylattices that can be chemically polymerized afford the investigator the opportunity to insert particular functional groups (e.g., methyl groups) or charged sites at predetermined positions, so that the effects of various surfaces on attached organisms can be tested. However, the widespread or prescribed use of one specific type of surface is both unlikely and unwarranted. Different organisms would be likely to react in different ways and could also change the nature of the substratum variably. In essence, the choice of a laboratory model system should depend on what the investigator wishes to learn. For studies of the adhesion process per se, it would be advisable to select an organism with a genetic system amenable to analysis. In this way specific mutants may be obtainable, and genetic coupled with biochemical, physiological, and morphological information can be obtained. For studies of particular ecosystems, it would be advisable to select the best laboratory system thought to mimic the natural state. In either case, the essential consideration is to standardize and characterize the substrata as completely as possible and to define carefully the parameters one wishes to evaluate.

For animate model systems, the investigator has equally little choice, inasmuch as these are inherently complex. Similar recommendations

TABLE 1 - Probable effect of bacterial cell state on cell activities and resistance to environmental stress[a].

Cell Activities	State of Cell: Free, Dispersed	Singly Attached or Monolayer	Multilayer Biofilm or Floc	Selected References
1. Nutrient uptake from bulk phase	E	E/D	D	—[d]
2. Frequency of cross-feeding	D	E/D	E	(7)
3. Frequency of genetic exchange	?	E/D	E/D	—
4. Insoluble polymer degradation	D	E	E	(Hoppe, this volume)
5. Metabolic sequences	D	E/D	E	(24)
6. Anaerobic growth in aerobic bulk phase	D	E/D	E	(White, this volume)
Environmental Stresses				
Abiotic				
1. Presence of toxic substances	S	P/S	P	(2)
2. Suboptimal pH	S	P/S	P/S	—
3. Desiccation	S	P/S	P	—
4. Thermal stress	S	P/S	P	(5)
Biotic				
1. Grazing by suspension feeders:				
- Macroparticle feeders[b]	P	S	S	(17)
- Microparticle feeders[c]	S	P	P	(6)
2. Grazing by "scrapers"	P	?	?	(15)
3. Attack by:				
- Bdellovibrio	S	P/S	P	(13)
- bacteriophage	S	P/S	P	(14)

[a]Symbols: E, enhanced; D, depressed; E/D, enhanced or depressed; P, protected; S, susceptible; P/S, protected or susceptible.
[b]Particles >1 μm diameter.
[c]Particles <1 μm diameter.
[d]Consensus of discussion group.

are made here as for the inanimate systems. The model will depend on the questions under investigation, but the system should be characterized as completely as possible.

SUMMARY

Attachment to a surface can undoubtedly affect the activity of microorganisms, although sometimes in ways that are not readily predictable based on our current knowledge. This reflects our poor understanding of microenvironments that exist or are created by attachment of cells to a surface, particularly complex surfaces such as plant or animal tissues. It also seems clear, however, that certain microbes have evolved rather spectacular strategies to exploit or escape from the attached state: strategies that are probably of great survival value in nature. In an effort to underscore deficiencies in our understanding of attachment and microbial activity, we have formulated a "wish list" of items that deserve attention for future research, as follows:

1. Better methods are needed to analyze microbial consortia without (or with minimum) perturbation of the system.
2. Because the nature of the cell surface is critical to the dynamics of attachment/detachment, more information is needed on the physiological and genetic modulation of cell surface structure and chemistry.
3. More information is needed on the physicochemical properties of prevalent adsorbants in biotopes.
4. Methods are needed that will discriminate the activities of attached versus unattached members of a microbial community.
5. A better understanding is needed of how a substratum affects attached microorganisms, i.e., by affecting the microbes directly? by affecting the cells' microenvironment? or both?
6. A better understanding is needed of microenvironments of attached cells, in terms of nutrient concentrations, diffusion gradients, size of boundary layers, etc.

REFERENCES

General

Beachey, E.H. 1980. Bacterial Adherence. London: Chapman and Hall.

Berkeley, R.W.C.; Lynch, J.M.; Melling, J.; Rutter, P.R.; and Vincent, B. 1980. Microbial Adhesion to Surfaces. Chichester: Ellis Horwood.

Bitton, G., and Marshall, K.C. 1980. Adsorption of Microorganisms to Surfaces. New York: John Wiley.

Characklis, W.G., and Cooksey, K.E. 1983. Biofilms and microbial fouling. In Advances in Applied Microbiology, ed. A.E. Laskin, vol. 29, pp. 93-138. New York: Academic Press.

Fletcher, M., and Marshall, K.C. 1983. Are solid surfaces of ecological significance to aquatic bacteria? In Advances in Microbial Ecology, ed. K.C. Marshall, vol. 6, pp. 199-236. New York: Plenum.

Marshall, K.C. 1976. Interfaces in Microbial Ecology. Cambridge, MA: Harvard University Press.

Specific

(1) Avnimelech, Y.; Troeger, B.W.; and Reed, L.W. 1982. Mutual flocculation of algae and clay: evidence and implications. Science 216: 63-65.

(2) Bitton, G., and Freihofer, V. 1978. Influence of extracellular polysaccharides on the toxicity of copper and cadmium toward Klebsiella aerogenes. Microb. Ecol. 4: 119-125.

(3) Bright, J.J., and Fletcher, M. 1983. Amino acid assimilation and electron transport system activity in attached and free-living marine bacteria. Appl. Envir. Microbiol. 45: 818-825.

(4) Fattom, A., and Shilo, M. 1984. Hydrophobicity as an adhesion mechanism of benthic cyanobacteria. Appl. Envir. Microbiol. 47: 135-143.

(5) Filip, Z. 1978. Effect of solid particles on the growth and endurance to heat stress of garbage compost microorganisms. Eur. J. Appl. Microbiol. Biotechnol. 6: 87-94.

(6) Güde, H. 1979. Grazing by protozoa as selection factor for activated sludge bacteria. Microb. Ecol. 5: 225-237.

(7) Haack, T.K., and McFeters, G.A. 1982. Nutritional relationships among microorganisms in an epilithic biofilm community. Microb. Ecol. 8: 115-126.

(8) Hamilton, W.A. 1983. Sulphate-reducing bacteria and the offshore oil industry. Trends Biotech. 1: 36-40.

(9) Heukelekian, H., and Heller, A. 1940. Relation between food concentration and surface for bacterial growth. J. Bacteriol. 40: 547-558.

(10) Hollibaugh, J.T., and Azam, F. 1983. Microbial degradation of

dissolved proteins in seawater. Limn. Ocean. 28: 1104-1116.

(11) Hoppe, H.-G. 1983. Significance of exoenzymatic activities in the ecology of brackish water: measurements by means of methylumbelliferyl-substrates. Mar. Ecol. - Prog. Ser. 11: 299-308.

(12) Kefford, B.; Kjelleberg, S.; and Marshall, K.C. 1982. Bacterial scavenging: utilization of fatty acids localized at a solid-liquid interface. Arch. Microbiol. 133: 257-260.

(13) Keya, S.O., and Alexander, M. 1975. Factors affecting growth of Bdellovibrio on Rhizobium. Arch. Microbiol. 103: 37-42.

(14) Martin, D.R. 1973. Mucoid variation in Pseudomonas aeruginosa induced by the action of phage. J. Med. Microbiol. 6: 111-118.

(15) Morrison, S.J., and White, D.C. 1980. Effects of grazing by estuarine gammaridean amphipods on the microbiota of allochthonous detritus. Appl. Envir. Microbiol. 40: 659-671.

(16) Poindexter, J.S. 1981. Oligotrophy: fast and famine existence. In Advances in Microbial Ecology, ed. M. Alexander, vol. 5, pp. 63-89. New York: Plenum.

(17) Porter, K.G.; Feig, Y.S.; and Vetter, E.F. 1983. Morphology, flow regimes, and filtering rates of Daphnia, Ceriodaphnia, and Bosmina fed natural bacteria. Oecologia 58: 156-163.

(18) Rittenberg, S.C., and Thomashow, M.F. 1979. Intraperiplasmic growth-life in a cozy environment. In Microbiology - 1979, ed. D. Schlessinger, pp. 80-85. Washington, DC: American Society for Microbiology.

(19) Rosenberg, E.; Kaplan, N.; Pines, O.; Rosenberg, M.; and Gutnik, D. 1983. Capsular polysaccharides interfere with adherence of Acinetobacter calcoaceticus to hydrocarbon. FEMS Microbiol. Lett. 17: 157-160.

(20) Shilo, M. 1980. Factors that affect distribution patterns of aquatic microorganisms. In Aquatic Microbial Ecology, eds. R. Colwell and J. Foster, pp. 5-11. Maryland Sea Grant Publication, University of Maryland, College Park, MD.

(21) Sims, R.C., and Little, L. 1973. Enhanced nitrification by addition of clinoptilolite to tertiary activated sludge units. Envir. Lett. 4: 27-34.

(22) Stanton, T.B., and Savage, D.C. 1984. Motility as a factor in bowel colonization by Roseburia cecicola, an obligately anaerobic bacterium from the mouse caecum. J. Gen. Microbiol. 130: 173-183.

(23) Timperley, D. 1981. The effect of Reynolds number and mean velocity of flow on the cleaning of in-place pipelines. In Fundamentals and Applications of Surface Phenomena Associated with Fouling and Cleaning in Food Processing, eds. B. Hallstrom, D.B. Lund, and C. Tragardh, pp. 402-412. Madison: University of Wisconsin.

(24) Whittenbury, R. 1978. Bacterial nutrition. In Essays in Microbiology, eds. J.R. Norris and M.H. Richmond, pp. 16/1-16/32. New York: John Wiley and Sons.

(25) Wolfe, R.S., and Higgins, I.J. 1979. Microbial biochemistry of methane - a study in contrasts. In International Review of Biochemistry, ed. J.R. Quayle, vol. 21, pp. 268-295. Baltimore: University Park Press.

Microbial Adhesion and Aggregation, ed. K.C. Marshall, pp. 223-232. Dahlem Konferenzen 1984. Berlin, Heidelberg, New York, Tokyo: Springer-Verlag.

Comparative Physiology of Attached and Free-living Bacteria

M. Fletcher
Dept. of Environmental Sciences, University of Warwick
Coventry CV4 7AL, England

Abstract. The environmental conditions at a solid-liquid interface differ from those in the bulk aqueous phase, and, accordingly, the physiological activity of bacteria attached to surfaces may differ from that of free living cells. There are three principal ways in which environmental conditions at a solid surface may influence the physiology of attached cells. First, nutrient concentration and/or accessibility may be different at the interface because of adsorption or irreversible binding of low molecular weight or macromolecular substrates. Second, processes, e.g., substrate transport and energy generation, which are sited in the cell membrane and which are central to all physiological processes, may be modified by elastic deformation of the cell envelope. Third, surfaces provide a site for colonization and the development of a bacterial biofilm, in which cells are embedded in a polymeric matrix. Such a colony microenvironment allows interactions between resident organisms, which frequently include a range of functional types, and affords protection from outside perturbations or lethal agents. Experimental measurements of the effects of solid surfaces on the activity of associated bacteria have varied considerably. The type of result obtained, ranging from promotion to inhibition of activity, has depended upon the type of activity measured, the organism, the substrate and/or its concentration, and the chemical composition of the substratum.

INTRODUCTION

One of the most important basic principles of experimental microbiology is that the physiology of a bacterium will be altered in response to changes in environmental conditions. Factors such as nutrient concentration, pH, temperature, and oxygen tension all have a profound influence on microbial metabolism. It should not be surprising then that bacteria attached to solid surfaces in liquid media may exhibit some differences in the rates, or balance, of various physiological processes, as compared with their counterparts dispersed in the bulk phase (cf. (3)). The microenvironment at the solid-liquid interface has properties which may be quite unlike those of the bulk phase, as conditions at the interface are determined, to some extent, by the properties of the solid substratum and by special physicochemical factors associated with surface phenomena and thermodynamic equilibria. From the point of view of a microbe, there are three principal ways in which attachment to a surface can alter its physiology by modifying environmental conditions; these are by a) influencing nutrient concentration and/or availability, b) modifying cell membrane-associated processes, e.g., substrate transport and energy generation, and c) providing a site for cell growth and colonization, thus allowing the development of a colony microenvironment and interactions between cells.

NUTRIENTS AND SURFACES

An interface tends to be thermodynamically unstable because of unsatisfied, potentially interactive surface molecules in both phases. As adsorption of solutes at the interface helps to achieve stability (by partially satisfying interactive surface sites), dissolved molecules tend to be adsorbed, thus concentrated, at a surface. When such adsorbed substances are nutrients, inhibitors, or substances affecting nutrient accessibility, then there may be a significant influence on the attached bacterium.

One type of solute adsorbed at the interface is low molecular weight species, which include a wide range of bacterial substrates, e.g., sugars, amino acids, and biogenic salts (NH_4^+, PO_4^{-3}, etc.), as well as inorganic ions (Ca^{2+}, Na^+, H^+, etc.). Such solutes are adsorbed and desorbed in an equilibrium, so that, although the overall concentration at the interface is higher than that in the bulk phase at any given moment, the time spent by any specific molecule at the interface is finite. The adsorption of ions, particularly H^+ and cofactors in enzyme reactions, can be quite important in environmental control of physiological processes. For example, the pH optima for glucose and succinate oxidation were found to be more than one pH unit higher for Micrococcus luteus cells attached to an anion exchange resin, as compared with suspended cells (7). On the other hand, with organic substrates of low molecular weight, the effect of surface adsorption on accessibility to attached cells is not clear. Theoretically, such adsorption should not offer a long-term advantage to the attached organism, for it would be the organism itself, whether attached or not, which would provide, via its uptake mechanism(s), the sink for the nutrient and thus maintain a diffusion gradient. However, two situations in which the surface may act in the long term as an accumulator of low molecular weight substrates occur a) when the substrate is amphipathic, i.e., surface active, or b) when the surface and substrate are oppositely charged. The utilization of fatty acids by bacteria has been shown to be facilitated when both were adsorbed at an interface (11). Similarly, bacterial nitrification has been promoted by the adsorption of the substrate NH_4^+ on zeolite (19).

However, bacterial utilization of such charged substrates requires that they be not so tenaciously bound by the substratum that they cannot be assimilated by the bacteria. For example, when certain amino acids were bound to clays, utilization by soil microorganisms became impossible (21).

Although nutrient adsorption and interaction with the substratum may be important in bacterial nutrient accumulation in some situations, a factor which is much more likely to be rate-limiting in many environments is transport of nutrient from the bulk phase to the boundary or viscous sublayer at the interface (3). This transport is facilitated enormously by water flow and circulation in the bulk phase. Thus the attached cell may be at an advantage simply by remaining on a stationary surface in a moving aqueous phase because of the constant replenishment of nutrients and removal of waste products.

Thus far, the nutrients considered have been low molecular weight substances which are usually reversibly adsorbed to surfaces and transported directly into the cell without prior breakdown. Macromolecular substrates differ, however, as they tend to be irreversibly bound by surfaces and must be broken down by extracellular or membrane-bound enzymes before uptake. Thus, utilization of adsorbed macromolecules will depend upon enzyme activity, which, in turn, may be influenced by the proximity of a solid surface. For example, binding of macromolecules to surfaces, e.g., clays, may make them inaccessible to enzymatic breakdown because of conformational changes or the binding or masking of the enzyme-labile sites (21). Furthermore, adsorption of the enzyme itself may either inhibit or enhance its activity (21). Also, other adsorbed macromolecules, e.g., humic substances, may inhibit enzymatic breakdown by binding the substrate or the enzyme (21).

There are other substances, apart from potential nutrients, which can also affect attached bacteria when the substance is either concentrated in, or removed from, the vicinity of the cells. These include inhibitors and substrate-capturing molecules, e.g., siderophores for iron transport.

EFFECT OF SURFACES ON CELL MEMBRANE-ASSOCIATED PROCESSES

Although it is an almost totally unexplored area, bacterial physiological processes may well be influenced directly by the proximity of a solid surface. There are at least two ways in which this could happen. First, the attractive forces which hold the cell at the surface could also produce some elastic deformation in the cell envelope. This deformation is easily observed with less rigid animal cells which become flattened when they are attached to a surface. Although it has not been observed with bacteria because of their rigid cell wall, it is still probable that some deformation occurs at a sub-microscopic level. Second, at charged surfaces, ion concentrations may affect stabilizing charge interactions in the cell envelope or the electric potential across the cell membrane. The consequent effects on physiological processes could be significant, as the bacterial membrane is the site of a number of vitally important enzymatic activities. Two functions which determine physiological activity and incorporate many of these enzymatic processes are energy generation, e.g., via electron transport or ATPases, and substrate transport, e.g., via phosphotransferase systems or those driven by ATP or proton-motive force. Is it not possible that some of the observed effects of solid surfaces on bacterial activity might be explained by induced changes in the cell membrane and consequent modifications in associated processes?

BIOFILM FORMATION

Once initial adhesion has occurred, attached cells may grow and produce copious amounts of highly hydrated exopolymers which form an intercellular matrix in a multilayer microcolony. The microenvironment in these thick bacterial layers, or biofilms, is quite different from that at a solid-liquid interface where only a few cells are attached, and the physiology of the residents will be accordingly modified. One very important feature of biofilms in natural environments is that they generally contain a wide range of functional types of organisms, including both producers, e.g., microalgae, and consumers, e.g., heterotrophic bacteria (5). Thus, the potential for interaction between the various components of the community is enormous, and presumably the organisms which become dominant are those which have found the biofilm microenvironment comparatively favorable to their physiological processes and growth. Factors which could favor nutrient utilization in biofilms are a) mutualistic or synergistic interactions between various community components, b) retention of extracellular enzymes near the cells by the polymer matrix, allowing their efficient use, and c) possible "capture" of nutrients in the bulk phase by the polymer matrix, which is frequently negatively charged. However, biofilm conditions may not necessarily promote nutrient assimilation, as the hydrated matrix acts as a gel diffusion barrier, thus retarding diffusion of nutrients and waste products (15). Clearly the biofilm can be an enormously complex system, much unlike the situation which occurs with attached monolayers of cells. It is extremely important to bear this distinction between the two types of attached cell in mind when considering the effects of solid surfaces on attached bacterial physiology and the possible mechanisms for such effects.

EXPERIMENTAL OBSERVATIONS

It is not possible to make a definitive statement about the effects of solid surfaces on the physiology of attached cells, as there is experimental evidence demonstrating that surfaces may have a variety of effects, ranging from promotion to inhibition of physiological activities. The type of result obtained depends in many cases on experimental design and can differ with the methods and variables employed.

First, the nature (increase or decrease) and extent of the effect of surfaces on physiological activity have depended on the type of activity measured. Some examples of recorded results are given in Table 1. The table is not meant to be a complete listing but gives some idea of the range of results which can be obtained with different methods.

Other measurements which have been used to evaluate differences in attached and suspended cell activity included evaluations of product or secondary metabolite formation (17) and morphological changes, such as fragmentation with continuous size reduction (9), spore formation, or flagellation.

Not only does the result obtained depend upon the type of activity being measured, but also upon a range of experimental conditions. The particular organism being used is important, as is its attachment ability and the consequent relationship between it and the surface, i.e., whether it is actually attached to the surface (18), or not associated (14), or only superficially associated with it. Environmental or experimental factors are considerably important, particularly the substrate (1, 21) and/or its concentration (10, 22) and the chemical composition of the substratum (2, 22). Furthermore, activity measurements of attached and free-living

TABLE 1 - Measured effects of solid surfaces on different types of bacterial activity.

Type of Activity Measured	Detected Response in Surface-associated Bacteria			Reference
	Increase	Decrease	Inconsistent[1]	
Growth	+			8, 13
Viability			+	9
Respiration:				
O_2 uptake			+	10
	+			20
CO_2 production		+		4
Tetrazolium staining			+	2
Heat generation			+	4
Substrate uptake			+	2
		+		4
	+			12
Substrate transformation		+		6

[1]Results differed (increase, decrease, or no effect) depending upon experimental conditions, e.g., organism used, substrate or its concentration, or the substratum composition.

cells may differ if a subpopulation of a heterogeneous bacterial suspension becomes selectively attached on the surface. Thus, the activities differ not because of attachment to a surface itself (e.g., because of proximity of a surface, microenvironmental conditions), but because the two subpopulations are inherently different. This could occur, for example, if physiological activity were required for firm attachment, as the more active cells would be more likely to attach (2).

Despite the confusion in this complex body of data, it is still possible to cite a feature common to most studies of attached bacterial activity. Most deal directly or indirectly with the ability to utilize a substrate or with the efficiency of its utilization. Many of the studies which illustrated an increase in activity when the bacteria were associated with the surfaces have done so only when low nutrient media were used (10, 15, 16, 22). Thus in low nutrient environments, surfaces may provide an advantage by assisting the capture and/or uptake of scarce nutrients.

Drawing from the foregoing discussion of ways in which surfaces can affect activity of attached cells, there are a number of mechanisms which could account for such surface-enhanced substrate utilization:

1) Scarce nutrients are concentrated at the surface through adsorption. This is indeed a commonly offered explanation, and although it is probably significant with macromolecular, charged, or surface active substrates, its relevance to uptake of other low molecular weight organics is not clear.

2) Substrate uptake mechanisms or respiratory efficiency may be affected by surface-induced modifications in cell membrane structure.

3) Substrate capture may be promoted by transport of substrate to the surfaces via flowing and circulating water masses.

4) In biofilms, higher bacterial numbers on a surface, as compared with the bulk phase, could be due to a relative increase in survival or sustentation of activity, rather than to an increase in growth. The polymeric intercellular matrix should offer some protection against lethal agents, e.g., biocides, phages, toxic metals, and unfavorable perturbations in the adjacent bulk phase.

5) Also in biofilms, numbers may be high because of mutualism between community members or facilitation of extracellular enzyme activity.

With so many possible explanations for observed differences in activity, it is rarely possible, particularly in a natural environment with its multiplicity of variables, to pinpoint with any confidence the mechanism responsible.

OVERVIEW AND OUTLOOK

Much of this discussion has emphasized the variety of results which have been obtained, the many possible explanations, and the impossibility of citing any one, or even few, mechanism(s) which is(are) responsible for the changes in physiological activity which occur when bacteria become associated with surfaces. My intention was not to convey pessimism - as much data has now been accumulated and a real start has been made in tackling the problem - but to urge caution. Although a few experiments have demonstrated the likely underlying mechanisms for observed differences in the physiology of attached and free-living cells (for example, pH (6, 20) and nutrient status (13) have been strongly implicated), it is still not possible to extrapolate such results to other situations where bacteria are attached to surfaces. The dominant factors on a rock in a stream should be quite unlike those in an immobilized-cell reactor. Yet we are now in a position to formulate questions which are ripe for investigation.

With bacterial monolayers, a question which really needs a definitive study is, to what extent the adsorption of low molecular weight substrates influences their subsequent uptake by attached bacteria? Similarly, the effect of adsorption on breakdown and utilization of macromolecular nutrients should be investigated. Also a question which is extremely important to our understanding of bacterial physiological responses is, to what extent can the substratum properties directly affect attached cell physiology by influencing membrane-associated processes? Moreover, could contact with a surface in this way "trigger" a physiological response, which then allows the attached cell to take better advantage of the conditions in its new microenvironment? An example of such a response could be the production of biofilm polymers which are produced and/or accumulated in copious amounts by biofilm bacteria. This leads to the question of the significance of these polymers: are they largely protective, or do they provide distinct advantages? There is no doubt that surfaces can, and often do, affect the physiology of associated bacteria. The question is now: what are the nature and underlying mechanisms for such effects?

REFERENCES

(1) Bell, C.R., and Albright, L.J. 1982. Attached and free-floating bacteria in a diverse selection of water bodies. Appl. Envir. Microbiol. 43: 1227-1237.

(2) Bright, J.J., and Fletcher, M. 1983. Amino acid assimilation and electron transport system activity in attached and free-living marine bacteria. Appl. Envir. Microbiol. 45: 818-825.

(3) Fletcher, M., and Marshall, K.C. 1983. Are solid surfaces of ecological significance to aquatic bacteria? In Advances in Microbial Ecology, ed. K.C. Marshall, vol. 6, pp. 199-236. New York: Plenum Press.

(4) Gordon, A.S.; Gerchakov, S.M.; and Millero, F.J. 1983. Effects of inorganic particles on metabolism by a periphytic marine bacterium. Appl. Envir. Microbiol. 45: 411-417.

(5) Haack, T.K., and McFeters, G.A. 1982. Nutritional relationships among microorganisms in an epilithic biofilm community. Microbial Ecol. 8: 115-126.

(6) Hattori, T., and Furusaka, C. 1960. Chemical activities of Escherichia coli adsorbed on a resin. J. Biochem. (Japan) 48: 831-837.

(7) Hattori, R., and Hattori, T. 1963. Effect of a liquid-solid interface on the life of micro-organisms. Ecol. Rev. 16: 64-70.

(8) Hattori, R., and Hattori, T. 1981. Growth rate and molar growth yield of Escherichia coli adsorbed on an anion-exchange resin. J. Gen. Appl. Microbiol. 27: 287-298.

(9) Humphrey, B.; Kjelleberg, S.; and Marshall, K.C. 1983. Responses of marine bacteria under starvation conditions at a solid-water interface. Appl. Envir. Microbiol. 45: 43-47.

(10) Jannasch, H.W., and Pritchard, P.H. 1972. The role of inert particulate matter in the activity of aquatic microorganisms. Mem. Ist. Ital. Idrobiol. 29(Suppl): 289-308.

(11) Kefford, B.; Kjelleberg, S.; and Marshall, K.C. 1982. Bacterial scavenging: utilization of fatty acids localized at a solid-liquid interface. Arch. Microbiol. 133: 257-260.

(12) Kirchman, D., and Mitchell, R. 1982. Contribution of particle-bound bacteria to total microheterotrophic activity in five ponds and two marshes. Appl. Envir. Microbiol. 43: 200-209.

(13) Kjelleberg, S.; Humphrey, B.A.; and Marshall, K.C. 1982. Effect of interfaces on small, starved marine bacteria. Appl. Envir. Microbiol. 43: 1166-1172.

(14) Laanbroek, H.J., and Geerligs, H.J. 1983. Influence of clay particles (illite) on substrate utilization by sulfate-reducing bacteria. Arch. Microbiol. 134: 161-163.

(15) Ladd, T.I.; Costerton, J.W.; and Geesey, G.G. 1979. Determination of the heterotrophic activity of epilithic microbial populations. In Native Aquatic Bacteria: Enumeration, Activity and Ecology, eds. J.W. Costerton and R.R. Colwell, pp. 180-195. Philadelphia: American Society for Testing and Materials.

(16) Marshall, K.C. 1976. Interfaces in Microbial Ecology. Cambridge, MA: Harvard University Press.

(17) Mattiasson, B., and Hahn-Hägerdal, B. 1982. Microenvironmental effects on metabolic behaviour of immobilized cells: a hypothesis. Eur. J. Appl. Microbiol. Biotechnol. 16: 52-55.

(18) Munch, J.C., and Ottow, J.C.G. 1982. Einfluss von Zellkontakt und Eisen (III) - Oxidform auf die bakterielle Eisenreduktion. Z. Pflanzenernähr. Bodenk. 145: 66-77.

(19) Sims, R.C., and Little, L. 1973. Enhanced nitrification by addition of clinoptilolite to tertiary activated sludge units. Envir. Lett. 4: 27-34.

(20) Stotzky, G. 1966. Influence of clay minerals on microorganisms II. Effect of various clay species, homoionic clays, and other particles on bacteria. Can. J. Microbiol. 12: 831-848.

(21) Stotzky, G., and Burns, R.G. 1982. The soil environment: clay-humus-microbe interactions. In Experimental Microbial Ecology, eds. R.G. Burns and J.H. Slater, pp. 105-133. Oxford: Blackwell Scientific.

(22) ZoBell, C.E. 1943. The effect of solid surfaces upon bacterial activity. J. Bacteriol. 46: 39-56.

Microbial Adhesion and Aggregation, ed. K.C. Marshall, pp. 233-249. Dahlem Konferenzen 1984. Berlin, Heidelberg, New York, Tokyo: Springer-Verlag.

Activities of Microorganisms Attached to Living Surfaces

D.C. Savage
Dept. of Microbiology, University of Illinois
Urbana, IL 61801, USA

Abstract. Bacteria associated with living surfaces may or may not multiply while attached to the living substratum. Whether or not they multiply, such bacteria must utilize nutrients present in the environment often in competition with the living host cells in the substratum and other microorganisms in an attached community. If they multiply while attached to a surface, then bacteria must conform in their reproductive strategies to conditions imposed on the habitat by the living tissue sharing it. Bacteria attached to living surfaces may penetrate into and through the surface into underlying host tissues, synthesize macromolecules that alter the function of the living substratum, or produce end products of metabolism that can be utilized as carbon and energy sources by the living cells in the substratum. Such products of metabolism may also function, along with nutritional competition, to regulate the population levels and localization of other microorganisms in attached communities. These processes are understood poorly at the molecular level.

INTRODUCTION

Representatives of species of cellular microorganisms of all major protist (42) classes (i.e., algae, protozoa, fungi, bacteria) are known to adhere to surfaces of other living creatures. Surfaces involved are those of organisms ranging from microbes to higher plants and mammals (including humans). While attached to surfaces, microbial cells may conduct all metabolic, synthetic, and reproductive functions necessary for colonizing the surface. Alternatively, they may not multiply but still may metabolize, synthesize, and secrete, or they may penetrate through the surface into subsurface tissues of the host organism. In this paper, some such activities of microorganisms associated with living surfaces will be discussed. However, the number of attached microorganisms, hosts, and activities that could be discussed is quite large. Accordingly, the discussion will be focussed in subject matter and scope.

The discussion will be focussed as follows: general points presented will be illustrated only with instances of bacteria adhering to the epithelial surfaces in mammals and (occasionally) higher plants. In addition, the activities discussed will be limited to some that involve interactions between the bacterium and the living substratum. Activities to be discussed will include the capacity of attached bacteria to penetrate through surfaces into epithelial cells or subepithelial tissues, the nutrition, secretory-excretory processes, and reproduction of such microorganisms. Attached bacteria are also undoubtedly engaged in transfer of genes from cell to cell. In addition, such bacteria "adhere" to the surface, i.e., one of their "activities" is to produce the mechanisms by which they adhere. These mechanisms of adhesion are discussed in detail by Jones and Dazzo (both this volume). Likewise, some aspects of genetic control of adhesion and genetic transfer are discussed by Silverman et al. (this volume). Therefore, those subjects will receive no further treatment in this paper.

For each activity discussed, the same format of presentation will be used. First, a brief background will be presented consisting of some information that, in the author's opinion, illustrates the basis for recent investigation. Second, some more or less critical comments will be made about some findings gained within the past few years on the subject and what research may be under way at this time. Finally, some questions will be raised about what aspects of the research may yield useful information to investigators in the future.

PENETRATION INTO CELLS OR SUBEPITHELIAL TISSUES

Background

Certain bacterial pathogens have long been known to penetrate into subepithelial spaces in mammals (42) and plants (7, 26) and to induce disease in the hosts while infecting the deeper tissues. Although of considerable importance, the activities of bacterial cells after they penetrate through an epithelium are not relevant to the purposes of this paper. Pertinent to the subject, however, are the activities of attached bacteria that lead to their penetrating into epithelial cells or subepithelial spaces.

Historically, most investigation in this area has been concerned with events ensuing after a bacterial pathogen penetrates epithelial surfaces (7, 26, 30, 42). In some cases, however, investigators have attempted to examine how certain invasive bacterial pathogens penetrate into or through epithelial cells. In one such study involving electron microscopic examination of an intestinal pathogen of mammals (45), the investigator noted that the bacterium involved (Salmonella typhimurium) induced degeneration of microvilli of intestinal epithelial cells when about 350Å from the membranes of those structures. Subsequently, the bacterial cells induced the epithelial cells to engulf them by a process now recognized to be endocytosis and to disgorge them into tissues below the epithelium by a process now recognized to be exocytosis (40). The bacteria also passed between epithelial cells, penetrating through tight junctions between cells and the basement membrane into the subepithelial tissues.

This study can be criticized; the animals involved in the study (guinea pigs) had been treated with opium to retard peristalitic and villous motility in their intestines and were given by gavage a quite large dosage of bacterial cells. Such treatment of the animals could have permitted the bacteria to interact with the intestinal epithelium in ways not characteristic of natural infections. Indeed, S. typhimurium probably invades into subepithelial tissues in the small intestines through the epithelium covering Peyer's patches, rather than through the absorptive epithelium of the intestine per se.

The epithelium covering the Peyer's patches contains cells, so-called microfold or M cells, that differ ultrastructurally from intestinal epithelial cells and are known to be actively endocytic (32). Such cells may be involved in transporting, via endocytic-exocytic processes, invasive bacterial pathogens such as S. typhimurium (22) and perhaps also nonpathogenic members of the normal intestinal microbiota (44) from the lumen of the intestine into the subepithelial tissues. Thus, intestinal enterocytes may be involved in only a minor way in such processes. Nevertheless, the early study with electron microscopy of the invasion of S. typhimurium into and through small bowel enterocytes (45) was seminal, by suggesting that endocytosis-exocytosis was involved in that invasion. Such processes induced by the pathogens (40) are now believed to mediate bacterial penetration into subepithelial spaces, not only of intestinal pathogens, but also of pathogens that invade into other tissues of the body, for example, Neisseria gonorrhoeae that penetrates into and through epithelial cells in the urogenital tract (21).

The electron microscopic study of bacterial invasion also illustrated (for the first time of which this author is aware) the capacity of the bacterial cells to induce dramatic ultrastructural changes in the membranes of mammalian epithelial cells without destroying the cells (45). As will be discussed in the next section, similar observations have been made more recently with indigenous and pathogenic bacteria that adhere to, but do not penetrate through, the microvillous membranes of intestinal epithelial cells.

Recent Findings

Within the past few years, research involving the microbiota indigenous to the gastrointestinal tract has revealed that strains of many bacterial species in mammals of several species adhere to epithelial surfaces in various regions of the tracts they colonize (35). Some of the attached bacteria do so by altering dramatically the ultrastructure of the microvillous membranes of enterocytes. A similar phenomenon has been observed with certain strains of enteropathogenic Escherichia coli (33). The membranes remain intact at the junctions of contact with the bacterial cells but undergo changes in architecture that have been interpreted to be sol-to-gel transformations of the membranes and underlying cytoplasm of the epithelial cells (41). In this process, the microvillous cores and terminal web disappear. These changes in the enterocyte membranes are induced by attached bacterial cells, in apparently much the same way S. typhimurium induces changes in the enterocyte membrane when it is penetrating into epithelial cells (45). In contrast to S. typhimurium, however, the indigenous bacteria and pathogenic E. coli strains do not penetrate into the enterocytes. Rather, they arrest the process of penetration at a point where an invagination of the enterocyte membrane essentially forms a firm attachment site for the bacterial cells (31).

The biochemistry of such processes remains unknown. The bacteria involved surely have on their surfaces, or excrete (see below), compounds that induce the architectural and biochemical changes observed. If the compounds are excreted, then they must be labile and active for only short periods; the changes in the enterocyte membranes occur only when the bacteria are closely associated or in direct contact with them. As is indicated by findings from recent work on the mechanisms of invasion of Shigella species (20), the putative substances are likely to be part of the outer surface of the bacterial cells. In some bacteria the capacity to induce the endocytosis-exocytosis phenomena is encoded by genes on plasmids (20, 22).

Future Research

Much investigation has been devoted recently to the mechanisms by which bacterial pathogens adhere to epithelial surfaces. Little attention has been given, however, to the specific mechanisms by which bacterial strains penetrate through epithelial membranes. Research on such problems should be facilitated by the discovery that the invasive characteristic is encoded or regulated in some bacterial pathogens by genes on plasmids. Attention should be given to the putative substance(s) produced by penetrating bacteria and to the receptor(s) on the epithelial cells that trigger the mechanism controlling endocytic phenomena in the cells (microfilaments, etc.). Indigenous and pathogenic bacteria that induce changes in the ultrastructure of epithelial cell membranes, without penetrating into the cells, should also be studied for the biochemistry and genetics of the activity, and for its effects on the absorptive and secretive functions of the epithelial cells.

NUTRITION OF BACTERIA ATTACHED TO LIVING SURFACES

Background

The nutrition of bacteria is usually studied with the microorganisms growing in pure batch or continuous cultures (42). In addition, the nutrition of a few bacterial strains has been studied while the organisms are growing in co-cultures with other strains, usually continuous culture (50). Comparatively little work has been done on the nutrition of bacteria attached to surfaces, however, especially surfaces of living animals and plants. Findings from research with bacteria attached to rigid, nonliving surfaces can be extrapolated only with care to bacterial strains metabolizing on living epithelial surfaces.

Some bacterial pathogens known to adhere to living surfaces in animals have long been known to have the capacity to hydrolyze macromolecules that occur in secretions on the surfaces they colonize. For example, Vibrio cholerae, which may adhere to the epithelium of the small intestine when causing cholera in man (13, 24), has been known from seminal research to produce glycosidases (in particular neuraminidase) that catalyze hydrolysis of polysaccharide moieties of glycoproteins and glycolipids found in mucins and mucous membranes (13, 24). Some other bacteria that colonize intestinal tracts are known to have similar capacities (34). As will be amplified shortly, such enzymes may be important in generating nutrients for bacteria on epithelial surfaces.

Recent Findings

Attention paid recently to the mechanisms by which bacteria adhere to epithelia has not been matched by efforts to determine how the bacteria gain energy and precursors for their synthetic processes while attached to surfaces. Presumably such effort has received little attention because doing research in the area is problematic, especially when a living epithelium is involved.

Some attached bacteria may derive their nutrition directly from the living substratum. For example, bacteria attached to the rumen wall hydrolyze sluffing squamous cells and presumably gain carbon and energy and other nutrients from the process (8). However, most attached bacteria probably gain their nutrients from secretions of the living substratum. For example, bacteria attached to rumen walls hydrolyze urea secreted by the host cells (47). Some bacterial pathogens such as V. cholerae are attracted chemotactically into mucous layers (16). As noted, these bacteria have a capacity to hydrolyze mucin constituents and may utilize the hydrolytic products as carbon, energy, and nitrogen sources. Similar circumstances may prevail with certain indigenous bacteria that colonize mucosal surfaces in intestines (35, 43) and bacteria that adhere to plant rootlets covered with a mucous-like gel (7).

In some bacteria in which the capacity to adhere to an epithelial surface is encoded on a plasmid (11), the plasmids encoding or regulating the adhesives also encode enzymes that catalyze the synthesis of a system specific for competing with the host for an essential nutrient (iron) (19). Soluble iron is a limiting nutrient in nature. As a consequence, living organisms make siderophores for chelating iron to facilitate its absorption into their cells (48). Certain bacteria make siderophores known as enterochelins which are synthesized by enzymes encoded by genes on the bacterial chromosomes (19, 48). Some plasmids have genes encoding a siderophore (aerobactin) of a chemical type different from enterochelins. Both systems are induced under conditions of limiting iron (19). Thus, the bacteria bearing the plasmid coding for synthesis of aerobactin are believed to have an augmented capacity, over that of strains lacking the plasmids, to compete with host tissues for iron (19).

Similar capacities for "nutrient trapping" (capture) may be important for bacteria to metabolize nutrients and synthesize macromolecules while attached to living surfaces. For example, because of the ion-exchange and molecular-sieve properties of their surface polymers (9), bacteria may benefit from nutrient gradients established by the host cells. Rather, the bacterial cells may find themselves in strong competition with the host for nutrients present in limited supply (such as iron). The iron chelators discussed above can compete with animal proteins that chelate iron (lactoferrin and transferrin), because they more avidly bind ferric ion (48). Other such nutrient-trapping systems may have evolved and be induced to function when bacteria are associated with living surfaces.

Competition for nutrients among bacterial strains in an indigenous community associated with a living surface may serve to control relative population levels of the various species in the community (15). Moreover, such competition may function to inhibit allochthonous (transient) species from establishing and multiplying in a community (15). However, such an hypothesis is difficult to test experimentally; secretory and excretory products of members of an attached community, especially as the products function in the local environment, may act as growth inhibitors to control relative population levels of the many types of bacteria in attached communities as well as prevent transients (pathogens) from establishing in such communities (1, 17, 18, 35). Thus, the effects of nutritional competition may be difficult to detect due to the influences of growth inhibitors.

Future Research

Little is known about the nutrition of bacteria associated with living surfaces, except what is known in general about bacterial nutrition as it is studied with organisms in cultures. Indeed, little is known even about whether bacterial cells associated with epithelial surfaces differ in nutrition and metabolism when they are attached to each other rather than living dispersed in a liquid environment. Therefore, creative experimentation is needed on almost any issue concerning nutrition of attached microorganisms. Issues that could be examined include the rates of diffusion of nutrients through thick layers of attached bacteria; the capacity of such bacteria to utilize as nutritional substrates endogenous substances produced by the host; how attached organisms compete, or otherwise interact, with each other and with host cells for nutrients; and any effects of such competition on the host cells.

SECRETORY AND EXCRETORY PROCESSES AND SURFACE COMPONENTS OF ATTACHED BACTERIA

Background

Interest in materials excreted and secreted by bacteria, and in macromolecular substances on bacterial surfaces, had its beginnings in the earliest days of microbiology (42). Only in the past two decades or so, however, have investigators been concerned about substances secreted or excreted by, or materials on the surface of, bacteria attached to living surfaces. The interest was stimulated by the discoveries that macromolecular substances harmful to the host were secreted by attached bacterial pathogens of several species (12, 14), that certain metabolic end products of indigenous bacteria are used as nutrients by animals (10), that some bacteria produce and secrete enzymes that catalyze hydrolysis or transformation of endogenous and exogenous compounds (49), and that macromolecular structures on the bacterial surfaces mediate their adhesion to other surfaces (35).

Recent Findings

Notable among secreted substances are the protein exotoxins produced by Corynbacterium diphteriae which adheres to pharyngeal epithelium while causing diphtheria in man (4), V. cholerae which, as discussed above, probably adheres to small bowel epithelium while causing cholera in man, and certain enteropathogenic strains of Escherichia coli that adhere to small bowel epithelia while causing watery diarrhea in humans and animals of other species. The influences on host cells of most of these toxic proteins are understood well at the molecular level (12, 14). Moreover, their molecular structure and mechanisms by which they are secreted by the bacterial pathogens are also reasonably well-defined (12). Such studies are routinely made with purified exotoxin or cells grown in liquid media, however, and not with cells associated with living surfaces. Thus, the mechanisms of synthesis and secretion of the proteins by attached bacteria are inferred from information gained from studies with cells living in culture media. Little is known, therefore, about whether the synthesis or secretion differs in attached cells from the processes in cultured cells, and about how the proteins move from the surface of the bacterial cell to that of the epithelial cell.

As do all microorganisms, attached bacteria produce in their energy-generating metabolism and excrete into their environment various reduced compounds, including short chain fatty acids (10, 25). As noted earlier, some such compounds (e.g., volatile organic acids) are known to be absorbed and function as carbon and energy sources for animal cells. Such compounds and lactic acid may function as well, in part by lowering local pH, in controlling the population levels of microorganisms in communities of complex composition on living surfaces (1, 17, 18, 35).

Some attached bacteria may also produce and secrete enzymes that catalyze many of the hydrolyses and transformations known to be performed by indigenous microorganisms, for example, deconjugation and transformation of certain steroids and other large molecules (49). Thus, metabolic products and other compounds altered by the enzymatic activity of attached bacteria are present in the interface (28) between living animal or plant surfaces and where those surfaces meet the environment. Presumably, therefore, such bacterial molecules could have strong influences on the physiology of the surfaces (37, 38, 46). Little evidence supports such an hypothesis at this time.

Some hydrolytic enzymes produced by certain bacterial species that adhere to epithelial surfaces in humans and some other animals catalyze the hydrolysis of immunoglobulins on the surfaces. Such enzymes, known as IgA-proteases, catalyze the hydrolysis of human IgA1 at a specific amino sequence near the so-called hinge region of the immunoglobulin molecule (23). IgA is present in high concentrations in human secretions and is regarded as a first immunological line of defense on epithelial surfaces (29). Bacteria producing enzymes that hydrolyze the molecule would presumably have some advantage in colonizing the epithelium. However, little direct evidence supports such an hypothesis.

Gram-negative and Gram-positive bacteria differ substantially in the types of macromolecules found on their surfaces (9, 39). Some molecules are present in envelope or capsular substances surrounding the bacterial cells (39). These can be complex polysaccharides or mixtures of macromolecules including polysaccharides, proteins, and other macromolecules (39). Some are proteinaceous projections such as fimbriae and flagella (33, 42, 43). Others are amphiphiles such as lipopolysaccharides (LPS) in Gram-negative bacteria and lipoteichoic acid (LTA) in Gram-positive bacteria. These amphiphiles may be inserted into the cell walls or membranes but have subcomponents that protrude into the surrounding environment (39). Some such macromolecules function as adhesives that bind the bacterial cells to epithelial substratum, processes which are amplified in the paper by Jones (this volume). However, some undoubtedly affect the function of host cells of the substratum.

Bacterial LPS (3) and teichoic acids (in particular lipoteichoic acid - LTA (27, 39)) are known to have the capacity to alter the functions of mammalian cell membranes. These molecules are known to be released into the surrounding environment by certain strains of some bacterial species (3, 27). If derived from attached bacteria, then the molecules could have direct access to epithelial cell membranes and thus could alter their function. Whether such circumstances prevail in any animals is not known. Nevertheless, germfree animals differ from animals with a microbiota in the enzymatic activities of their intestinal epithelial cells (37, 38, 46).

Future Research

Much recent research involves study of the precise mechanisms of the effects on animal cells of bacterial protein toxins. Little effort has yet been expended, however, on how bacterial synthetic functions may be modulated or enhanced in the environment on the epithelial surface. Likewise, virtually nothing is known about how surface molecules such as amphiphiles may affect epithelial cell membranes. Finally, and perhaps most importantly for the lifelong health of mammals, the mechanisms by which bacterial macromolecules or metabolic end products affect the functions of epithelial and other host cells are virtually unexplored. These areas all involve the interface between an animal and its environment and have implications in the health and quality of life. Therefore, they deserve research attention.

REPRODUCTION OF BACTERIA ATTACHED TO LIVING SURFACES

Background

As with many of the topics discussed in this paper, much is known about bacterial reproduction as it takes place in broth cultures or on the surfaces of agar media (42). Thus, much can be inferred about reproduction of bacteria associated with living surfaces. Moreover, research on the reproduction of bacteria attached to inert surfaces (e.g., Caulobacter species, others) gives guidance about reproductive mechanisms that have evolved so that bacteria can successfully colonize surfaces (42). Nevertheless, little of the information available serves as a complete model for reproductive niches filled by bacteria associated with living surfaces.

Depending upon the circumstances, bacteria may or may not multiply while attached to an epithelial surface. Some strains may multiply at some remote site and somehow find their way to a surface; others may multiply on the surface. Published evidence is often not informative on this matter; few investigators attempt to determine whether the attached bacteria in which they have interest multiply on the epithelium.

In mammals, living surfaces usually consist of macromolecules in dynamic states on and in membranes (36, 38). Epithelial cells are constantly shedding from the surface and being replaced by new cells. Thus, microorganisms attached to mammalian epithelia must have reproductive strategies that allow them to maintain their populations in dynamic three-dimensional habitats formed in secretions on the surface of the membranes of living cells.

Recent Findings

The gastrointestinal ecosystem of laboratory rodents (primarily rats and mice) can serve as an illustration in microcosm of various reproductive strategies that have evolved in bacteria associated with mammalian epithelial surfaces (35). The stomachs of such animals contain two incompletely separated compartments lined with two epithelia of quite different architecture and function. The compartment to the left in the stomachs of the animals is lined with stratified squamous epithelium, that to the right with columnar epithelium. The intestines of the animals are also lined with columnar epithelium. The columnar epithelial cells have microvilli on their lumenal surfaces, the membranes of which are covered with a glycocalyx consisting of glycoproteins and probably also glycolipids. Each epithelium is constantly replaced in a normal process of turnover, the squamous cells by desquamation, the columnar cells by migration from pits in the mucosa, the crypts of Lieberkuhn (where mitosis takes place), to extrusion zones on the epithelial surface. Columnar epithelium is often covered with mucous gel. The squamous epithelium may be coated with glycoproteins from secretions. In normal, healthy, well-fed animals (not housed germfree), the epithelium in each of the major regions (the two stomach "compartments," the small intestine, and the cecum and colon) has associated with it a microbial community. The community in each region has a microbial composition that is unique to the particular ecology of the region. Some or all members of those communities may be attached to the epithelium.

The stratified squamous epithelium of the stomach normally hosts a community of attached lactic acid bacteria composed principally of lactobacilli and streptococci. The columnar epithelium of the stomach often serves as a substratum for an attached layer of yeasts (Candida pintolopesii). The columnar epithelium of the lower two thirds of the small intestine has associated with it a community of filamentous bacteria, each of which adhere tightly by one end to the microvillous membranes. The columnar epithelium of the cecum and colon hosts bacterial communities of complex composition, composed of anaerobic bacteria of many genera and species, and some filamentous and helical-shaped prokaryotes of undefined taxonomy. The latter communities do not adhere strongly, if at all, to the epithelial surface.

Each of these communities has a somewhat unique reproductive strategy for maintaining itself on its epithelial surface. The lactic acid bacteria multiply by binary fission and maintain a rather thick layer covering the epithelium, with bacterial cells attached to the squamous cells and to each other. When the squamous cells sluff from the surface, the bacterial layer may simply multiply to cover the denuded surface. A similar phenomenon may occur with the yeast, except that that organism multiplies by budding and maintains an essentially continuous layer in the mucous gel on the surface by growing deep in crypts and foveal pits where the pH may be quite low. Likewise, the bacteria colonizing the epithelium of the cecum and colon cover the epithelium with a continuous and thick layer in mucous gel. In that layer, however, the bacteria involved multiply in most cases by binary fission; some of them divide deep in the crypts of Lieberkuhn. Many of them are motile and undoubtedly use that motility to move about the surface to new epithelium generated in the crypts of Lieberkuhn. Some of these bacteria can form endospores which may ensure survival when they are removed from host surfaces into hostile environments.

In striking contrast to those strategies are the elaborate and complex reproductive systems of the filaments attached to the epithelium of the small intestine (35, 36). In this region the epithelium covers finger- or leaf-shaped projections (villi) extending from the mucosa into the lumen. The epithelium of these structures is produced by mitosis in the crypts and migrates out of those pits at the bases of the villi along the lumenal villous surface until they are extruded into the intestinal lumen at the tips of the villi. The trip from crypt to villous tip can require as little as two days in some animals (37). In this region of the tract as well, the lumenal contents can move along the intestine at considerable velocity. Possibly as a consequence, the bacterial filaments adhere tightly to the epithelium with one end (a "holdfast") in a "socket" in the membrane of an epithelial cell. In addition, the bacteria and epithelium are covered with viscous mucous gel (5).

In this ecological circumstance, the filaments multiply by at least three processes: a) binary fission leading to chain elongation (6), b) narrowing and breakage of the filament at differentiated sites to produce new holdfasts that must somehow move to the epithelial surface and initiate new filaments (2), and c) producing per segment two intracellular prokaryotic bodies that can move out of ruptured "mother-cells" to the epithelium and form holdfasts and initiate new filaments (6). These mechanisms may have evolved so that the bacteria can colonize "new" epithelial cells migrating along the villi from the crypts. Such enterocyte migration would carry away attached filaments at the extrusion zones. Thus the survival of the microbial population would depend upon its capacity to colonize new epithelium produced. In times of stress (starvation of the animal), some of the filaments can also make endospores.

Future Research

As with some other matters discussed earlier, many problems remain to be studied about reproduction of attached bacteria. One fundamental question is whether or not a particular attached bacterium multiplies at all on its living substratum or multiplies elsewhere, somehow makes its way to the epithelium, and adheres to it. Such a case may be more typical of bacteria that invade into or through an epithelium than of those that remain attached to the surface. Once it is established that a particular microorganism multiplies while attached to an epithelium, then the strategy of a given reproductive process could be analyzed in reference to the anatomy and physiology of the living substratum, any overlying secretions, and bacterial metabolic end products or antibodies that may function in population control in attached communities of complex composition (15). Mathematical models as analyzed in computer simulation may be particularly useful in such complex analyses (15). Such efforts would be worthwhile, because even though a given bacterium may be able to adhere to a given surface, it will not be able to colonize that surface unless it can reproduce in the nutritional and environmental conditions prevailing in the habitat.

Acknowledgements. The author expresses appreciation to J.K.W. Hansen and C. Paris who prepared the manuscript by word processing.

REFERENCES

(1) Barrow, P.A.; Brooker, B.E.; Fuller, R.; and Newport, M.J. 1980. The attachment of bacteria to the gastric epithelium of the pig and its importance in the microecology of the intestine. J. Appl. Bacteriol. 48: 147-154.

(2) Blumershine, R.V., and Savage, D.C. 1978. Filamentous microbes indigenous to the murine small bowel: a scanning electron microscopic study of their morphology and attachment to the epithelium. Microb. Ecol. 4: 95-103.

(3) Bradley, S.G. 1979. Cellular and molecular mechanisms of action of bacterial endotoxins. Ann. Rev. Microbiol. 33: 67-94.

(4) Cromartie, W.J. 1941. Infection of normal and passively immunized

chick embryos with Corynebacterium diphtheriae. Am. J. Path. 17: 411-420.

(5) Davis, C.P. 1976. Preservation of gastrointestinal bacteria and their microenvironmental associations of rats by freezing. Appl. Envir. Microbiol. 31: 304-312.

(6) Davis, C.P., and Savage, D.C. 1974. Habitat, succession, attachment and morphology of segmented, filamentous microbes indigenous to the murine gastrointestinal tract. Infec. Immun. 10: 948-956.

(7) Dazzo, F.B. 1980. Adsorption of microorganisms to roots and other plant surfaces. In Adsorption of Microorganisms to Surfaces, eds. G. Bitton and K.C. Marshall, pp. 253-316. New York: John Wiley.

(8) Dinsdale, D.; Cheng, K.-J.; Wallace, R.J.; and Goodlad, R.A. 1980. Digestion of epithelial tissue of the rumen wall by adherent bacteria in infused and conventionally fed sheep. Appl. Envir. Microbiol. 39: 1059-1066.

(9) Dudman, W.F. 1977. The role of surface polysaccharides in natural environments. In Surface Carbohydrates of the Prokaryotic Cell, ed. I. Sutherland, pp. 357-414. London: Academic Press.

(10) Elsden, S.R.; Hitchcock, M.W.S.; Marshall, R.A.; and Phillipson, A.T. 1946. Volatile acid in the digesta of ruminants and other animals. J. Exp. Biol. 22: 191-202.

(11) Elwell, L.P., and Shipley, P.L. 1980. Plasmid-mediated factors associated with virulence of bacteria to animals. Ann. Rev. Microbiol. 34: 465-496.

(12) Field, M. 1979. Modes of action of enterotoxins from Vibrio cholerae and Escherichia coli. Rev. Infec. Dis. 1: 918-925.

(13) Finkelstein, R.A.; Boesman-Finkelstein, M.; and Holt, P. 1983. Vibrio cholerae hemagglutinin/lectin/protease hydrolyzes fibronectin and ovomucin: F.M. Burnet revisited. Proc. Natl. Acad. Sci. USA 80: 1092-1095.

(14) Finkelstein, R.A., and LoSpalluto, J.J. 1969. Pathogenesis of experimental cholera. Preparation and isolation of choleragen and choleragenoid. J. Exp. Med. 130: 185-202.

(15) Freter, R.; Brickner, H.; Fekete, J.; Vickerman, M.M.; and Carey, K.E. 1983. Survival and implantation of Escherichia coli in the intestinal tract. Infec. Immun. 39: 686-703.

(16) Freter, R.; O'Brien, P.C.M.; and Macsai, M.S. 1981. Role of chemotaxis in the association of motile bacteria with intestinal mucosa: in vivo studies. Infec. Immun. 34: 234-240.

(17) Fuller, R. 1977. The importance of lactobacilli in maintaining normal microbial balance in the crop. Br. J. Poult. Sci. 18: 85-94.

(18) Fuller, R.; Houghton, S.B.; and Brooker, B.E. 1981. Attachment of Streptococcus faecium to the duodenal epithelium of the chicken and its importance in colonization of the small intestine. Appl. Envir. Microbiol. 41: 1433-1441.

(19) Griffiths, E.; Rogers, H.J.; and Bullen, J.J. 1980. Iron, plasmids and infection. Nature 284: 508-509.

(20) Hale, T.L.; Sansonetti, P.J.; Schad, P.A.; Austin, S.; and Formal, S.B. 1983. Characterization of virulence plasmids and plasmid-associated outer membrane proteins in Shigella flexneri, Shigella sonnei, and Escherichia coli. Infec. Immun. 40: 340-350.

(21) Johnson, A. 1981. The pathogenesis of gonorrhoea. J. Infection 3: 299-308.

(22) Jones, G.W.; Robert, D.K.; Svinarich, D.M.; and Whitfield, H.J. 1982. Association of adhesive, invasive, and virulent phenotypes of Salmonella typhimurium with autonomous 60-megadalton plasmids. Infec. Immun. 38: 476-486.

(23) Kornfeld, S.J., and Plaut, A.G. 1981. Secretory immunity and the bacterial IgA proteases. Rev. Infec. Dis. 3: 521-534.

(24) Lankford, C.E. 1960. Factors of virulence of Vibrio cholerae. Ann. NY Acad. Sci. 88: 1203-1212.

(25) Lee, A., and Gemmell, E. 1972. Changes in the mouse intestinal microflora during weaning: role of volatile fatty acids. Infec. Immun. 5: 1-7.

(26) Lippincott, J.A., and Lippincott, B.B. 1980. Microbial adherence in plants. In Bacterial Adherence, ed. E.H. Beachey, pp. 375-398. London: Chapman and Hall.

(27) Markham, J.L.; Knox, K.W.; Wicken, A.J.; and Hewett, M.J. 1975. Formation of extracellular lipoteichoic acid by oral streptococci and lactobacilli. Infec. Immun. 12: 378-386.

(28) Marshall, K.C. 1976. Interfaces in Microbial Ecology. Cambridge,

MA: Harvard University Press.

(29) McNabb, P.C., and Tomasi, T.B. 1981. Host defense mechanisms at mucosal surfaces. Ann. Rev. Microbiol. 35: 477-496.

(30) Nester, E.W., and Kosuge, T. 1981. Plasmids specifying plant hyperplasias. Ann. Rev. Microbiol. 35: 531-565.

(31) Neutra, M.R. 1980. Prokaryotic-eukaryotic cell junctions: attachment of spirochetes and flagellated bacteria to primate large intestinal cells. J. Ultra. Res. 70: 186-203.

(32) Owen, R.L., and Nemanic, P. 1978. Antigen processing structures of the mammalian intestinal tract: an SEM study of lymphoepithelial organs. In Scanning Electron Microscopy/1978, vol. II, pp. 367-378. AMF O'Hare, IL: SEM Inc.

(33) Pearce, W.A., and Buchanan, T.M. 1980. Structure and cell-membrane - binding properties of bacteria fimbriae. In Bacterial Adherence, ed. E.H. Beachey, pp. 289-344. London: Chapman and Hall.

(34) Salyers, A.A.; West, S.E.H.; Vercellotti, J.R.; and Wilkins, T.D. 1977. Fermentation of mucins and plant polysaccharides by anaerobic bacteria from the human colon. Appl. Envir. Microbiol. 34: 529-533.

(35) Savage, D.C. 1980. Adherence of normal flora to mucosal surfaces. In Bacterial Adherence, ed. E.H. Beachey, pp. 33-59. London: Chapman and Hall.

(36) Savage, D.C. 1983. Morphological diversity among members of the gastrointestinal microflora. Int. Rev. Cyt. 82: 305-334.

(37) Savage, D.C.; Siegel, J.E.; Snellen, J.E.; and Whitt, D.D. 1981. Transit time of epithelial cells in the small intestines of germfree mice and ex-germfree mice associated with indigenous microorganisms. Appl. Envir. Microbiol. 42: 996-1001.

(38) Savage, D.C., and Whitt, D.D. 1982. Influence of the indigenous microbiota on amounts of protein, DNA, and alkaline phosphatase activity extractable from epithelial cells of the small intestines of mice. Infec. Immun. 37: 539-549.

(39) Schockman, G.D., and Wicken, A.J., eds. 1981. Chemistry and Biological Activities of Bacterial Surface Amphiphiles. New York: Academic Press.

(40) Silverstein, S.C.; Steinman, R.M.; and Cohn, Z.A. 1977. Endocytosis. Ann. Rev. Biochem. 46: 669-722.

(41) Snellen, J.E., and Savage, D.C. 1978. Freeze-fracture study of the filamentous, segmented microorganism attached to the murine small bowel. J. Bacteriol. 134: 1099-1107.

(42) Stanier, R.Y.; Adelberg, E.A.; and Ingraham, J.L. 1976. The Microbial World, 4th ed. Englewood Cliffs, NJ: Prentice Hall.

(43) Stanton, T.B., and Savage, D.C. 1983. Colonization of gnotobiotic mice of Roseburia cecicola, a motile, obligately anaerobic bacterium from murine ceca. Appl. Envir. Microbiol. 45: 1677-1684.

(44) Steffen, E.K., and Berg, R.D. 1983. Relationship between cecal population levels of indigenous bacteria and translocation to the mesenteric lymph nodes. Infec. Immun. 39: 1252-1259.

(45) Takeuchi, A. 1967. Electron microscopic studies of experimental salmonella infection. Am. J. Path. 50: 109-136.

(46) Umesaki, Y.; Tohyama, K.; and Mutai, M. 1982. Biosynthesis of microvillus membrane-associated glycoproteins of small intestinal epithelial cells in germfree and conventionalized mice. J. Biochem. 92: 373-379.

(47) Wallace, R.J.; Cheng, K.-J.; Dinsdale, D.; and Orskov, E.R. 1979. An independent microbial flora of the epithelium and its role in the ecomicrobiology of the rumen. Nature 279: 424-426.

(48) Weinberg, E.D. 1974. Iron and susceptibility to infectious disease. Science 184: 952-956.

(49) Winter, J., and Bokkenheuser, V.D. 1979. Bacterial metabolism of corticoids with particular reference to the 21-dehydroxylation. J. Biol. Chem. 254: 2626-2629.

(50) Wolin, M.J. 1975. Interactions between the bacterial species of the rumen. In Digestion and Metabolism in the Ruminant, eds. I.W. McDonald and A.C.I. Warner, pp. 134-148. Sydney: The University of New England Publishing Unit.

Microbial Adhesion and Aggregation, ed. K.C. Marshall, pp. 251-282. Dahlem Konferenzen 1984. Berlin, Heidelberg, New York, Tokyo: Springer-Verlag.

Utilization of Substrates and Transformation of Solid Substrata

Z. Filip* and T. Hattori**
*Institute for Water, Soil, and Air Hygiene
German Federal Health Office
6070 Langen, F.R. Germany
**Institute for Agricultural Research
Tohoku University, Sendai, Japan

Abstract. Solid adsorbents which play a role in microbial adhesion influence the growth and metabolic activity of microorganisms and utilization of substrates. Usually, different adsorbents enhance the specific growth rate and molar growth yield in pure and mixed cultures of bacteria, actinomycetes, and fungi. In the dependence of the microorganisms tested, of adsorbents used, and of cultivation conditions maintained, stronger glycolytic or oxidative utilization of a carbon source can be observed. Formation of humic substances is generally increased. The mechanism of the effect of adsorbents on microbial utilization of substrates is very complex and includes the adsorption of cells, substrates, enzymes and other microbial metabolites, as well as degradation products. Electron transfer from cell to solid surface and the hydrophobicity of the solid surface may be involved. Different biotic and abiotic materials serve as solid substrata in microbial adhesion, and they can be transformed by microorganisms. Both superficial and substantial transformation may occur in which exoenzymes and other metabolic products are involved. Further analytical studies are necessary to elucidate the relationship between microorganisms and surfaces and how they influence microbial growth, utilization of different substrates, and transformation of solid substrata.

INTRODUCTION

The environmental effects influencing microorganisms in their natural terrestrial and aquatic ecosystems are extremely complex, and they include actual biotop factors (biotic and abiotic) and climatic and topographic factors. The number of special studies which have been published on the effect of their various components, e.g., temperature, humidity, organic matter content and quality, atmosphere, pH, rH, osmotic pressure, etc., on microorganisms is very large. The present paper attempts to evaluate some existing facts on the significance of solid adsorbents as an important natural ecological factor influencing the main manifestations of the physiological activity of microorganisms, i.e., microbial growth and transformation (utilization) of substrata. To illustrate the authors' specific points of view, references cited in this paper have been restricted primarily to works from their own or their co-authors' laboratories. Nevertheless, excellent studies by other workers in this field should be mentioned here (e.g. (56, 57)). When adhered on surfaces of different substrata, microorganisms can affect them in several ways which include nutrient relationships, superficial deterioration, degradation, and structural changes. Some examples of such effects are also given in this paper.

GROWTH AND METABOLIC ACTIVITY

Solid adsorbents, such as inorganic and organic colloids, are the site of the main physicochemical relations in the environments. The colloid particles of some adsorbents (e.g., clays) probably played an important role even at the beginning of life on the Earth by contributing to protein formation (52). Consequently, the solid adsorbents may have also influenced the first well organized forms of life, microorganisms. However, it is not for historical reasons that we should study the interactions of microorganisms and solid adsorbents; rather, because microorganisms are in close contact with solid adsorbents in soils, waters, and sediments, and microbial growth and metabolic activity can be strongly influenced by adsorbents.

Microbial Humification

Due to their importance in natural environments, the chemistry, biochemistry, and biology of terrestrial and aquatic humic substances are receiving increased attention (40). Inasmuch as humic substances are mostly products of microbial activity (30), adsorbents influencing microorganisms may also exert an effect on the formation of humic substances. It was found in model experiments that the humification of protein substrate in the presence of bentonite was in inverse proportion to the degree of its mineralization. A mixed culture of soil microorganisms in intensively aerated liquid medium formed the most humic substances in treatments without clay, whereas mineralization was higher in treatments with bentonite. Substances on the humic acid type had greater optical density in the latter medium, however, a lower $Q_{4/6}$ value was recorded. In a sand-bentonite mixture, mineralization fell as the clay concentration was raised, and the amount and optical density of humic substances rose at the same time (14, 17). No correlation of either mineralization or humification processes to clay content was found in a mixture of loamy, sandy soil, or of loamy soil when bentonite was added, but more humic acids were found firmly sorbed by the mineral portion of the soil in the bentonite treatments (15, 17).

Though the formation of humic substances in natural environments is controlled primarily by mixed microbial populations, in some microhabitats, however, several species of microorganisms capable of forming humic-type substances may temporarily or permanently gain the upper hand. They include certain microscopic fungi (45), of which Epicoccum nigrum and Stachybotrys chartarum were cultivated in a nutrient solution containing adsorbents (montmorillonite, kaolinite, and quartz sand) (25, 26). In the adsorbent treatments, particularly with montmorillonite, higher synthesis of humic substances was found (Table 1) as well as greater biomass formation and a higher economic coefficient.

In order to show more clearly the extent to which the energy source, glucose, was converted to humic acid-type substances relative to cellular substance, the ratios of total humic acid formed to glucose consumed and the ratios of total humic acid to total cell substance synthetized were calculated (Table 2). After twenty and thirty days of incubation, the presence of montmorillonite greatly increased these ratios.

The catalytic action of adsorbents on the polymerization of humic precursors was emphasized by Scheffer and Kroll (54), Kyuma and Kawaguchi (43), and others. In respirometric and spectrophotometric experiments carried out by Filip et al. (24), however, the oxygen uptake of 5-methylpyrogallol, 2, 3, 4-trihydroxybenzoic acid, 2, 3, 5-trihydroxytoluene, as well as the mixtures of these phenols with protocatechuic acid, caffeic acid, and orcine, was not at all or only to a small extent influenced by the presence of montmorillonite in the water suspension buffered at pH 7 or 8. Furthermore, montmorillonite and kaolinite, when saturated with H^+, K^+, Ba^{2+}, Al^{3+}, or Fe^{3+} ions and added to the pyrogallol solution buffered at pH 6.0, influenced neither the speed of the color changes nor the intensity of the brown color, evaluated as an extinction at 300 nm.

TABLE 1 - Formation of humic substances in stationary cultures of *E. nigrum* with different concentrations of montmorillonite (reproduced from (26)).

	Grams of Ash-free and Dry Humic Acid in 1 l of Culture Medium							
	Montmorillonite (% W/v)							
	0	0.25	0.5	1	0	0.25	0.5	1
Humic Acid	20 days				30 days			
From Culture Solution (HA1)	0	0	0.24	0.88	0.12	0.38	0.57	0.98
From Cells Extracted with NaOH (HA2)	0	0.66	0.80	0.32	0.10	0.73	0.78	0.47
From Cells After Acid Hydrolysis (HA3)	0	0.54	0.45	0.02	0	0.15	0	0.18
Total	0	1.20	1.49	1.22	0.22	1.76	1.35	1.63
	Grams Carbon/l in Supernatant of HA2							
Fulvic Acid	0.87	0.70	0.49	0.33	0.95	0.48	0.44	0.23

TABLE 2 - Ratio of total humic acid x 100 (Table 1) to biomass of E. nigrum and to glucose consumed in stationary cultures (reproduced from (26)).

	20 Days		30 Days	
Montmorillonite Additions (% w/v)	g of HA (tot.) x 100 per 1g Cells	per 1g Glucose	g of HA (tot.) x 100 per 1g Cells	per 1g Glucose
Control	0	0	4.1	1.5
0.25	18.2	4.7	17.5	4.2
0.5	15.7	5.8	18.0	4.5
1.0	15.8	4.1	29.6	5.4

In model experiments, the sorption to clays of some phenolic acids, usually produced during the first stages of fungal growth, could be measured (24). An expansion of the montmorillonite interlayer spacing from the original 12 Å to 18 Å was observed during the first ten days of cultivation. From the twentieth to thirtieth day, the montmorillonite interlayer spacing contracted again to 12 Å. These changes should also be recognized if the clay mineral was in dialysis tubes, so that a sorption of some low molecular weight substances, which could be phenolics, could be supposed. The sorption to clays of such substances, which can act as growth inhibitors, could enhance fungal growth in the first stages of cultivation. Later, when exchanged for ammonium released during cell autolysis, the same substance could support the dark polymer formation because they polymerize auto-oxidatively in water solutions above pH 7.

The observations made indicate that, by affecting the growth and metabolic activity of microorganisms, adsorbents can indirectly influence the formation of humic substances.

Mixed Culture Studies

Examination of a soil suspension under the microscope usually shows microorganisms forming larger or smaller aggregates with mineral adsorbents, mostly clays (17). The influence of clays on soil microorganisms is very complex and includes trophogenic relationships, cation exchange capacity, sorptive activity, and osmotic and other physicochemical effects in microbial environments. Different effects

TABLE 3 - Effect of bentonite on CO_2 release from sand cultures inoculated with a suspension of soil microorganisms and enriched in glucose and mineral nitrogen (mg CO_2 h^{-1} per 100 g sand) (reproduced from (19)).

Treatment	Bentonite (%)			
	0	0.12	0.25	0.50
Control	0.38	0.21	0.21	0.21
$(NH_4)_2SO_4$ added	0.29	0.27	0.33	0.33
Glucose added	2.63	4.73	4.19	10.96
Glucose + $(NH_4)_2SO_4$ added (C:N=10:1)	7.38	7.12	7.70	16.56

of clay minerals on soil microorganisms and also the influence of microorganisms on clay minerals have been discussed by Filip (18, 19, 23).

Increased cell counts of several nutritional groups of soil microorganisms were generally noted after addition of an adsorbent (e.g., bentonite) to sand or soil samples (17). Usually the highest clay amendment (0.5% = 200 q ha^{-1}) was most effective. Respiratory activity of soil microbiota, however, did not differ substantially in the amended or unamended samples. Immediately after addition of some readily utilizable carbon or nitrogen sources, the microbial activity increased quickly in clay-amended samples (Table 3). An adsorbent (bentonite) also enhanced ammonification activity of soil microorganisms as reported by Filip (16, 19) and as shown in Table 4.

Different adsorbents support substrate utilization of a complex microbiota even under undesirable stress conditions. The curves in Fig. 1 demonstrate a high metabolic activity of compost microorganisms exposed to a heat shock at 70°C in the cultures containing inorganic and organic solid particles. At five days (the third day of the 70°C period), 42 mg CO_2 day^{-1} was evolved in controls, but 380 mg CO_2 day^{-1} was observed in the cultures with clay. Glass beads and humic acids were less effective; nevertheless, the cultures containing these particles produced about five- to sevenfold more CO_2 than the controls (21).

TABLE 4 - Effect of bentonite on ammonification activity of soil microorganisms in sand cultures enriched in casein (0.5%) (mg% NH^{+}_{4}-N) (reproduced from (19)).

Variant	Bentonite (%) 0	0.12	0.25	0.50
1) Control	2.3	3.1	3.9	6.5
2) Peptone added	136.4	152.0	156.7	181.8
3) = 2) - 1)	134.2	148.9	152.8	175.3

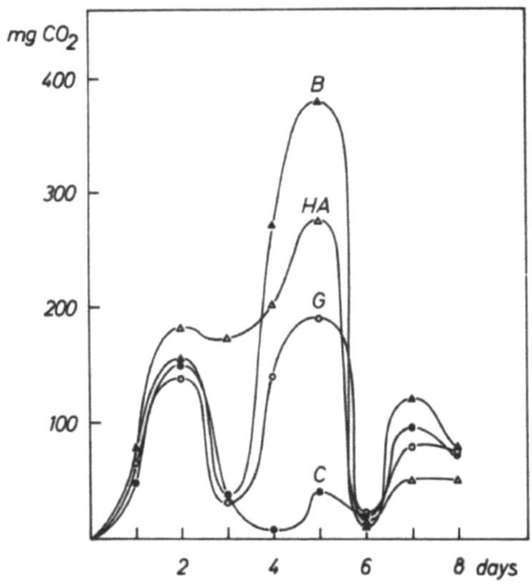

FIG. 1 - Daily CO_2 evolution in 30°C submerged cultures of compost microorganisms exposed to a heat shock at 70°C between the third and fifth days. C = control; G = glass beads (10% w/v); B = bentonite (0.5% w/v); HA = humic acid (0.5% w/v) (reproduced from (21)).

If polymers serve as a substrate for microorganisms, the degradation process is due to the activity of microbial exoenzymes. Some hydrolytic enzymes such as esterase and histidase can sequentially break intermolecular bonds in polyurethane, e.g., as shown by Filip (20). In the tests with adsorbents (kaolinite, bentonite), however, ring frequencies of isocyanuric acid in the polyurethane remained intact. Also, the loss in weight of different polyurethanes incubated in garbage landfill leakage water rich in microorganisms and with clays added was either lower than in controls or not detected. This obvious inhibitory effect of adsorbents (clays) can be explained by the sorption and decrease in activity of microbial hydrolases. Many examples of such enzyme inactivation through adsorbents are discussed by Filip (19) and Burns (9).

Pure Culture Studies

In our experiments we have attempted to determine whether microbial activity in culture is influenced by the presence of solid adsorbents such as clays or other solid particles.

Montmorillonite, for example, when added to cultures of Saccharomyces cerevisiae, enhanced the biomass formation in aerobic (shake and stationary) cultures, but not under anaerobic conditions (Table 5). The growth increase over the controls was up to 90% (28).

In these experiments, the glucose concentration in the nutrient solution was 4%, i.e., 204 mmoles per liter. The amounts of glucose consumed and the corresponding amounts of ethanol formed are given in Table 6. The glucose consumption was enhanced by montmorillonite both in shake and stationary cultures, but not under anaerobic conditions. Also, an acceleration in the formation of ethanol could be observed only in

TABLE 5 - Amount of biomass of S. cerevisiae in culture solution (g l^{-1}) (reproduced from (28)).

Cultivation	Shake		Stationary		Anaerobic Cultures	
in Days	c[a]	m[a]	c	m	c	m
1	0.47	0.89	-	-	0.33	0.29
2	1.20	1.88	-	-		
3	1.47	2.45	1.10	1.62	1.07	1.03
4	1.75	2.95	1.47	2.10	1.21	1.28
5	2.45	3.34	1.75	2.90	1.19	1.29
6	2.56	3.08	2.08	3.12	1.16	1.23
7	2.42	2.87	2.15	3.14		

[a]c = Control, m = 0.5% w/v Montmorillonite

TABLE 6 - Glucose consumption and ethanol formation in S. cerevisiae culture solution (reproduced from (28)).

Cultivation	Shake		Stationary		Anaerobic Cultures	
in Days	c	m	c	m	c	m
3	99[a]	149	53	74	113	105
	87[b]	159	67	96	166	154
4	128	173	85	103	170	178
	135	189	111	137	270	278
5	158	201	122	156	189	194
	174	237	146	185	310	316
6	187	201	180	201	195	201
	220	224	207	237	322	332
7	202	203	203	203	-	-
	220	215	241	257		

[a]Glucose consumption in mmole per liter
[b]Ethanol in mmole per liter
c and m as in Table 5

aerobic cultures. Under aerobic conditions, the ratio between ethanol formed and glucose consumed showed a somewhat higher formation of ethanol per unit of consumed glucose in the presence of montmorillonite and indicated, therefore, a somewhat higher rate of glycolysis in the presence of the clay minerals.

A calculation of the "economic coefficient" (biomass x 100 divided by the amount of glucose) showed an increase of this coefficient in the presence of montmorillonite in the aerobic cultures and indicated a more efficient conversion of glucose into biomass.

Respirometric experiments in Warburg-vessels showed a decrease in O_2 consumption with increasing montmorillonite concentration, but CO_2 evolution did not change correspondingly (Table 7). Therefore, the CO_2/O_2 ratio increases, which indicates a higher rate of glycolysis of the cells

TABLE 7 - O_2 uptake and CO_2 release of S. cerevisiae with montmorillonite (reproduced from (28)).

Amount of montmorillonite (mg/vessel)	mmole O_2/2 h	mmole CO_2/2 h	CO_2/O_2	mmole Ethanol/ vessel/2 h
0	24.8	59.0	2.38	31.1
2.5	24.4	60.8	2.49	30.3
5	23.6	60.4	2.56	31.3
10	23.4	60.6	2.59	33.1
20	22.4	59.0	2.64	35.3

in the presence of montmorillonite. Warburg experiments made under anaerobic conditions showed no effect of montmorillonite on the CO_2 release and the formation of ethanol. Short time experiments with ^{14}C-labeled glucose established a higher incorporation of substrate into the biomass in the presence of montmorillonite.

A higher incorporation of substrate into the cell polymer fraction was established also by use of ethanol labeled either in the C_1- or C_2-carbon by ^{14}C. Ethanol is introduced via alcohol dehydrogenase into the Krebs cycle, which functions both as a pathway for the oxidation of acetyl-CoA and as a source of cell synthetic intermediates. From the relative contribution of the C_1- and C_2-carbon to the respiratory CO_2, it is possible to calculate approximately how much of the ethanol carbons are used for oxidation or for synthetic processes. The experiments with montmorillonite added showed an acceleration of the O_2 uptake, but also an increase in the incorporation of the methyl carbon into cell polymers. Also, the carbinol carbon was incorporated to a somewhat higher extent, which may indicate a higher flow through the glyoxylate cycle (29).

In well aerated culture solutions, Ca-montmorillonite at 0.25% concentration markedly accelerated and increased growth, glucose consumption, and CO_2 evolution by various Streptomyces, Micromonospora, and Nocardia species (44). Clay in dialysis tubing was almost as effective as that suspended directly in the medium. Used clay in dialysis tubing

TABLE 8 - Economic coefficient values of the cell synthesis by Streptomyces violaceus-niger in the presence and absence of clay (reproduced from (44)).

Incubation Days	Control	Montmorillonite
3	27.6	29.5
6	13.6	23.2
9	10.3	16.3
15	7.3	11.2
21	5.3	10.6

was a little less effective than fresh clay. Economic coefficient values indicated a greater biomass synthesis per unit glucose consumed for the cultures containing clay (Table 8).

Tests were also made to compare the growth effect of Ca-montmorillonite with that of Ca-humate and $CaCO_3$ (Table 9). Finely powdered $CaCO_3$ slightly improved the growth of S. violaceus-niger and also that of Nocardia globurula. The Ca-humate at half the concentration of the clay but with more than double the exchange capacity was almost as effective as the clay in increasing biomass formation. When Ca-humate was supplemented with sufficient Fe^{2+}, Mn^{2+}, Cu^{2+}, and Zn^{2+} to supply 5 ppm Fe and Mn and 1 ppm Cu and Zn to the culture solutions, its effect on the growth of S. violaceus-niger was not altered.

The marked stimulatory effect of montmorillonite could be explained on the basis of adsorption of one or more metabolic inhibitors. Support for this theory is indicated by the observations that clay, even when closed in dialysis tubing which allows transport of molecules low in weight, generally improved growth after a short lag period. Clay once used in experiments was not as effective as fresh clay. This view is further supported by the fact that when maximum biomass was present in the control cultures, there was considerable glucose remaining in the culture solution, whereas in the cultures with clay, the glucose was generally all consumed at the time of maximum biomass production.

TABLE 9 - Influence of montmorillonite, Ca-humate, and $CaCO_3$ on growth of S. violaceus-niger (reproduced from (44)).

Incubation Days	Control	Weight of Cells, Dry and Ash-free ($g.l^{-1}$) Mont. 0.25%	Ca-humate 0.12%	CaCO 0.12%
3	0.8	2.2	1.6	1.0
6	1.6	3.9	3.7	2.0
9	1.2	3.0	2.5	2.0
15	0.9	2.6	2.2	1.7
21	0.8	2.3	1.8	1.5

The stimulatory effect of the humic acid could also be explained by its ability to adsorb inhibitory molecules as suggested by Waksman and Iyer (59) and Badura (7). However, it is possible that small amounts of stimulatory organic molecules may be released from the humates. It is also possible that the humates supply needed trace elements. If these were a factor in the experiments discussed here, sufficient Fe, Zn, Cu, and Mn were apparently supplied from the original humate, as supplementation with these ions did not further increase growth.

An accelerated biomass formation in the presence of clay was observed with Aspergillus niger which also showed enhanced citric acid formation and, later, consumption. With several penicillia the growth was inhibited or not influenced in stationary cultures, but enhanced in the better aerated shake cultures (28).

A different response to the presence of an adsorbent in the culture of several bacteria was also observed. The respiratory activity of Pseudomonas pyocyanea, Bacterium prodigiosum, Bacillus megaterium, and Staphylococcus aureus was 1.5 - 3-fold lower if bentonite was present. There was no influence observed with Sarcina lutea, whereas glucose utilization increased up to tenfold and oxygen consumption up to 21-fold in Azotobacter chroococcum cultures with clay added (60).

MECHANISMS

Growth on an Anion-exchange Resin

Factors affecting the growth of bacterial cells adhered to a surface may be varied and, more or less, complicated. Kinetic studies of Escherichia coli cells adhered on an anion-exchange resin, Dowex-1 in the chloride form, have contributed to a deeper understanding of these factors.

Distribution or arrangement of newborn cells derived from adhered cells. In the case of the growth of bacterial cells suspended freely in a liquid medium, both parent and daughter cells distribute randomly in the medium. But, in the case of adhered cells, parent cells are immobilized on the surface and newborn daughter cells are destined to be either immobilized on the surface as are the parent cells or released into the bulk liquid medium.

The experiment described in Fig. 2 was designed to obtain information on the distribution of daughter cells. Resin particles loaded with different amounts of E. coli cells per gram of resin were incubated under stationary conditions in a glucose-minerals medium. After various lag periods, the turbidity due to cells in the medium began to increase independently of the cell density on the resin particles. Such a linear increase in the turbidity was not observed when a carbon or nitrogen source was omitted

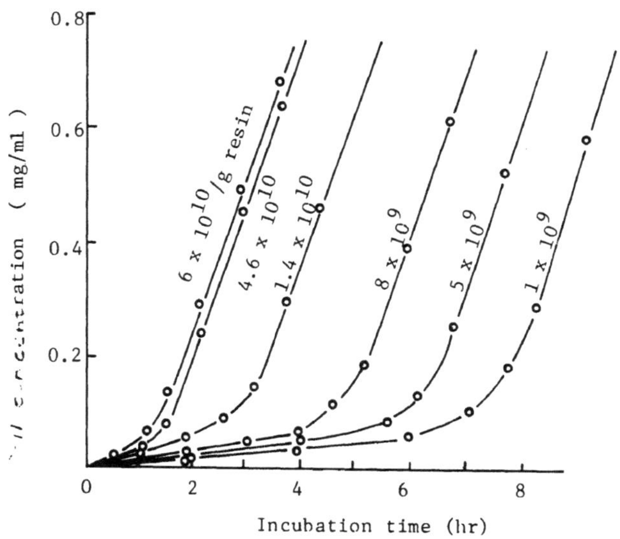

FIG. 2 - Relation between the duration before cell release and initial density of adhered cells. 1.5 g of resin with different densities of adhered cells were incubated at 30°C and the increase in cell concentration in the liquid part of the culture was followed.

or when chloramphenicol (50 µg/ml) was added and, consequently, the linear increase is considered to be closely related to the growth of adhered cells (35).

The results can be interpreted by a model in which, in the case of the growth of adhered cells, newborn cells would be immobilized on the surface until the surface area is fully covered with these cells and, thereafter, the cells would be released into the liquid medium. The critical value of cell density on the surface was estimated at ca. 5×10^{10} cells/g resin, or 8.8×10^7 cm^2 resin surface, in the case of a batch culture. In the case of a continuous flow culture, similar detachment of newborn cells was observed, and the critical density was estimated at ca. 1.3×10^{10} cells/g resin, or 2.3×10^7 cells/cm^2 resin surface. These values of the critical density correspond to a cell density when cells cover the surface in a monolayer, a point which is supported by electron micrographs of carbon replicas of the cell-loaded resin particles (36). It is noteworthy that adhered cells in a continuous flow medium showed a synchronous growth when they were transferred into a new medium and cultured under stationary conditions (31).

Growth equation for adhered cells in a continuous flow system. The growth of cells adhered on the resin particles may be formulated as follows (31): when adhered cells were incubated in a continuous flow system at high growth rates (D = 1), cell concentration in the effluent increased to a steady state level and the concentration of a limiting substrate also became steady, thus we have

$$\frac{dX}{dt} = \mu X + \mu_a X_a - DX, \qquad (1)$$

where X is cell concentration in the liquid phase of culture (mg dry wt/ml), X_a the amount of adhered cells represented in terms of liquid phase volume of the culture (mg dry wt/ml), μ the specific growth rate of free cells (hr^{-1}), μ_a the specific growth rate of adhered cells (hr^{-1}), and D the dilution rate (hr^{-1}). The value of μ under a condition of D = 1 is estimated from the slope of logarithmic plots of washout curve when free cells are incubated without adhered cells; the value was 0.5 hr^{-1} over a wide range of glucose concentrations (4.29 mM to 22.70 mM), indicating that the maximal growth rate was realized at these concentrations. The value of μ_a can be estimated from the cell concentration of the effluent (X) in a steady state in which cell density is almost the same as the initial one. That is, by putting

TABLE 10 - Growth rate of adhered cells in relation to the amount of resin and substrate concentration.

Resin (g)	X [a] (mg dry wt/ml)	X_a [b] (mg dry wt/ml)	μ_a [c] (hr^{-1})	S [d] (mMd)
10	0.026	0.008	1.560	19.40
20	0.0155	0.016	0.465	9.18
20	0.0285	0.016	0.855	19.66
50	0.014	0.0416	0.168	13.22
60	0.013	0.050	0.130	9.20
50	0.029	0.0416	0.348	22.08
80	0.007	0.066	0.052	4.66
80	0.0125	0.066	0.093	9.60
75	0.026	0.0625	0.208	21.08
95	0.033	0.0791	0.208	21.20
75	0.030	0.0625	0.240	21.28
75	0.030	0.0625	0.240	21.28

[a] Steady level cell concentration in liquid phase.

[b] Amount of adhered cells in terms of cell concentration, the value of which was obtained by dividing amount of adhered cells (0.3 mg dry wt/g resin) by the culture volume.

[c] Growth rate of adhered cells given by Eq. 2, in which $\mu = 0.5$, $D = 1$.

[d] Steady level substrate.

$\frac{dX}{dt} = 0$ in Eq. 1, we have

$$\mu_a = (D-\mu)\, X/X_a. \qquad (2)$$

Values of μ_a estimated under a condition of $D = 1\ hr^{-1}$, where $\mu = 0.5\ hr^{-1}$, are given in Table 10.

It is noteworthy that the value of μ_a varies widely not only with substrate concentration but also with the amount of adhered cells in the system (X_a), where the amount of adhered cells is changed by the amount of cell-loaded resin and, thus, the cell density on the surface of each resin particle is kept constant. As in the case of free cells, we can expect the following relation between μ_a and substrate concentration (S):

$$\mu_a = \mu_{max,a} S/(K_{s,a} + S), \tag{3}$$

where $\mu_{max,a}$ and $K_{s,a}$ are the maximum growth rate of adhered cells and the appropriate constant, respectively. Equation 3 can be rewritten

$$\frac{1}{\mu_a} = \frac{K_{s,a}}{\mu_{max,a}} \frac{1}{S} + \frac{1}{\mu_{max,a}} . \tag{4}$$

Using Eq. 4 we obtained Lineweaver-Burk type plots as depicted in Fig. 3. With different amounts of adhered cells we have a different linear relationship which is similar to a competitive type of inhibition in an enzyme reaction. Thus the value of $K_{s,a}$ varied with the amount of adhered cells or cell-loaded resin particles, although $\mu_{max,a}$ did not.

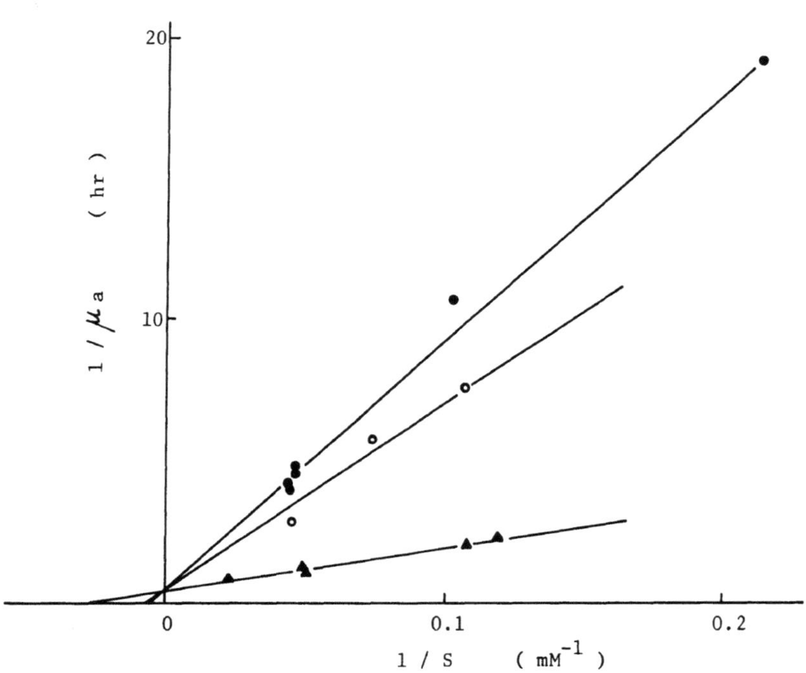

FIG. 3 - Reciprocal plots of μ_a against glucose concentration. The amounts of cell-loaded resin used are: ●, 75-95 g; ○, 50-65 g; ▲, 20 g.

Another point to be noticed in the kinetics of the growth of adhered cells is that value of $K_{s,a}$ was 0.3 - 1.5 x 10^{-1} M when glucose concentration was 4.98 - 22.70 mM. The value is too high compared with the value of free cells (2 x 10^{-5} M) obtained by Monod (47).

These remarkable results can be interpreted by assuming that, because the density of growing cells on the surface is very high and the cells take up glucose very vigorously, the rate of glucose consumption will exceed its diffusion to the resin surface, which results in the formation of a glucose-depleted sphere surrounding the cell-loaded resin particles. The increase in the value of $K_{s,a}$ in proportion to the amount of cell-loaded resin particles will reflect the enhancement of glucose depletion by overlapping of these spheres.

This assumption seems incompatible with the fact that, in the case of glucose oxidation by nongrowing E. coli cells, the relation between O_2 uptake rate and glucose concentration was nearly the same with free and adhered cells (34) and the value of K_s was estimated to be between 10^{-4} and 10^{-5} M, which approximately coincides with the value obtained by Monod in the case of growing cells. This fact indicates that no glucose-depleted sphere was formed in this case. Therefore, our question is, why is a glucose-depleted sphere formed in the case of growing adhered cells and not in the case of nongrowing adhered ones? Anwering this question may be possible by consideration of the difference in glucose consumption rates between growing and nongrowing adhered cells; 5-50 x 10^{-6} mole/mg dry cells/hr for the former, and 0.03 x 10^{-6} mole/mg dry cells/hr for the latter (the latter estimation was made from O_2 uptake in the case of glucose oxidation by assuming a complete oxidation of the substrate).

The maximum specific growth rate of adhered cells ($\mu_{max,-a}$) was estimated at 2.0 hr^{-1} and that of free cells (μ_{max}) at 0.49 hr^{-1}. The result is very impressive compared with the fact that when nongrowing cells were used, the oxidation rate of glucose by attached cells was 4.6% of that by free cells at pH 6.0 (38).

Growth yield of adhered cells. Growth yield of adhered cells in the continuous flow system can be estimated by using the following equations. During the growth of adhered cells in a continuous flow system, Eq. 1 applies to the change of cell concentration in the liquid phase of the culture and the following equation to the change of substrate concentration in the culture:

$$\frac{dS}{dt} = DS_R - DS - \frac{\mu X}{Y} - \frac{\mu_a X_a}{Y_a} , \tag{5}$$

where S and S_R are the concentrations of substrate of liquid phases in the culture vessel and the reservoir, respectively, and Y and Y_a the growth yields of free and adhered cells, respectively. Under steady state conditions, dX/dt = 0 in Eq. 1 and dS/dt = 0 in Eq. 5. Thus we have Eq. 2, and

$$\frac{1}{Y_a} = \frac{D}{D - \mu} \frac{S_R - S}{X} - \frac{\mu}{D - \mu} \frac{1}{Y} . \tag{6}$$

Values of μ and Y are estimated from the results of washout experiments with free cells; the former is from the logarithmic plots of cells washed out and the latter from the ratio of the amount of cells in a volume of the effluent to the amount of glucose consumed in that volume (S_R - S) x V. The value of (S_R - S)/X can be obtained from plots in Fig. 4. Thus the value of Y_a can be estimated from these values using Eq. 6.

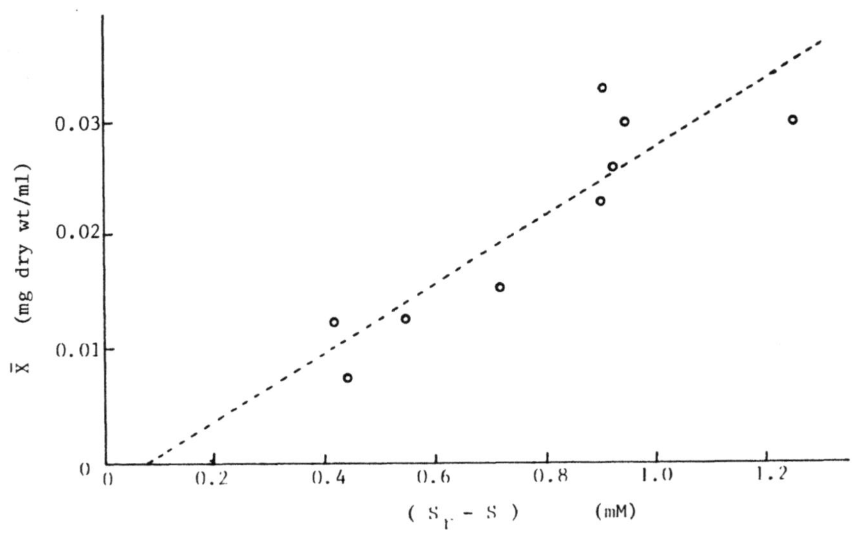

FIG. 4 - Steady level cell concentration, $\bar{X}$, in relation to glucose consumption, S_R - S. The dotted line is a regression of $\bar{X}$ on S_R - S. $\bar{X} = 0.0026 + 0.0306\ (S_R - S)$, r = 0.886.

We estimated the value of Y_a at 50.03 g dry wt/mole glucose when Y = 22 g dry wt/mole glucose. It is notable that the value of Y_a is about two times larger than that of Y.

Effects of cell adhesion onto the surface of the resin particles on bacterial physiology. We have shown in earlier reports (33, 38, 39) that chemical activities of nongrowing cells of E. coli, Azotobacter agile, Pseudomonas fluorescens, and Micrococcus luteus were also affected by cell adhesion onto the resin surface. Rates of oxidation of organic substances were markedly depressed by adhesion. A relation between pH and activity was shifted toward the alkaline side by about one unit in every case when cells were attached. Activity substrate concentration was also shifted towards high concentration by the adhesion in the case of anionic substrate but not in the case of neutral or amphoteric substrate. Activation energies for substrate oxidation and growth were observed to be markedly varied by cell adhesion in the case of E. coli. It was also reported that lag time before linear oxidation of lactose or xylose by E. coli or of succinate by A. agile was markedly shortened by adhesion. It was also found that, in the case of B. subtilis, kinetics of both cell proliferation and spore formation were greatly affected by cell adhesion (32).

Undoubtedly, adhesion of bacterial cells onto the surface of the resin will result in various modifications in bacterial metabolism or physiology. Among these modifications, the central problem may be why cell adhesion acts to depress chemical activities of nongrowing cells on one hand, and on the other hand, to stimulate cell growth, giving a higher efficiency of utilization of carbon and/or energy sources. Further analytical studies are necessary to elucidate mechanisms of these effects at the molecular level.

Electron Transfer

Effects of solid surfaces in bacterial activities involve a variety of known and unknown factors, and transfer of electrons from bacterial cells to solid surfaces may be among such factors (37). During the metabolism of bacterial cells, electrons are removed from various substances and, under usual experimental conditions, these electrons are passed through a series of intermediate carriers until the last carrier of the series reacts in its reduced state with oxygen or other terminal electron acceptors. But we can suppose that, in a heterogenous environment such as soil and mud, for example, some part of the electrons thus liberated may be transferred to solid surfaces as an electron circuit that is earthed and, in such cases, solid surfaces may play a role in the metabolism itself.

Allen et al. (3) observed the production of an electric current by E. coli cells during the metabolism of organic compounds using H-type electrochemical cells containing smooth platinum cylindrical electrodes separated by an ion-exchange membrane. However, Allen observed negligible current when bacterial cells were prevented from coming into contact with the anode by insertion of a dialysis membrane separator (1, 3). Thus, the current is considered to result from the bacterial cells making intimate contact with, and transferring potentially available electrons to, the electrode surface. Presumably, electrons are present on the cell wall surface and discharged to the electrode from the portion of the organism making contact with the electrode. Since the cell membrane located inside the cell wall is the site where electrons are liberated most vigorously through oxidation reactions, the production of electric current is expected to be larger if spheroplasts of the organism are used instead of intact cells. Allen showed that this is the case. Spheroplasts produced several times the coulombic output than did an intact cell control (2).

Recently Morisaki (48) examined the production of electric current by E. coli using a two-electrode system in which each electrode consisted of a platinum anode and silver peroxide cathode with an anion-exchange membrane as a separator; one electrode was covered with a cellulose dialysis membrane at the anode to prevent contact of the bacterial cells with the electrode (reference electrode), and the other was not covered (determination electrode). He noticed that the production of electric current was diminished markedly in proportion to the amount of an anion-exchange resin, Dowex-1 (200 - 400 mesh), which was added to the reaction mixture, although the rate of substrate (glucose) uptake by bacterial cells was enhanced by the addition of the resin. When resin particles were eliminated from the reaction mixture, electric current recovered to that of the original level. These results were interpreted by assuming that electrons on the surface of bacterial cells can be transferred to the resin particles by contact with them and, thus, a competition for electrons occurs between the electrode and the resin particles.

Another important observation reported by Morisaki is that, by an addition of the resin particles into an E. coli-substrate buffer system without the electrode system, rates of substrate (glucose) uptake and carbon dioxide production were stimulated, but the production of lactic and succinic acids was diminished (Fig. 5). The results of these experiments suggest that glucose metabolism of E. coli cells became more oxidative by an addition of the resin particles. The effect may be induced by transferring electrons from bacterial cells to resin particles.

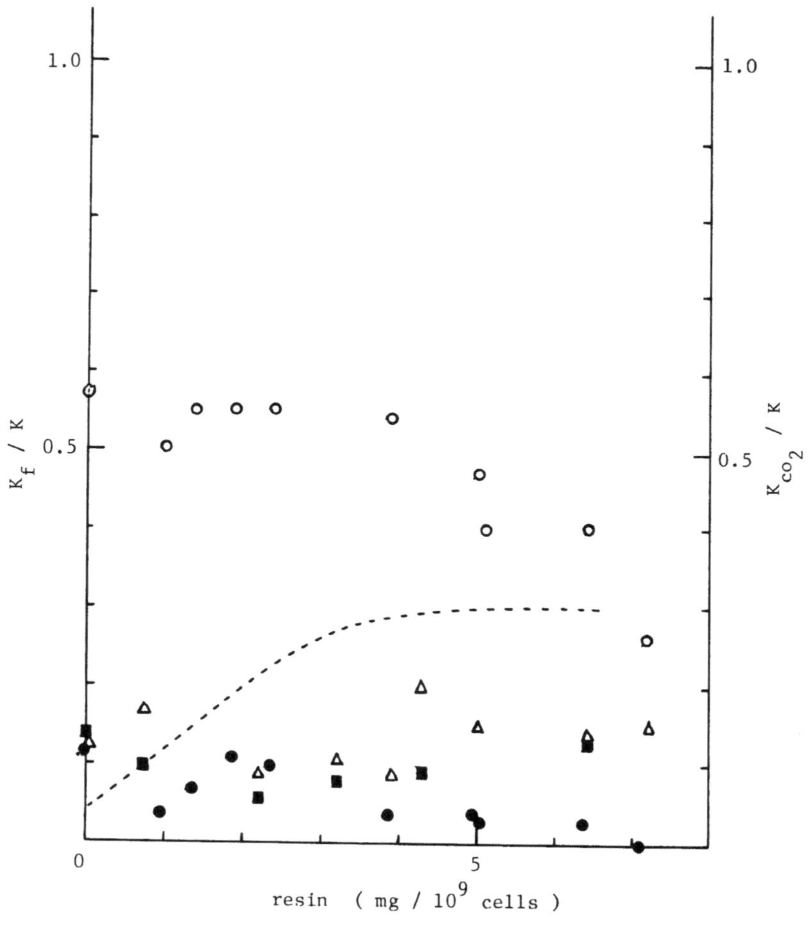

FIG. 5 - Effect of resin on the rates of fermentative product formation of CO_2 production. Kf/K is the ratio of the rate constant of product formation to that of glucose uptake; Kco_2/K is the ratio of the rate constant of CO_2 production to that of glucose uptake; both were corrected for the number of carbons in each product. Dotted line, Kco_2/K; O, lactic acid; •, succinic acid; Δ, acetic acid; ▪, ethanol. Ethanol was evaporated by a first-order reaction at the rate constant of 2.5 x 10^{-3} min^{-1}.

Effect of Hydrophobic Surfaces

In the study of the interaction between bacterial cells and solid surfaces, the hydrophobicity is considered an important factor and is discussed energetically in terms of surface tension. Researchers' attention is directed toward the effect of the hydrophobicity resulting in a closer interaction or adhesion between cells and solid surfaces. However, evidence was recently presented suggesting that the hydrophobicity of a surface may affect the metabolism of E. coli cells through a transient contact of cells with a surface.

Morisaki (49) examined glucose oxidation by E. coli cells with and without an addition of various solid particles including an anion-exchange resin (Dowex-1), a cation-exchange resin (Dowex-50), styrene-divinylbenzene copolymer, acrylic polymer, polytetrafluoroethylene, kaolinite, montmorillonite, silicic acid, celite, and alumina. Uptake rates of glucose and molecular oxygen by cells were, more or less, affected by an addition of solid particles without appreciable cell adhesion. He noticed that rates of these reactions were affected toward different directions in most cases; when the glucose uptake rate was diminished, the oxygen uptake rate was enhanced. This fact was formulated by introducing the following index, the respiratory ratio (R):

$$R = \frac{O_2 \text{ taken up (mole)}}{\text{glucose taken up (mole)} \times 6} \times 100.$$

Some examples are depicted in Fig. 6. Morisaki's formulation of this influence is that solid particles affect the metabolism of cells so that the ratio of glucose metabolized by respiration to the total amount of glucose taken up by cells is in inverse proportion to the rate of glucose uptake. This means that solid surfaces affect not only an individual activity but also some regulation mechanisms of cell metabolism. The author showed that the effect of solid particles does not always directly involve electron transfer as described above. Styrene-divinyl-benzene copolymer which was most effective in influencing the R value either did not affect or hardly affected the production of electric current; on the other hand, an anion-exchange resin, Dowex-1, and kaolinite markedly depressed the current production and affected the R value only slightly (33). Morisaki (unpublished) noticed that substances effective in influencing the R value are hydrophobic, and those not effective are less hydrophobic or even hydrophilic. Thus, it was proposed that surface tension may be an important factor affecting the relationship between glucose uptake rate and the respiratory ratio (R). This idea should be

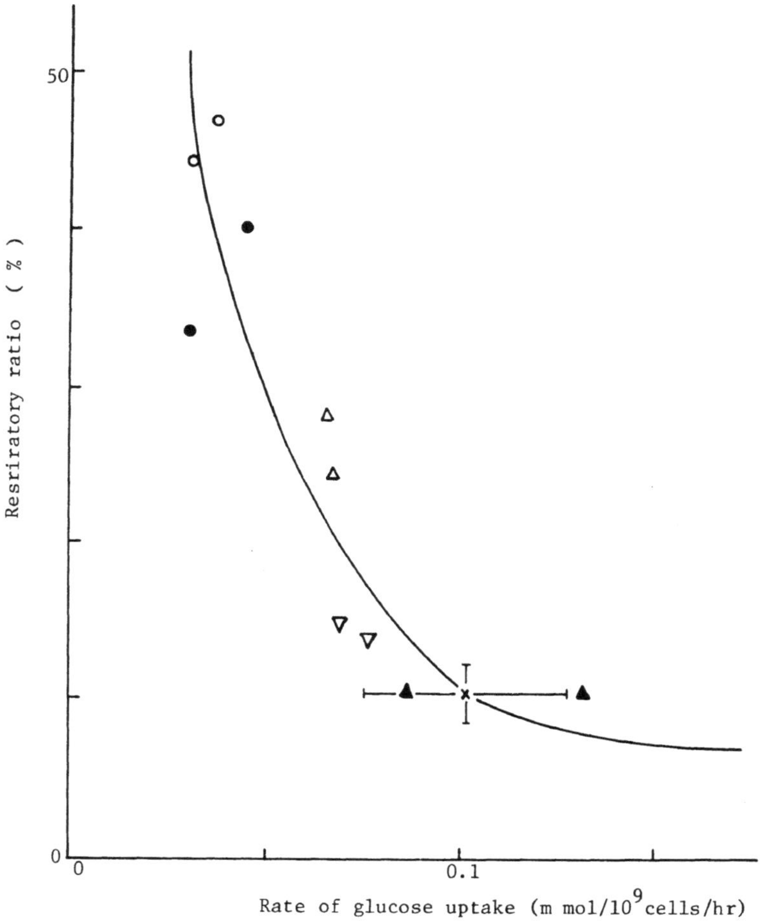

FIG. 6 - Relation between glucose uptake rate and respiratory ratio. O, styrene-divinylbenzene copolymer; •, polytetrafluoroethylene; Δ, pyrophyllite; ∇, anion-exchange resin; ▲, kaolinite; x, control, with the mean and standard errors which were obtained from 49 trials (49).

tested by further analytical studies.

TRANSFORMATION OF SOLID SUBSTRATA

In the previous sections of this paper we have pointed out different ways of how microbial metabolism is affected by adhesion of microorganisms or by the presence of adsorbents either in cultivation media or in natural microbial habitats. Another point of interest is, how can microorganisms

TABLE 11 - Some examples of substrata for microbial adhesion.

Substrata						
of biotic origin			of abiotic origin			
animate	inanimate		natural		anthropogenic	
	org.	inorg.	org.	inorg.	org.	inorg.
animal tissue	rigid (wood)	crusts (calcareous, chitinous)				metals
plant tissue	semi-rigid (litter, plant products)		carbonaceous	minerals		glass
		lithified algal mats	fossils	ores	plastics	ceramics
microbial cells	soft (meat, fruits)	teeth				artif. stones

affect solid substrata at the surface to which they adhere? This question cannot be answered easily because both microorganisms and substrata are manifold and their specific properties can differ greatly. Table 11 summarizes some substrata of importance for microbial adhesion.

Some microorganisms can utilize certain solid substrata as substrates and in such instances microbial exoenzymes or other metabolic products are involved. However, substrata can also be altered without serving as a nutrient source for attached microorganisms.

Not infrequently, animal tissues of different composition serve as substrata which can be colonized by different microorganisms, especially by bacteria. Mechanisms of those interactions as well as resulting microbial activity have been described by Jones and Savage (both this volume). Due to the activity of microbial exoenzymes, macromolecules on the surface of animal cells can be hydrolyzed, ultrastructural changes which cause cell degeneration can be initiated, and even microbial penetration into the substratum may occur. With the participation of specific bacterial metabolites, pathological effects may be evoked in animal cells.

Inanimate tissues such as stored raw meat can be colonized superficially by about 10^8 - 10^9 cm^{-2} of aerobic or anaerobic bacteria (Pseudomonas sp., Lactobacillus sp., and others) which utilize low molecular weight, soluble components present on the surface and cause meat spoilage (27). Streptococci and actinomycetes predominate in attached microbial deposits on teeth (51). Both dental caries and periodontal inflammation can be caused by such dental plaque. Acid production by plaque bacteria has been strongly implicated in the development of dental decay. However, factors responsible for gum disease have yet to be identified (53). Several aspects concerning mechanisms of adhesion to inanimate surfaces are discussed by Kjellberg (this volume).

Different microorganisms seek their own ecological niche on plant hosts, and observed attachment supports this suggestion. About $10^5 mm^{-2}$ or 10^6-10^8 g^{-1} of microbes have been reported as colonizing plant tissues (41). Among these epiphytic microorganisms, pectolytic and cellulolytic bacteria and fungi which can cause enzymatic damage to plant cells frequently occur. The roots come into contact with microorganisms more often than other parts of plants. As shown by Dazzo (this volume), not only plant nutrition and symbiosis, but also morphogenesis and disease development may be affected by microbial adhesion to plant root surfaces.

Microecological studies often show adhesion phenomena occurring among different microorganisms. For example, several Gram-negative organisms have been found to adhere to surfaces of various Actinomyces species and to streptococci (55). Nesbitt et al. (50) reported bacterial colonization of the fungus Phytophtora cinnamoni. Correlated with the increase in bacterial numbers, there was also an increase in hyphal lysis. The authors conclude that many of the bacteria residing in the "hyphasphere" may exude extracellular metabolites which cause breakdown of hyphal cytoplasm.

Microorganisms which adhere on the surface of wood and timber are often involved in biodegradation of these materials. However, whereas Basidiomycetes produce decay in various forms (e.g., brown rots, white rots), molds do not cause evident weakening of the wood (10). Soft rot decay brought about by Ascomycetes and Fungi Imperfecti produces superficial softening often leading to erosion of the surface. According to Fazzani et al. (13), bacteria colonizing timber in aquatic environments cause an increase in permeability of the wood, due to the ray cells and pit membranes being attacked. They also attack fibers, tracheids, and vessels, which results in a decrease in structural strength of the timber.

Among abiotic substrata, minerals play a most important role as substrata for microbial adhesion. Their chemical composition, exchange capacity, sorptive capacity (associated in some cases with a change in volume), and catalytic activity can all be of importance in their effect on the activity of microorganisms (18). On the other hand, as minerals are composed of biogenic elements, they are not biologically inert. Different microorganisms may bring about destruction, mineralysis, and weathering of minerals through the action of the mucus envelope of the cells or the action of various (especially acid) metabolites (4, 42). Sushkina and Tsiyurupa (58), in their comprehensive presentation of pedologic aspects, emphasize the role of proactinomycetes and, especially, of mycobacteria in both mineral destruction and the subsequent formation of new soil aggregates. Boyle and Voigt (8) report that the biotite alteration by Asperigillus niger and mixed population of soil microorganisms result in the release of potassium and other ions from mineral lattices. Similar alterations are brought about by oxalic, citric, and other organic acids. Cyanobacteria and lichens participate in the weathering and alteration of rocks, and the most active species are those whose thallus adheres best to the mineral substratum (6, 11). On the other hand, mobile aluminum has been reported to deposit on Metallogenium cells which contribute to the crystallization of biogenic hydroxides and their transformation to gibbsite (5).

Eckhardt (12) reports the weathering of a sandstone monument as being affected by microorganisms. The production of acids by microorganisms living on the stone seems to be the most important agent of deterioration. Oxalic and citric acids were detected both in the cultivation medium and in crust material from the monument.

Microbial corrosion of metals is another example of the both ecologially and economically important process related to the adhesion of microorganisms. Miller and King (46) discuss several aspects of this topic in detail. In summary, microbial metal corrosion is due to a) uptake of nutrient (including oxygen) by microbial growth adhering to metal surfaces, b) liberation of corrosive metabolites or end products of fermentative growth (i.e., organic acids), c) production of sulfuric acid by Thiobacillus sp., and d) interference with the cathodic process in oxygen-free conditions by the sulfate-reducing bacteria.

Extensive growth of microorganisms, especially of microscopic fungi, also occurs on the surface of plastics and may result in deterioration and/or degradation of these substrata. Mechanical disruption of fabrics by growing fungal hyphae, nutrient uptake from non-polymerized substances remaining on the surface of the plastics, and hydrolytic degradation by microbial exoenzymes may also be involved as shown by our investigations (20, 22).

In addition to the necessary ecological observations, further biochemically oriented studies are required in order to elucidate in more detail several aspects of microbial transformatin of solid substrata which are of major scientific, technological, and practical importance.

REFERENCES

(1) Allen, M.J. 1966. The electrochemical aspect of some biochemical systems. VII. The current generating site in metabolizing E. coli systems. Electr. Act. 11: 7-13.

(2) Allen, M.J. 1968. Cellular electrophysiology. XVIII. Coulokinetic behaviour of E. coli spheroplasts. Electr. Act. 15: 1565-1568.

(3) Allen, M.J.; Bouwen, R.J.; Nicholson, M.; and Vasta, B.M. 1963. The electrochemical systems. III. A new approach to investigation of electrical energy producing reactions in biological systems. Electr. Act. 8: 991-995.

(4) Aristovskaya, T.V. 1971. The role of microorganisms in transformation of primary and secondary minerals. Fourth Meeting of the Soviet Soil Society, Trans. II, I: 238 (in Russian).

(5) Aristovskaya, T.V.; Zykina, L.V.; and Sokolova, T.A. 1983. On the possibility of biogenic formation of aluminium hydroxide minerals in soils. Pochvovedenye (Moscow), Nr. 9, pp. 67-73 (in Russian).

(6) Ascaso, C.; Galvan, J.; and Rodrigues-Pascal, C. 1982. The weathering of calcareous rocks by lichens. Pedobiologia 24: 219-229.

(7) Badura, L. 1965. On the mechanism of the "stimulating" influence of Na-humate upon the process of alcoholic fermentation and multiplication of yeast. Act. Soc. Bot. 34: 287.

(8) Boyle, J.R., and Voigt, G.K. 1973. Biological weathering of silicate minerals. Implication for tree nutrition and soil genesis. Plant Soil 38: 191-201.

(9) Burns, R.G. 1980. Microbial adhesion to soil surfaces: consequences for growth and enzyme activities. In Microbial Adhesion to Surfaces,

eds. R.C.W. Berkeley, J.M. Lynch, J. Melling, P.R. Rutter, and B. Vincent, pp. 249-262. Chichester, England: Ellis Horwood.

(10) Cavey, J.K. 1975. Isolation and characterization of wood inhabiting fungi. In Microbial Aspects of the Deterioration of Materials, eds. D.W. Lovelock and R.J. GIlbert, pp. 23-38. London, New York, San Francisco: Academic Press.

(11) Danin, A., and Gerson, R. 1983. Weathering patterns on hard limestone and dolomite by endolithic lichens and cyanobacteria: Supporting evidence for eolian contribution to terra rossa soil. Soil Sci. 136: 213-217.

(12) Eckhardt, F.E.W. 1978. Microorganisms and weathering of the sandstone monument. In Environmental Biogeochemistry and Geomicrobiology, ed. W.E. Krumbein, vol. 2, pp. 675-686. Ann Arbor, MI: Ann Arbor Science Publishers.

(13) Fazzani, K.; Furtado, S.E.J.; Eaton, R.A.; and Jones, E.B.G. 1975. Biodeterioration of timber in aquatic environments. In Microbial Aspects of the Deterioration of Materials, eds. D.W. Lovelock and R.J. Gilbert, pp. 39-58. London, New York, San Francisco: Academic Press.

(14) Filip, Z. 1968. Development of microorganisms and humus substances formation in media with different content of bentonite. Pochvovedenye (Moscow), Nr. 9, pp. 55-61 (in Russian).

(15) Filip, Z. 1969. A characteristic of humic substances in a soil incubated with additions of bentonite. Rostlin. Vyr. (Prague) 15: 377-390 (in Czech).

(16) Filip, Z. 1969. Die Ammonisierungsaktivität der Bodenmikroflora in einer Umwelt mit verschiedenem Bentonitgehalt. Zbl. Bakteriol. II Abt. 123: 616-621.

(17) Filip, Z. 1970. Über die Beeinflussung der Bodenmikroorganismen, der Huminstoffbildung und der Krümelung von Bodenproben durch Bentonit. Landbau. Völ. 20: 91-96.

(18) Filip, Z. 1973. Clay minerals as a factor influencing the biochemical activity of soil microorganisms. Fol. Microbiol. 18: 56-74.

(19) Filip, Z. 1975. Wechselbeziehungen zwischen Mikroorganismen und Tonmineralien und ihre Auswirkung auf die Bodendynamik. Habilitationsschrift, University of Giessen.

(20) Filip, Z. 1978a. Decomposition of polyurethane in a garbage landfill leakage water and by soil microorganisms. Eur. J. Appl. Microbiol. 5: 225-231.

(21) Filip, Z. 1978b. Effect of solid particles on the growth and endurance to heat stress of garbage compost microorganisms. Eur. J. Appl. Microbiol. 6: 87-94.

(22) Filip, Z. 1979. Polyurethane as the sole nutrient source for Aspergillus niger and Cladosporium herbarum. Eur. J. Appl. Microbiol. 7: 277-280.

(23) Filip, Z. 1979. Wechselwirkungen von Mikroorganismen und Tonmineralien - eine Übersicht. Z. Pflanz. B. 142: 375-386.

(24) Filip, Z.; Flaig, W.; and Rietz, E. 1977. Oxidation of some phenolic substances as influenced by clay minerals. In Soil Organic Matter Studies, vol. II, pp. 91-96. Vienna: International Atomic Energy Agency.

(25) Filip, Z.; Haider, K.; and Martin, J.P. 1972a. Influence of clay minerals on growth and metabolic activity of Epicoccum nigrum and Stachybotrys chartarum. Soil Biol. Biochem. 4: 135-145.

(26) Filip, Z.; Haider, K.; and Martin, J.P. 1972b. Influence of clay minerals on the formation of humic substances by Epicoccum nigrum and Stachybotrys chartarum. Soil Biol. Biochem. 4: 147-154.

(27) Gill, C.O., and Newton, K.G. 1978. Ecology of meat spoilage at chill temperatures. In Microbial Ecology, eds. M.W. Loutit and J.A.R. Miles, pp. 443-447. Berlin, Heidelberg, New York: Springer-Verlag.

(28) Haider, K.; Filip, Z.; and Martin, J.P. 1970. Einfluss von Montmorillonit auf die Bildung von Biomasse und Stoffwechselprodukten durch einige Mikroorganismen. Arch. Mikrob. 73: 201-215.

(29) Haider, K.; Filip, Z.; and Martin, J.P. 1971. Action of clay minerals on different metabolic processes of soil fungi. In Studies About Humus, Transactions of the International Symposium "Humus et Planta V." Prague: UVTI.

(30) Haider, K.; Martin, J.P.; and Filip, Z. 1975. Humus biochemistry. In Soil Biochemistry, eds. E.A. Paul and A.D. McLaren, vol. 4, pp. 195-244. New York: Dekker.

(31) Hattori, R. 1972. Growth of E. coli on the surface of anion-exchange

resin in continous flow system. J. Gen. Appl. Microbiol. 18: 319-330.

(32) Hattori, R. 1976. Growth and spore formation of B. subtilis adsorbed on an anion-exchange resin. J. Gen. Appl. Microbiol. 22: 215-226.

(33) Hattori, R., and Hattori, T. 1963. Effect of a liquid-solid surface on the life of microorganisms. Ecol. Rev. 16: 63-70.

(34) Hattori, R., and Hattori, T. 1981. Growth rate and molecular growth yield of E. coli adsorbed on anion-exchange resin. J. Gen. Appl. Microbiol. 27: 287-293.

(35) Hattori, R.; Hattori, T.; and Furusaka, C. 1972a. Growth of bacteria on the surface of anion-exchange resin. I. Experiment with batch culture. J. Gen. Appl. Microbiol. 18: 271-283.

(36) Hattori, R.; Hattori, T.; and Furusaka, C. 1972b. Growth of bacteria on the surface of anion-exchange resin. II. Electron microscopic observation of adsorbed cells on the resin surface by carbon replica method. J. Gen. Appl. Microbiol. 18: 285-293.

(37) Hattori, T. 1973. The Microbial Life in the Soil, pp. 215-216. New York: Dekker.

(38) Hattori, T., and Furusaka, C. 1960. Chemical activity of E. coli adsorbed on a resin. J. Biochem. (Tokyo) 48: 831-837.

(39) Hattori, T., and Furusaka, C. 1961. Chemical activities of A. agile adsorbed on a resin. J. Biochem. (Tokyo) 50: 312-315.

(40) Josephson, J. 1982. Humic substances. Envir. Sci. Technol. 16: 20A-24A.

(41) Káš, V. 1966. Mikroorganismen im Boden, pp. 182-189. Wittenberg Lutherstadt: Ziemsen Verlag.

(42) Kutuzova, R.S. 1971. The role of microorganisms in the weathering of silica compounds in acidic and alkalinic soils. Fourth Meeting of the Soviet Soil Science Society, Trans. II, I: 240 (in Russian).

(43) Kyuma, K., and Kawaguchi, K. 1964. Oxidative changes of polyphenols as influenced by allophane. Soil Sci. Soc. (Japan) 28: 371.

(44) Martin, J.P.; Filip, Z.; and Haider, K. 1976. Efect of montmorillonite and humate on growth and metabolic activity of some actinomycetes.

Soil Biol. Biochem. 8: 409-413.

(45) Meyer, F.H. 1970. Abbau von Pilzmycel im Boden. Z. Pflanz. B. 127: 193-199.

(46) Miller, J.A.D., and King, R.A. 1975. Biodeterioration of metals. In Microbial Aspects of the Deterioration of Materials, eds. R.J. Gilbert and D.W. Lovelock, pp. 83-104. London, New York, San Francisco: Academic Press.

(47) Monod, J. 1942. Recherches sur la croissance des cultures bacteriennes. Paris: Hermann Press.

(48) Morisaki, H. 1982. The electric current from E. coli and the effect of resin on it. J. Gen. Appl. Microbiol. 28: 73-86.

(49) Morisaki, H. 1983. Effect of solid-liquid interface on metabolic activity of E. coli. J. Gen. Appl. Microbiol. 29: 195-204.

(50) Nesbitt, H.J.; Malajczuk, N.; and Glenn, A.R. 1978. Bacterial colonization of Phytophtora cinnamoni Rands. In Microbial Ecology, eds. M.W. Loutit and J.A.R. Miles, pp. 371-375. Berlin, Heidelberg, New York: Springer-Verlag.

(51) Ørstavik, D. 1980. Salivary factors in initial plaque formation. In Microbial Adhesion to Surfaces, eds. R.C.W. Berkeley, J.M. Lynch, J. Melling, P.R. Rutter, and B. Vincent, pp. 407-423. Chichester, England: Ellis Horwood.

(52) Paecht-Horowitz, M. 1973. Die Entstehung des Lebens. Angew. Chem. 85: 422-430.

(53) Rútter, P. 1979. The accumulation of organisms on the teeth. In Adhesion of Microorganisms to Surfaces, eds. D.C. Ellwood, J. Melling, and P. Rutter, pp. 139-164. London, New York, San Francisco: Academic Press.

(54) Scheffer, F., and Kroll, W. 1960. Die Bedeutung nicht-metallischer Oxyde im organischen Stoffkreislauf des Bodens unter besonderer Berücksichtigung des katalytischen Einflusses der Kieselsäure auf Huminsäureauf- und abbaureaktionen. Agrochimica 4: 97.

(55) Slots, J., and Gibbons, R.J. 1978. Attachment of Bacteriodes melaninogenicus subsp. asaccharolyticus to oral surfaces and its possible role in colonization of the mouth and periodontal pockets. Infec. Immun. 19: 254-264.

(56) Stotzky, G. 1974. Activity, ecology, and population dynamics of microorganisms in soil. In Microbial Ecology, eds. A.I. Laskin and H. Lechevaliere, pp. 57-135. Cleveland, OH: CRC Press.

(57) Stotzky, G. 1980. Surface interaction between clay minerals and microbes, viruses, and soluble organics, and the probable importance of these interactions to the ecology of microbes in soil. In Microbial Adhesion to Surfaces, eds. R.C.W. Berkeley, J.M. Lynch, J. Melling, P.R. Rutter, and B. Vincent, pp. 231-247. Chichester, England: Ellis Horwood.

(58) Sushkina, N.N., and Tsiyurupa, I.G. 1973. Microflora and the Primary Soil Forming Process. Moscow: Izdatelstvo Moskovskogo Univerziteta (in Russian).

(59) Waksman, S., and Iyer, K.R.N. 1932. Contribution to our knowledge of the chemical nature and origin of humus. II. The influence of "synthesized" humus compounds and of "natural" humus upon soil microbial processes. Soil Sci. 34: 43.

(60) Zvyagintsev, D.G. 1973. Interactions of Microorganisms and Solid Surfaces. Moscow: Izdatelstvo Moskovskogo Univerziteta (in Russian).

Microbial Adhesion and Aggregation, ed. K.C. Marshall, pp. 283-301. Dahlem Konferenzen 1984. Berlin, Heidelberg, New York, Tokyo: Springer-Verlag.

Attachment of Bacteria: Advantage or Disadvantage for Survival in the Aquatic Environment

H.-G. Hoppe
Institut für Meereskunde
2300 Kiel, F.R. Germany

Abstract. This review is based on the evaluation of observations of attached bacterial numbers and activities from the aquatic biosphere. The term "activity" is discussed with respect to these bacteria, and it is suggested that measurements of extracellular activities should be applied as a most relevant means to determine the ecological role of attached bacteria. The portion of attached bacteria can vary from a few % to 94% of total bacteria abundance in different aquatic regions depending on particle abundance, particle composition, and nutrient conditions in the water phase. Characteristics of uptake and respiration of dissolved organic compounds do not necessarily exhibit an advantage for attached bacteria in comparison to free-living ones. However, preliminary experiments have shown that attached bacteria are provided with special extracellular enzymatic faculties concerning V_{max} and K_m of selected enzymes. In sediment systems, where mechanical stress can be a dominant factor, attachment to protected areas is a necessity for the survival of bacteria.

INTRODUCTION

The attachment of bacteria to particles or, more generally, interfaces has frequently been observed in ecological surveys of aquatic habitats. The role of attached bacteria in sea- and inland waters has been discussed in different terms, all of which have a bearing on their chance of survival in the aquatic environment: a) the implication of particle-bound bacteria in the aquatic food web on the grazing activity of microzooplankton, b) the metabolic activity of attached bacteria in comparison with that of free-living bacteria, c) the attachment of bacteria in relation to dissolved organic nutrients and particle concentration, and d) the effect of attached bacteria on the balance of the size-composition of organic substrates and particles in the water phase.

Results of field observation and experiments designed to approach these problems are still widely controversial, and thus, it remains an open question as to whether the attached mode of life is a positive strategy for the survival of bacteria in aquatic biotopes. Naturally the survivors among attached bacteria may benefit from the conditions of their microenvironment, and therefore, the problem must be viewed within the context of a comparison with free-living bacteria, i.e., which mode of life offers the better chance for survival under constant or changing circumstances?

The task has been enunciated a few years ago by Wangersky (38), who claimed that the rate of nutrient regeneration is a function of the number of actively metabolizing bacteria, which in turn appears to be dependent upon the particle content of the water. However, in more recent investigations it has been found that many inorganic and even organic particles in the water phase are not heavily colonized by bacteria. In a variety of "normal" aquatic biotopes, no more than 10% of the bacteria were found to be attached to surfaces (2, 22, 43), and also free-living bacteria exhibited a considerable metabolic activity (13). Of course, the interpretation of the terms, "particle" or "solid surface," may lead to an underestimation of the attached bacterial fraction due to the presence of organic aggregates of extremely fragile nature, which are destroyed during the sampling procedure (18). Activity measurements of the microorganisms will certainly provide an impression of the growth conditions in a particular biotope and, thus, the chances for survival in a more or less nutrient-depleted environment. Marshall (25) interprets the attraction of microorganisms to interfaces as the escape of these microorganisms from the nutritionally-deficient aqueous phase. The heterotrophic uptake of attached bacteria expressed as "uptake per cell" was found to be about four times greater than that of free-living ones (22). On the other hand, Bell and Albright (3) observed that only an increase in amino acid uptake was significantly associated with the attached fraction, whereas glucose uptake was primarily mediated by the free-floating fraction. Gordon et al. (10) detected no enhancements of bacterial activity by organic particles. In fact, the respiration of attached bacteria was diminished in comparison with that of free bacteria.

Measurements of the extracellular enzymatic properties of attached bacteria may lead to a better understanding of nutrient availability and survival chances for and of these microorganisms, their contribution to the organic substrate composition, and nutrient turnover in the aquatic environment. According to Khailov and Finenko (20), macromolecular components are more effectively adsorbed by surfaces than are low molecular weight components, and this adsorbed material is hydrolyzed and incorporated at a high rate by the attached bacteria. Whether a fraction of the hydrolyzed products escape incorporation and, consequently, contribute to the pool of easily degradable organic substances in the water phase - a benefit to the free-living bacteria - remains uncertain.

TECHNIQUES APPLIED FOR THE STUDY OF BACTERIAL ATTACHMENT

There are many forms of discontinuities in aquatic ecosystems (e.g., air bubbles; thermoclines; haloclines; animal, plant, or detritus surfaces; air-water or water-sediment interfaces; oxic-anoxic interfaces) (18) in which the enrichment or attachment of bacteria has been observed. The sampling of water and materials from these environments without the destruction of the natural structure of its components is a prevailing and unsolved problem. For example, it is hardly conceivable that delicate organic lamellae could survive the mechanical stress of inflow into an evacuated ZoBell bottle. Similar difficulties may occur during the isolation of particulate material in laboratory experiments, when filtration or centrifugation processes are involved. For the in situ estimation of attached bacterial abundance, the use of special free-floating sedimentation traps which collect larger fast-sinking particles may completely alter the commonly accepted opinion regarding quantities of particle-bound bacteria (8). Urgently required for in situ studies of the fine scale distribution of bacteria in natural waters are methods conserving particulate and bacterial structure and allowing the microscopical observation of microzonation.

Several techniques have been adopted to visualize attached bacteria and to measure their metabolic activity. Besides SEM-techniques, which give an insight into the mechanisms of attachment and the local distribution of bacteria on particles (e.g., (39)), epifluorescence staining methods are widely used to quantify the number and biomass of attached bacteria (43). A commonly employed dye for staining is acridine orange. DAPI (4'6-diamidino-2-phenylindole) provides a better differentiation between bacteria and detritus (30). Autoradiographic techniques have proven to be very useful in the detection of actively metabolizing bacteria and qualification of the incorporation of radiotracers originating from dissolved and particulate organic substrates (14) (Fig. 1a, b).

In some special cases the INT (2-p-iodophenyl-3-p-nitrophenyl-5-phenyl tetrozoliumchloride) vital stain is also adequate in providing data on the activity of attached bacteria with respect to their respiratory potential (15).

In a biological sense the contemporary term, "activity," has assumed very different meanings, and with respect to attached bacteria, it should therefore be regarded with a certain amount of caution. Which

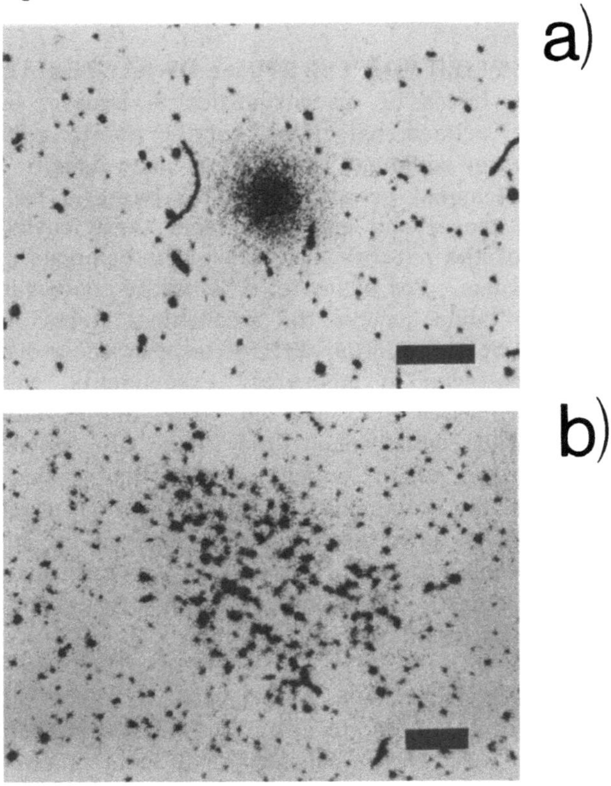

FIG. 1 - Microautoradiographic studies of bacterial attachment, bars indicate 6 μm. a) Natural free-living bacteria of different shape, in the center a disintegrating particle showing diffusion of the organic ^{3}H-label in the thin water layer between filter and X-ray film. b) Aggregation of ^{3}H-labeled bacteria on a large but thin organic detrital lamella. Many of the attached bacteria are bigger than the free-living ones around.

measurement of activity gives the most accurate impression of physiological condition and of the bacterial contribution to the functioning of the respective biotope? Heterotrophic uptake of dissolved radioactively labeled organic compounds and respiration have frequently been employed to study the activity of bacteria in general and particularly of attached bacteria. This kind of measurement yields an estimate of the enrichment of particulate matter via bacterial biomass, and thus, it is of importance for food chain studies. Furthermore, if adsorbed macromolecular organic matter or detritus is a substrate for attached bacteria, the decomposition of these materials may produce a very interesting pattern of activity. Through measurements of bacterial extracellular enzymatic potentials in conjunction with uptake measurements of the products of enzymatic breakdown, the importance of such substrates as a nutritional source for bacteria as well as their input into the pool of low molecular dissolved organic matter may be determined. Preliminary investigations of this nature have been published by several authors (16, 17, 29, 33).

QUANTIFICATION OF BACTERIAL ATTACHMENT IN NATURAL WATERS

The quantification of attached bacteria and the determination of the ratio between attached and free-living bacteria are only infrequently reported in the literature. This may be due to the fact that the counting of bacteria with epifluorescence techniques is a laborious task, and the uncertainties involved in the recognition of attached bacteria are considerable. Nevertheless, some such estimations obtained through comparable techniques have been made for water and aquatic ecosystems of various natures (Table 1).

Before the development of epifluorescence techniques, which also enabled investigators to detect reliably minute free-living bacteria, the predominating opinion was that marine bacteria were present mainly in aggregates or attached to detritus particles (32). Wiebe and Pomeroy (41) and Zimmermann (42) were the first investigators offering evidence indicating that only a few percent of total bacterial abundance in oceanic and brackish water (Baltic Sea) could be attributed to attached bacteria. Comparable results from more recent investigations were obtained from different types of ponds and marshes (22) from the Hudson River Plume during a diatom spring bloom (7) and from coastal kelp environments (24). However, there are also a few recent publications regarding various marine and inland water biotopes, which seem to correspond with the old theory (3, 4). It is, indeed, difficult to find an explanation for the variations in this pattern. Cammen and Walker (4) found a strong

TABLE 1 - Numbers of bacteria and portion of attached bacteria from different sources (x 10^6 ml^{-1} water or x 10^6 cm^{-3} wet sediment).

Source	Total number of bacteria	Number of attached bacteria	Attached bacteria %	Reference
Brackish marsh (Phragmites)	2.7		3.1	22
Nobska pond (2‰ salinity)	4.4		6.0	
Ice house pond (freshwater)	1.8		2.1	
Mill pond	12		1.3	
Salt pond (3-18‰ salinity)	6.8		1.3	
Antarctic Ocean			<20	41
Duplin River			<30	
Kiel Bight, sand in 12m depth	682 - 2301		49 - 64	40
Bay of Fundy lower Bay	0.2- 1) 4.0	0.0 - 0.5		4
upper Bay (extreme value)		2	8 - 94	
Hudson River	1.06 1)	0.2	20	7
Plume (33‰)	0.92	0.17		
coastal spring	1.00 1)	0.10	10	
bloom (diatoms)	0.82	0.09		
Kelp beds (South Africa)			4.5 - 5.8	24

TABLE 1 (cont.)

Source	Total number of bacteria	Number of attached bacteria	Attached bacteria %	Reference
Beach "St. Peter Ording," sand				31
1 cm depth	116	106	>90	
30 cm depth	125	121		
Estuarine and inland waters of British Columbia (average values)	0.36 1)	0.35	44	3
	0.16 2)	0.097 2)	53	
	0.092 3)	0.061 3)	57	
Kiel Bight (brackish water)				42
2m depth			2.8–3.8	
above ground			3.7 - 8.1	
Kiel Canal	2.1		24	42
Elbe River (Brockdorf)	9.1		51	
Elbe River (Lauenburg)	2.8		14	
Kiel Fjord (brackish water)				9
March	30000 4)	19000 4)	63	
May	77560 4)	25460 4)	33	
November	14325 4)	1100 4)	7	

1) free-living bacteria instead of total number of bacteria.
2) numbers of free-living active bacteria and numbers of attached active bacteria, based on amino acid uptake activity.
3) the same as 2), but numbers based on glucose uptake activity.
4) numbers of viable bacteria grown on ZoBell-agar (saprophytes per ml); bacteria grown from a 1 µm filtrate were assumed to be free-living bacteria.

correlation with suspended particulate matter, the content of which was very high in some areas of the Bay of Fundy (5 to 20 g l^{-1}). However, a larger load of particulate matter is also characteristic for brackish water tropical lagoons, and the portion of attached bacteria in these waters does not exceed 20% (17). In this case and others it may be assumed that in (POM-rich) water the content of dissolved organic nutrients is considerable, and thus, there would be no significant nutritional advantage for attached bacteria. In most coastal and estuarine waters about 50% of the particles are observed to be free from bacteria, about 50% contain a few bacteria, and only some individual particles are heavily colonized (11). Zimmermann (42) estimated that in the Kiel Fjord only 4% of the particles present exhibited a dense bacterial population comparable to a bacterial microcolony. Apparently these particles either have a more favorable composition of nutrients than the less populated ones, or they possess special capacities for nutrient sorption (5). The origin of such material is still a subject for speculation, and there may be different mechanisms involved in its formation. Studies on the submicroscopical fine structure of some particles may disclose that they are composed of colloidal compounds originating from dissolved organic substances. Other particles demonstrate similarities with the released contents of phytoplankton cells following cellular disruption, and a third type resembles residues of destroyed fecal pellets (42). The concentration of these types of particles in combination with favorable hydrographic and aging conditions will certainly promote a relatively high percentage of attached bacteria compared with free bacteria. However, studies of chemical particle composition with respect to investigations of bacterial abundance and activities are rare. It is, hence, difficult to uncover the explanations for the varying degrees of bacterial attachment corresponding to different water bodies and seasons (12).

ATTACHMENT AND SURVIVAL OF BACTERIA IN SPECIAL AQUATIC BIOTOPES

Within the aquatic environment there exist limited biotopes which offer much better nutritional bases for attached bacteria than the bulk of the water phase or in which attachment is even a necessity for survival. Such natural biotopes for the first category are fields of benthic macroalgae, blooms of certain planktonic microalgae, neuston, and hydrographic stratifications. Different kinds of sand and sediments or the fecal pellet microenvironment are constituents of the second.

Biodegradation of macrophyte (seagrass) debris and its implication in the food chain has been investigated by Velimirov et al. (37). They found as many as 4 x 10^4 bacterial cells attached to $1mm^2$ of the brown surfaces of leaves. Decreasing particle sizes were paralleled by decreasing C/N ratios and increasing O_2 uptake (max. values 10 mg O_2 $g^{-1}h^{-1}$) for the particle size fraction 0.1 - 1 mm. Changes in sections of the leaves from green to brown were accompanied by a rapid leaching of soluble sugars resulting in particles with a carbon concentration of 24-33%, which comprised mainly structural carbohydrates. A worthy investigation would be to determine whether bacterial attachment and production continues after depletion of soluble sugars, and whether the exoenzymatic capacity of the attached bacteria is large enough to decompose residues of the particles while still in aqueous suspension.

In Baltic Sea waters the component of attached bacteria increased up to 60% of total bacteria numbers during the occurrence of blue-green algal surface agglomeration in late summer. These spiral-shaped Nodularia spumigena masses form a unique microbiotope for bacteria in a nutrient-poor environment. Even in late stages of decay some sections of cells within the long filaments retain their vitality as demonstrated by INT-staining for respiration and autoradiography for photosynthetic capacities (15). The nodules formed by the algal filaments are, therefore, characterized by living organic matter as well as detrital organic matter. Grazing on the nodules by animals is obviously inefficient; whether this is due to toxic properties of the Nodularia filaments is an open question. However, there is some evidence that the organic substances of the nodules are introduced to the food chain via the "bacterial loop" (1). Attached bacteria play a dominant role in this decomposition process. In a Nodularia agglomerate of 0.25 cm^3, which contained about 4400 algal filaments, 7.5 x 10^8 attached bacteria were observed and these represented 3.8 - 7.6 µg C. Taking into account the distribution of nodules in the water column, it was calculated that the number of attached bacteria in 1 l of Nodularia nodule-containing water was equivalent to the total bacterial content of 1.5 to 3 m^3 of surrounding water. Within the Nodularia agglomerates a special food chain based on the constant input of organic substances by the photosynthetically active parts of the filaments was established. Attached bacteria benefitted from the relatively high exudation* ratios of living cells (ca. 58% of primary production) and quite obviously from the decomposition of the dead or "unhealthy" parts of the filaments. Free and sessile ciliates are associated in great numbers

*Exudation measurements from "unhealthy" cells may include a considerable amount of lysis products.

with the algal flocks and have been observed microscopically to feed on attached bacteria. The ciliates themselves are a component of the diet of rotifers, which are present in considerable abundance in and around the flocks. These organisms seem to be responsible for the export of organic matter from the Nodularia flock microenvironment.

INT-staining revealed that the colonization of filaments by bacteria was initiated in unhealthy sections of the cellular strand. When these sections were overcrowded, the healthy cells were also covered with bacteria, although in most filaments there remained at least a few noninfected healthy cells.

The ambivalence of microbial attachment with respect to survival chances in the natural habitat is clearly recognizable in the Nodularia microbiotope. High abundances of grazers such as flagellates and ciliates limit the abundance of bacteria and their individual life span. This is counterbalanced by enhanced reproduction of the bacteria due to favorable nutritional conditions in the agglomerates. In such an environment fast-growing bacterial populations will dominate as long as their nutrient supply is sufficient. These considerations are in agreement with Kato's (19) definitions of the characteristics of bacteria. Kato categorized free-living bacteria as "K-strategists" with high efficiency in utilization of resources in short supply and attached bacteria as "T-strategists" which are fast-growing and capable of rapid colonization of unoccupied habitats.

In sand and other sediments exposed to strong water currents or wave action, the advantages of attached bacteria become most obvious. In regard to sand grains, the degree of roundness seems to be extremely important for the attachment of bacteria, since bacteria were most abundant in niches not exposed to mechanical stress (Fig. 2). In these niches or "caves," detritus was also trapped and served as a nutritional foundation for the bacteria (40). During the ultrasonic handling of samples, the energy required to liberate bacteria from the sand grains varied for different types of bacteria. Some bacteria were firmly attached to rough surface structures and were completely resistant to ultrasonic measures.

Weise (39) determined a ration of 90% for attached to unattached bacteria in a tidal beach area of the North Sea coast, whereas only 50% of the bacteria were attached to sandy sediments in the Kiel Bight (Baltic Sea region at a depth of 14 m). Furthermore, although there was a highly significant correlation (r = 0.99) between bacteria and grain size distribution in the wave action zone of this beach area, in the sediment of the Kiel Bight this relationship was not clearly evident. Concentrations of dissolved organic substances in the interstitial water of Kiel Bight sediments were about four times higher than in the investigated tidal zone, due to dissimilarities in water agitation and nutrient input. The difference in the proportion of attached bacteria in the two environments may, therefore, be a consequence of a combination of the two factors - mechanical stress and nutrient supply. However, it should be noted that the advantage of attached bacteria for survival on sand grains in the wave action zone is not attachment per se, but attachment in special protected ecological sites within their habitat.

METABOLIC PROPERTIES OF BACTERIA ATTACHED TO ORGANIC OR INORGANIC SURFACES

In aquatic marine biotopes, a large portion of bacteria, often more than half of the total population, has been found to be in a state of metabolic "dormancy" or starvation (13, 27, 35). Nevertheless, autoradiographic observations revealed that most attached bacteria took up dissolved organic nutrients in detectable quantities. Despite their relatively small number, the importance of attached bacteria in substrate turnover and in the aquatic food chain may be considerable, provided that they are associated with certain special metabolic properties. The first and most obvious approach is readily pursuable in the methodical sense, namely, the determination of heterotrophic uptake patterns of attached bacteria.

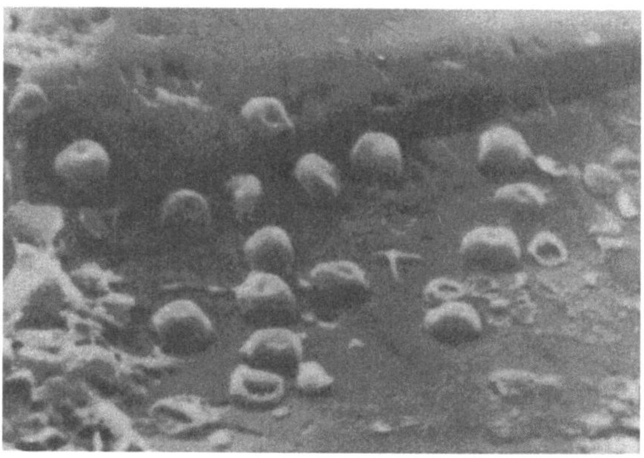

FIG. 2 - SEM studies of bacterial attachment, bar indicates 1 µm. Coccoid bacteria attached to a sheltered area of a sand grain (SEM photograph by W. Weise).

The disadvantage here is that such measurements remain controversial, as already mentioned in the introduction. Using grain density microautoradiography, Paerl and Merkel (28) found a higher relative uptake rate of $^{32}PO_4$ per attached bacterial cell than for a free-living one in different (mainly lake) biotopes. Because differential filtration was avoided in their methodology and the influence of background and adsorption was eliminated statistically, the results of this "direct" approach are reliable. The specific cellular uptake activities varied seasonally as well as locally, and, therefore, consistent direct correlations between biomass and activities of attached vs. unattached bacteria could not be assessed.

Working with bacteria of natural aquatic communities, it is especially difficult to associate measured activities with one or the other bacterial group. Arbitrary systems have, hence, been developed to determine growth characteristics of attached bacteria in specific circumstances. The effects of kaolin particles on Aeromonas sp. were investigated by Sugita et al. (36). Using different concentrations of histidine as a nutritional substrate, they found a higher growth rate in the presence than in the absence of kaolin, which they attributed to bacterial attachment and a consequent decrease of K_s. The latter was interpreted as a result of "physiological changes" in attached bacterial cells. Employing glucose and glutamic acid as substrates, Gordon et al. (10) studied the effects of hydroxyapatite particles on the metabolism of marine bacteria (Vibrio alginolyticus). Their results do not support the hypothesis that bacterial activity is stimulated by the presence of inorganic surfaces. Respiration of attached bacteria was even lower than that of free bacteria. These findings resemble the low rates of respiration in sediments, where attached bacteria are the dominant fraction (26). As a possible explanation for this phenomenon, a methodological problem concerning respiration (21) and the equilibrium of $C/^{14}C$ in the internal pool of organic substances has been discussed. Readily available pools for respiratory requirements may be greater in well nourished sedimentary bacteria, and, thus, it takes longer to establish continuous radioactive labeling. In some respects, especially regarding nutrition, prevailing conditions for bacteria attached to particulate materials in the water column may be comparable to those of sedimentary bacteria. Lower respiration ratios may consequently be related to the same factor.

Morphological changes of attached bacteria which have an influence on their metabolic capacities have been reported by Kjelleberg et al. (23). A copiotrophic marine Vibrio sp. revealed additional survival mechanisms when attached to interfaces lacking sufficient amounts of nutrients. In such a situation, bacterial cells became smaller, thereby favorably altering their surface/volume ratio and, furthermore, permitting a greater packing density.

However, in light of these facts, the question arises as to why so many natural interfaces are not colonized by bacteria. Studies on the prevention of marine fouling may give some indications for probable mechanisms inhibiting bacterial attachment. For example, instead of using highly toxic heavy metals, Chet et al. (6) employed organic compounds at nontoxic concentrations (acrylamide, benzoic acid, tannic acid). It was found that a negative chemotactic response enabled motile bacteria to flee from unfavorable environments. Naturally occurring substances produced by living organisms and adsorbed to surfaces may produce the same effect as demonstrated by the seaweed, Sargassum, which produces tannins, preventing the formation of epiphytic bacterial populations at the tips of its branches.

By means of organic compounds coupled to a fluorescence tracer, preliminary experiments on extracellular enzymatic activities of bacteria indicated that at high substrate concentrations the liberation of easily degradable, low molecular weight organic matter from these substrates was much higher than the uptake of the liberated products by the natural bacterial population (16). These findings were confirmed by direct measurements of amino acids and peptide concentrations with the fluorescamin method. Similar to the query regarding the factors involved in substrate uptake, the question now becomes, "what are the major determinants affecting extracellular enzymatic activities?" From fractionated filtration experiments and bacterial culture studies, it has been concluded that bacteria play a dominant role. In light of our knowledge concerning the adsorption of macromolecular organic compounds at interfaces, the development of extracellular enzymatic activities by attached bacteria affording the exploitation of these nutrients might well be expected.

A comparison of extracellular enzymatic capacities of an < 8 μm filtrate with that of a threefold concentration of the > 8 μm particles of the same water sample revealed that extracellular enzymatic activity in the concentrated sample was disproportionately large. This was particularly true for V_{max}, which was considerably larger than suggested by the bacterial number or bacterial biomass ratio (Table 2). In contrast, K_m was lower in the sample enriched with particles. However, this effect could only be observed for one of the two enzymatic systems investigated, namely, for glucosaminidase. No significant changes could be found for β-glucosidase. Hence, in one case the same pattern for attached bacteria was observed as previously described by Sugita et al. (36) for the growth rate of attached bacteria on histidine. The example is, of course, valid only when bacteria alone are producers of the extracellular enzymes in question and when enzymes adsorbed on surfaces lose their activity. There is some evidence from the literature that extracellular enzymatic activity originates mainly from bacteria and that it remains cell-associated. The concentration of free dissolved enzymes which might become adsorbed to surfaces is low (16, 33).

TABLE 2 - Specific extracellular enzymatic activity of bacteria calculated from an enzyme cinetical approach with methylumbelliferyl-N-acetyl-β-D-glucosaminid as a substrate.

	Total bacteria number ($x10^6 ml^{-1}$)		V_{max} glucosaminidase activity	
	Free	Attached	$\mu g\ C\ l^{-1} h^{-1}$	$ng\ Cx10^{-7} h^{-1} cell^{-1}$
unfiltered seawater	5.76	0.55	0.71	1.12
< 8 µm filtrate of seawater	5.73*	-*	0.56	0.97**
> 8 µm concentrated particle fraction of seawater	7.81	1.73	1.12	2.01***

* In the sample of investigation, bacteria attached to the < 8 µm particles were below the detection limit.

** V_{max} predominantly for free-living bacteria cells.

*** V_{max} only for attached bacteria cells.

CONCLUSION AND HYPOTHESIS

Paerl and Merkel (28) posed the question: "Given the conclusion that particles provide a favorable environment for bacteria, why then are not more free-living bacteria prone to find an attachment site?" They assume that particles "act as a key limiting factor to the establishment," and better microbial growth should be the result of the greater abundance of particles. However, in the natural aquatic environment only a small portion of particles present is normally colonized to any considerable degree by bacteria, and therefore, it seems that the nature of particles and their adsorption capacities must be decisive. Many particles and interfaces may not provide the conditions necessary for the successful competition of attached bacteria with bacteria in the water phase due to specific advantages associated with the latter. These advantages are the reduction in both predation (34) and inhibitory substances, more individual space, and the availability of dissolved organic substances ready for incorporation by bacteria. The composition of the prevailing microbial population with respect to their substrate affinities may also be of importance, because bacteria well adapted to low concentrations of nutrients in the water will not significantly benefit from the attached mode of life.

If free-living bacteria benefit most from dissolved small organic molecules, then their activity will also be dependent upon it. This type of nutrient is supplied mainly through the excretion of photosynthetic products of phytoplankton and macroalgae, from animals, and from the decomposition of macromolecular and particulate organic materials via extracellular enzymes. Concerning the latter, our preliminary investigations lend some support to the assumption that from the products of extracellular enzymatic breakdown, only a small fraction is immediately incorporated by bacteria, whereas the main component serves as an input into the pool of dissolved small molecules (16). This is at least true when there is a large supply of substances demanding extracellular enzymatic breakdown prior to their incorporation. Of course, little is known about the sizes of macromolecular pools in aquatic regions. Organic particles, however, will certainly function as centers of formation of low molecular organic compounds, if the argumentation up till now is accepted. It appears likely that attached bacteria have developed extracellular enzymatic capacities enabling the decomposition of their organic substrate and the substances adsorbed on it. According to Traube's law (20), especially high molecular weight compounds (MW between 10,000 and 200,000) exhibit a large degree of surface activity. Their inclusion into the heterotrophic food chain may, therefore, be mediated by two mechanisms: adsorption (physicochemical) and subsequent biodegradation by bacteria adhering to detritus (20). In preliminary experiments, attached bacteria exhibited higher extracellular enzymatic capacities for the breakdown of methylumbelliferyl-glucosamine and its naturally occurring substrate analogues than did free-living bacteria. Furthermore, the resulting products, which are not immediately consumed by the attached bacteria, provide a nutritional benefit for the free-living bacteria.

This dynamic process, albeit hypothetical in character, may be a step forward in defining the "key limiting factor of the establishment" determining bacterial attachment as expressed by Paerl and Merkel (28). The role of attached bacteria in the aquatic environment would then be twofold: a) the establishment of an equilibrium between pools of differently sized molecules, and b) the production of nutrients for attached bacterial maintenance and, to some extent, for that of their free-living counterparts. In essence, the advantage of attached bacteria, with respect to nutrition, is that competitors are restricted to only a small number of bacteria and that special extracellular enzymatic capacities of a few individual species may promote an increase in survival chances for the dense population as a whole.

REFERENCES

(1) Azam, F.; Field, J.G.; Gray, J.S.; Meyer-Reil, L.-A.; and Thingstad, F. 1983. The ecological role of water-column microbes in the sea. Mar. Ecol.-Prog. Ser. 10: 257-263.

(2) Azam, F., and Hodson, R.E. 1977. Size distribution and activity of marine microheterotrophs. Limn. Ocean. 22: 492-501.

(3) Bell, C.R., and Albright, L.J. 1982. Attached and free-floating bacteria in a diverse selection of water bodies. Appl. Envir. Microbiol. 43: 1227-1237.

(4) Cammen, L.M., and Walker, J.A. 1982. Distribution and activity of attached and free-living suspended bacteria in the Bay of Fundy. Can. J. Fish. Aquat. Sci. 39: 1655-1663.

(5) Chave, K.E. 1970. Carbonate-organic interaction in sea water. In Organic Matter in Natural Waters, ed. D.W. Hood, pp. 373-386. University of Alaska Press.

(6) Chet, I.; Asketh, P.; and Mitchell, R. 1975. Repulsion of bacteria from marine surfaces. Appl. Microbiol. 30: 1043-1045.

(7) Ducklow, H.W., and Kirchman, D.L. 1983. Bacterial dynamics and distribution during a spring diatom bloom in the Hudson River plume, USA. J. Plankton Res. 5: 333-355.

(8) Fellows, D.A.; Karl, D.M.; and Knauer, G.A. 1981. Large particle fluxes and the vertical transport of living carbon in the upper 1500 m of the northeast Pacific Ocean. Deep-Sea Res. 28A: 921-936.

(9) Gocke, K. 1975. Untersuchungen über die Aufnahme von gelöster

Glukose unter natürlichen Verhältnissen durch größenfraktioniertes Nano- und Ultrananoplankton. Kiel. Meeresforsch. 31: 87-94.

(10) Gordon, A.S.; Gerchakov, S.M.; and Millero, F.J. 1983. Effects of inorganic particles on metabolism by a periphytic marine bacterium. Appl. Envir. Microbiol. 45: 411-417.

(11) Hanson, R.B., and Wiebe, W.J. 1977. Heterotrophic activity associated with particulate size fractions in a Spatina alterniflora salt-marsh estuary, Sapelo Island, Georgia, USA, and the continental shelf waters. Mar. Biol. 42: 321-330.

(12) Hargrave, B.T. 1972. Aerobic decomposition of sediment and detritus as a function of particle surface area and organic content. Limn. Ocean. 17: 583-596.

(13) Hoppe, H.-G. 1976. Determination and properties of actively metabolizing heterotrophic bacteria in the sea, investigated by means of micro-autoradiography. Mar. Biol. 36: 291-302.

(14) Hoppe, H.-G. 1977. Analysis of actively metabolizing bacterial populations. In Microbial Ecology of a Brackish Water Environment, ed. G. Rheinheimer, pp. 179-197. Berlin: Springer-Verlag.

(15) Hoppe, H.-G. 1981. Blue-green algae agglomeration in surface water: a microbiotope of high bacterial activity. Kiel. Meeresforsch. 5: 291-303.

(16) Hoppe, H.-G. 1983. Significance of exoenzymatic activities in the ecology of brackish water: measurements by means of methylumbelliferyl-substrates. Mar. Ecol.-Prog. Ser. 11: 299-308.

(17) Hoppe, H.-G.; Gocke, K.; Zamorano, D.; and Zimmermann, R. 1983. Degradation of macromolecular organic compounds in a tropical lagoon (Cíenaga Grande, Columbia) and its ecological significance. Int. Rev. ges. Hydrobiol. 68: 811-824.

(18) Karl, D.M. 1982. Microbial transformation of organic matter at ocean interfaces: a review and prospectus. EOS 63: 138-140.

(19) Kato, K. 1984. A concept on the structure and function of bacterial community in aquatic ecosystems. Verh. Internat. Verein Limnol. 22, in press.

(20) Khailov, K.M., and Finenko, Z.Z. 1970. Organic macromolecular compounds dissolved in sea-water and their inclusion into food chains. In Food Chains, ed. J.H. Steele, pp. 6-18. Edinburgh: Oliver and

Boyd.

(21) King, G.M., and Berman, T. 1984. Potential effects of isotopic dilution on apparent respiration in ^{14}C-heterotrophy experiments. Mar. Ecol.-Prog. Ser., in press.

(22) Kirchman, D., and Mitchell, R. 1982. Contribution of particle-bound bacteria to total microheterotrophic activity in five ponds and two marshes. Appl. Envir. Microbiol. 43: 200-209.

(23) Kjelleberg, S.; Humphrey, B.A.; and Marshall, K.C. 1982. Effect of interfaces on small, starved marine bacteria. Appl. Envir. Microbiol. 43: 1166-1172.

(24) Linley, E.A.S., and Field, J.G. 1982. The nature and ecological significance of bacterial aggregation in a near-shore upwelling ecosystem. Estuarine, Coastal Shelf Sci. 14: 1-11.

(25) Marshall, K.C. 1979. Growth at interfaces. In Strategies of Microbial Life in Extreme Environments, ed. M. Shilo, pp. 281-290. Dahlem Konferenzen. Weinheim, New York: Verlag Chemie.

(26) Meyer-Reil, L.A.; Dawson, R.; Liebezeit, G.; and Tiedge, H. 1978. Fluctuations and interactions of bacterial activity in sandy beach sediments and overlying waters. Mar. Biol. 48: 161-171.

(27) Morita, R.Y. 1982. Starvation-survival of heterotrophs in the marine environment. In Advances in Microbial Ecology, ed. K.C. Marshall, vol. 6, pp. 171-198. New York: Plenum Publishing Corporation.

(28) Paerl, H.W., and Merkel, S.M. 1982. Differential phosphorous assimilation in attached vs. unattached microorganisms. Arch. Hydrobiol. 93: 125-143.

(29) Pancholy, S.K., and Lynd, J.Q. 1972. Quantitative fluorescence analysis of soil lipase activity. Soil Biol. Biochem. 4: 257-259.

(30) Porter, K.G., and Feig, Y.S. 1980. The use of DAPI for identifying and counting aquatic bacteria. Limn. Ocean. 25: 943-948.

(31) Rheinheimer, G. 1981. Investigations on the role of bacteria in the food web of the Western Baltic. Kiel. Meeresforsch. Spec. Publ. 5: 284-290.

(32) Seki, H. 1970. Microbial biomass on particulate organic matter in seawater of the euphotic zone. Appl. Microbiol. 19: 960-962.

(33) Somville, M., and Billen, G. 1983. A method for determining exoproteasic activity in natural waters. Limn. Ocean. 28: 190-193.

(34) Sorokin, Y. 1970. Formation of aggregates by marine bacteria. Oceanography 192: 905-907 (translated from Russian).

(35) Stevenson, L.H. 1978. A case for bacteria dormancy in aquatic systems. Microbial Ecol. 4: 127-133.

(36) Sugita, H.; Ishida, Y.; and Kadota, H. 1979. Kinetic analysis of promotive effects of Kaolin particles on growth of an aquatic bacterium. Bull. Jap. Soc. Sci. Fish. 45: 1381-1383.

(37) Velimirov, B.; Ott, J.A.; and Novak, R. 1981. Microorganisms on macrophyte debris: biodegradation and its implication in the food web. Kiel. Meeresforsch. Spec. Publ. 5: 333-344.

(38) Wangersky, P.J. 1977. The role of particulate matter in the productivity of surface waters. Helg. W. Meer. 30: 546-564.

(39) Weise, W. 1975. Fluoreszenz- und Raster-Elektronenmikroskopische Untersuchungen über die Bakterienbesiedlung von marinen Sandsedimenten. Diploma Thesis, University of Kiel, F.R. Germany.

(40) Weise, W., and Rheinheimer, G. 1979. Fluoreszenzmikroskopische Untersuchungen über die Bakterienbesiedlung mariner Sandsedimente. Botan. Marin. 22: 99-106.

(41) Wiebe, W.J., and Pomeroy, L.R. 1972. Microorganisms and their association with aggregates and detritus in the sea: a microscopic study. Mem. Ist. Ital. Idrobiol. 29: 325-352.

(42) Zimmermann, R. 1975. Entwicklung und Anwendung von fluoreszenz- und rasterelektronenmikroskopischen Methoden zur Ermittlung der Bakterienmenge in Wasserproben. Thesis, University of Kiel, F.R. Germany.

(43) Zimmermann, R. 1977. Estimation of bacterial number and biomass by epifluorescence microscopy and scanning electron microscopy. In Microbial Ecology of a Brackish Water Environment, ed. G. Rheinheimer, pp. 103-120. Berlin: Springer-Verlag.

Standing, left to right:
Brian Vincent, Bernard Atkinson, Hans Reichenbach, Bernhard Schink, Tony Rose, David Garrod

Seated, left to right:
Peter Hirsch, David Jenkins, Gode Calleja, Byron Johnson, Peter Wilderer

Microbial Adhesion and Aggregation, ed. K.C. Marshall, pp. 303-321. Dahlem Konferenzen 1984. Berlin, Heidelberg, New York, Tokyo: Springer-Verlag.

Aggregation
Group Report

G.B. Calleja, Rapporteur
B. Atkinson
D.R. Garrod
P. Hirsch
D. Jenkins
B.F. Johnson
H. Reichenbach
A.H. Rose
B. Schink
B. Vincent
P.A. Wilderer

INTRODUCTION

Definitions

For the simple expediency of accommodating the interests of every member of this group, we define a microbial aggregate as a collection of microbial cells in intimate contact. This definition, agreed upon by the majority, avoids any connotation of mechanism as to origin and genesis of the collection. It is not in agreement with the more restrictive definition put forward by some of us: a gathering followed by a binding together. An aggregate, as treated here, is a structure - the product of a process. Aggregation is one of the processes that may lead to an aggregate. Thus, a microcolony on a solid surface falls into this category; likewise, a biofilm, as used by the other groups. However, a biofilm needs a substratum to which cells attach, whereas a floc most often does not. Moreover, flocs are transportable entities, whereas biofilms are more or less stationary.

Experimental Systems

Even in its restricted definition, the subject of microbial aggregation is large enough to accommodate a variety of well studied systems (1, 3, 4, 6-10, 12-15, 17, 19-23, 26). Only a few of these systems were represented in this workshop. These include three of the oldest and best studied microbial aggregation systems: flocculation of brewing yeast, aggregation in cellular slime molds, and aggregation in myxobacteria. Also represented were sex-directed flocculation of fission yeast, artificially provoked flocculation of industrially important microorganisms (brewing yeast, Aspergillus niger, actinomycetes, etc.), sludge-floc formation in waste treatment, and a variety of naturally occurring microcolonies and consortia, including rosettes or stars found in a variety of species. Unrepresented in our discussions were a large number of well studied systems: mating-aggregate formation during conjugation in Escherichia coli; pellicle formation in fimbriate bacteria; agglutination associated with bacterial transformation; formation of dental plaque and aggregation of oral bacteria; cell aggregation in Streptococcus faecalis; sexual agglutination in Hansenula wingei, Saccharomyces cerevisiae, and other yeasts; sexual agglutination in Chlamydomonas and other algae; mating reaction in Blepharisma, Paramecium, and other protozoa; formation of synemmata, coremia, and sclerotia in many species of filamentous organisms.

Expanded in its scope, the subject of microbial aggregation becomes a collection of phenomena that embraces many population arrangements (1, 7, 9, 13, 20). Indeed, the term becomes synonymous with enhanced concentration of a microbial population. The collection of cells in contact is nothing more than a structure which has a higher cell-population density than has the immediately surrounding space, be that two-dimensional or three-dimensional.

Classification of Aggregates

An aggregate can be natural or artificial. It is natural if its genesis is part of the natural history of the organism, whether that natural history occurs in the laboratory or in nature. It is artificial if it is provoked by man-made substances or if it occurs under nonphysiological conditions. In water and wastewater treatment, for instance, one may employ such techniques as chemical flocculation as well as aggregation by naturally occurring microorganisms.

Aggregation may be active or passive. An active process denotes a coming together, i.e., the gathered cells must be originally dispersed. The gathering may be directed, as by chemotaxis, or stochastic, as by fluid motion. Passive aggregation arises from the failure of progeny to separate materially from each other.

An interaction among cells of two or more types may be termed heterotypic. The interaction may involve some interdependence, unilateral or bilateral; such an arrangement may be called a consortium. An interaction among cells of one type is homotypic.

Coverage of Discussion and Report

This report covers the following topics: structures and functions, methods of characterizing an aggregate, mechanisms of movement and of contact, cell-cell communication, abiotic environmental influences, and the implications of being aggregated or dispersed.

STRUCTURES AND FUNCTIONS

Structures

Because the phenomenon of being aggregated is observed in a variety of microbes, a variety of aggregated structures is expected and observed. This morphological variety, in turn, is reflected in the variety of terms used for the communal structures: pseudoplasmodium, grex, or slug for cellular slime molds; mating aggregate for E. coli; floc for yeasts and for activated-sludge aggregates; pellet for Aspergillus niger; coremium, sclerotium, synnema, strand, and stroma for filamentous fungi and actinomycetes; film or pellicle for fimbriate bacteria; consortium for heterotypic interactions; clump for many different organisms; star and packet for microcolonies, etc. At the macroscopic level, the aggregated structures may take the form of fruiting structures that look like umbrellas, mounds, columns, mats, etc., or they may come in the form of hats, balls, stars, flakes, etc.

The aggregation number can range from a few cells to billions of cells. The lower limit must be a number large enough so as to exclude the dispersed state. This minimum number must be operationally defined, based on the aggregation numbers of what may be considered non-aggregated cells in a given system.

Activated-sludge flocs exist in an environment where there is a wide range of physical, chemical, and biological conditions. The aeration basin may have conditions of high fluid shear so that for flocs to remain intact they must possess levels of structure beyond that of polymer-type bridging. Large flocs often contain a superstructure of filamentous organisms that reinforces flocs and directs the way in which these flocs grow. Flocs devoid of filaments usually are approximately spherical and small. Flocs containing filamentous organisms assume irregular, nonspherical shapes. Frequently these filaments become covered with a dense population of rod-shaped bacteria positioned head-on. Within such flocs there is also a population of unattached, freely moving organisms, such as protozoa, that contributes to the overall floc metabolism. Protozoa and metazoa may also be attached.

The arrangement of cells in an aggregate may be random, or it may be dependent on localized hydrophobicity, localized presence of polymer (capsules, slimes, pili, fimbriae, flagella) or cell geometry. Only a few examples can be cited from the great variety of structural arrangements that have been observed in nature or in cultures. In the case of Caulobacter, three-dimensional rosettes are formed where the offspring are produced at the periphery. Hyphomicrobium, Planctomyces, Pseudomonas, and other bacteria show the same arrangement. Migrating cells in myxobacteria and cellular slime mold aggregates reorient themselves with their long axes in the direction of movement. It has been demonstrated that different cell-surface molecules (glycoproteins) are responsible for end-to-end and side-to-side adhesion in aggregation streams of cellular slime molds. Where fibrils (pili, fimbriae, flagella) are involved, the primary intercellular contacts lead to a loose arrangement of the participating cells, the intercellular distances varying up to several cell diameters. In contrast to fibril-wall or fibril-fibril interactions, wall-wall contacts, as those of the yeast Hansenula wingei, result in an arrangement so compact as to cause mural deformation.

Mixed aggregates (consortia) of bacteria may owe their cell arrangement to localized elaboration of polymers (Hirsch, this volume). Chloroplana vacuolata is a flat, band-shaped consortium of green photosynthetic bacteria where one type (species) without gas vesicles alternates with similarly shaped cells of another species that contain gas vesicles. Barrel-shaped consortia of Chlorochromatium aggregatum consist of a colorless, central, spindle-shaped cell which is surrounded by one or a few layers of Chlorobium cells, arranged with their long axes parallel to the long axis of the central organism. In the case of Peloploca undulata, filaments of rod-shaped bacteria with gas vesicles are surrounded by filaments of similar size of other bacteria without gas vesicles. The whole consortium has the shape and structure of a twisted rope bundle. Individual cells in microbial microcolonies may be arranged either regularly or irregularly, depending on the cell shape, specialized sites of adhesion, and motility.

It is evident that the aggregated structures encountered are greatly influenced by whether there are morphogenetic, developmental, or sexual events that take place after the primary event of aggregation, as well as by the genesis of the cell collection. Differentiation of the members of the collection could have arisen from cell reproduction, and thus, the final shape of the colony will be primarily determined by the mode of budding or fission. Cell plates formed by Thiopedia rosea, Lampropedia sp., or Merismopedia arise from polymer formation at the division site which prevents complete cell separation.

Functions

In a number of examples, most notably in several yeast systems, heat-killed cells do aggregate, provided they have become competent prior to their being killed. In these particular instances, the binding activity has been isolated from the rest of the cellular functions.

In living aggregated systems, however, cells in the aggregate may or may not actually respire, metabolize, or grow as much as dispersed cells in the immediate neighborhood. Many aggregative systems of cellular slime molds, algae, and yeasts, for instance, require nongrowing cells. In yeasts, competence to aggregate is catabolite-repressible. Thus, enzymic activities may not be due to the aggregative state but to the pre-aggregative condition. In the cellular slime molds, certain enzymes are enriched or diminished, and specific gene activation has been demonstrated.

When the aggregative state is a prelude to development or sexual activity, grossly identifiable functions, such as copulation, meiosis, sporulation, and fruiting-body formation, seem to be assumed exclusively by the aggregated cells. Consequently, at the molecular level, all functions associated with these developmental post-aggregation events must take place. An example is mutual erosion of apposed cell walls to form a single continuous wall from originally two closed walls. This must involve simultaneous enzyme-catalyzed hydrolysis and synthesis of carbohydrate polymers at the site of fusion.

Cell lysis may also occur during aggregation, especially among those cells in the core of an aggregate. There are at least three cases reported and studied: autolysis in myxobacteria, conjugation-induced lysis in fission yeast, and lethal zygosis in E. coli.

CHARACTERIZING AN AGGREGATE

Colloid Theory Considered

Some very simple microbial aggregation systems can be viewed as colloidal systems that may be fruitfully probed with the analytical tools of colloid science (25). The quantification of aggregation must ultimately be expressed in terms of energy expended. There are at least four interdependent properties which may be quantified in the context of aggregation: a) strength, b) morphology, c) extent, and d) rate.

Strength

Strength may be assessed directly using rheological techniques (e.g., cone/plate, Couette-type rheometers). Basically, one measures the shear stress (τ) as a function of shear rate (γ). In Fig. 1, the slope of the curve (initially flocculated system) is given by

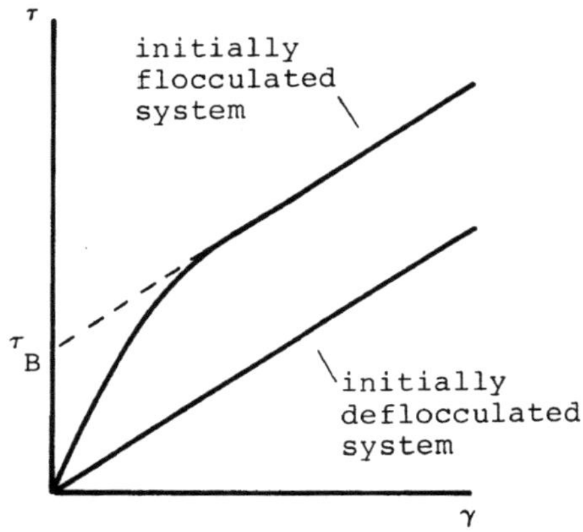

FIG. 1 - Shear stress (τ) as a function of shear rate (γ).

$$\frac{d\tau}{d\gamma} = \eta \,, \tag{1}$$

where = viscosity, and

$$\tau_B = \tau_v + \tau_E \,, \tag{2}$$

where τ_B = Bingham yield value, τ_v = hydrodynamic contribution (energy dissipated in viscous losses), and τ_E = interaction contribution (energy dissipated in breaking interactions). Also

$$\tau_B = N.E, \tag{3}$$

where N = total number of contacts and E = contact energy. Hence, E may be obtained from τ_B if N can be determined or assumed (e.g., total deflocculation).

Morphology

For a system which is composed of spherical, monodispersed particles, the "compactness" of the flocs which form depends on E. Compactness may be assessed in terms, for example, of the average coordination number (z) of a particle inside a floc and/or the concentration (c_f) of particles in the floc. In general, c_f and z decrease as E increases. This is also reflected in the sediment volume: the higher E is, the more volume the sediment occupies. This may be rationalized as follows.

Consider what happens when a singlet meets a doublet in the early stages of aggregation (Fig. 2). Thermodynamically B is preferred to A (because three contacts are formed rather than two), but hydrodynamically A is preferred (less water has to be "pushed out of the way," to put it crudely). Hence, initially A wins. Now if E is low, then the particles will roll around one another to produce B. However, the tendency to do this will reduce as E increases (akin to a surface friction effect). Hence, flocs are more linear (chain-like), the higher E is, compared to the more compact flocs at lower E.

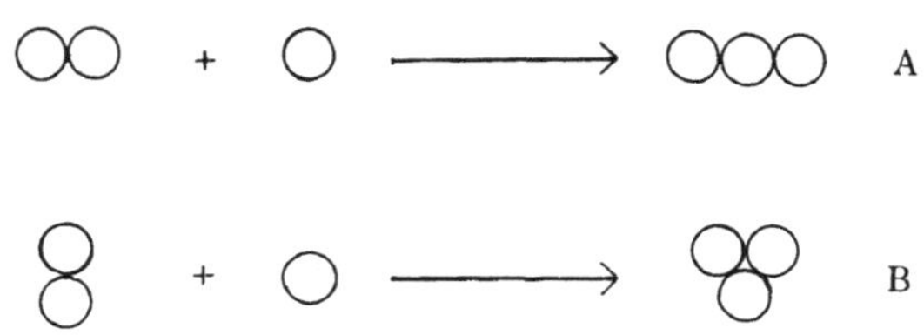

FIG. 2 - Early events in aggregation.

Extent

For a system in which flocculation is weak and reversible (E not more than ~5kT), the system at equilibrium will effectively "phase separate" (Fig. 3). The concentrations of particles in the two phases, at equilibrium, are related by the following equation:

$$c_d = c_f \exp\left(\frac{-zE}{2kT}\right), \quad (4)$$

where k = Boltzmann constant and T = absolute temperature. (There is the number 2 in the denominator because there are two particles per contact.) Hence, E can be determined if c_f and z are known (or can be approximated).

FIG. 3 - "Phase separation" during flocculation.

Rate

For a system undergoing flocculation into a weak energy trough (E), without having to overcome any energy barrier, the rate constant for flocculation ($\underline{k}$) is given (semi-empirically) by the following equation:

$$\underline{k} = \underline{k}_0(1 - \exp{-E/kT}). \quad (5)$$

Then, for E = 0, $\underline{k}$ = 0: no flocculation; for $E \longrightarrow \infty$, $\underline{k} \longrightarrow \underline{k}_0$: rapid (irreversible) flocculation. The term $\underline{k}_0$ may be calculated theoretically:

$$\underline{k}_0 = 8\pi aD, \quad (6)$$

where D = diffusion coefficient of the particles. But

$$D = \frac{kT}{6\pi\eta a}, \quad (7)$$

where η = viscosity of the continuous phase. Therefore,

$$\underline{k}_0 = \frac{4kT}{3\eta}, \quad (8)$$

that is, $\underline{k}_0$ is dependent only on T and η. (N.B.: for water at 25°C, $\underline{k}_0 \sim 5 \times 10^{-13}$ cm^3 s^{-1}.) Hence, if $\underline{k}$ can be measured (e.g., by light scattering), E may again be obtained.

Other Measurable Characteristics

The group reviewed critically the methods currently used for describing in quantitative terms the microbial aggregates that have to date received attention. Measuring the size of an aggregate can be a relatively easy task when that aggregate is compact, as with grex formation in Dictyostelium discoideum. Simple visual or microscopic methods can then be employed. Considerably greater problems arise when attempts are made to find dimensions of less compact aggregates, such as yeast-cell aggregates and aggregates encountered in water and wastewater treatment processes. Preparation of such materials for microscopy, be it light or electron microscopy, presents many problems. Indeed, such is the ease with which components may leave or enter the aggregate that size measurements may have little meaning.

Closely associated with the problems of measuring aggregate size are those of establishing aggregate shape. Shape has a less rigorous meaning than size. The problems are greater, the more loosely associated are the components in the aggregate. For such aggregates, methods need to be refined for fixing aggregates without causing appreciable distortion of the morphology of the aggregate. To date, very little information has become available on the shape of loosely aggregated microorganisms.

Cell number in aggregates - and by difference the proportion of a cell population in aggregates - is easily determined. Many workers prefer light-scattering methods. However, others prefer methods involving counting of cells by light microscopy. A method widely accepted for aggregates of yeast cells involves measuring first the number of free cells in suspension, and then the total number of cells after aggregates have been chemically or enzymically dispersed (5). The number or the proportion of cells in aggregates can then be determined. For the method to be applied more widely, particularly among the recently studied species of bacteria which are found as aggregates, more needs to be known of the chemicals and enzymes which will disperse aggregates.

Similar problems are encountered when attempts are made to determine the number, types, and distribution of components in aggregates. The success of such attempts especially hinges around the availability of chemicals and enzymes that will disperse aggregates. With some of the recently discovered aggregated bacteria, problems will also be encountered in devising media that will support growth of the disaggregated bacterial components.

MECHANISMS

Movement and Competence

In passive aggregation, growth of an aggregate is due to the increase in the number of cells by continual cell division, without complete separation, of an initially isolated cell and its progeny. When aggregation is achieved exclusively by way of an active process, an aggregate can arise only if there is movement so as to allow cells to come together. Chemotactic signals, as in motile organisms, may be used to direct cells to an aggregation center. In nonmotile organisms that are cultured in a turbulent milieu, chemotactic signals will probably not work, nor are they necessary. Random collision due to fluid motion is sufficient.

In inducible aggregation systems, the development of competence to aggregate must be a prerequisite. Inducible here means that there is a period in the life cycle of an aggregative organism in which the cells are dispersed and unable to aggregate among themselves. Changes in the characteristics of the cell surface and/or appendages must take place to allow the cells to attach to each other.

Contact Sites

The forces that keep cells together may simply be ionic interactions, as has been proposed for flocculation of brewing yeast (Rose, this volume). For such a system, colloid theory may be sufficient to explain aggregation. Where specificity of interaction is undeniably demonstrated, hydrogen bonds, strengthened by hydrophobic interactions, are favored by most investigators. Where developmental events, such as sex or parasexual and nonsexual fusion of cell walls, are post-aggregative events, the weak bonds mentioned above are subsequently supplanted by covalent bonds. Such is the case of the mating reaction in a number of eukaryotic protists.

Fimbriae, flagella, or fibrillar materials have been proposed to act as if they were grappling hooks (as electron microscopy would suggest), allowing further and more intimate contacts to be made later. It is obvious that when two cells approach each other, the superficial components that stick out farthest would be the first to touch.

When the interaction between cell surfaces becomes covalent, as when mural fusion occurs, then the cell-cell interaction should no longer be susceptible to disruption by changes in ionic strength, by high temperatures and other hydrogen-bond-breaking agents, and by surfactants, or even by enzymes that hydrolyze binding polymers.

Our understanding of aggregative phenomena ranges from simple observation of consortia of microorganisms which are attached to each other in patterned assemblies to systems where putative adhesion molecules have been identified or isolated. Ultimately, analysis of all of these systems will require the latter type of procedure.

Some systems, such as the flocculation of brewing yeast, should be readily accessible to both biophysical and biochemical analysis, because flocculation can be brought about under laboratory conditions and because large quantities of material are available. Analysis will be much more difficult when these organisms can be obtained only in small quantities, when the organisms are not readily grown in the laboratory, or when the systems are complex and poorly defined as wastewater flocs. Mechanisms which have been found in experimentally more amenable systems may provide a clue for initiating studies of the complex systems.

In Dictyostelium, molecular preparation of the cell surface for chemotaxis and adhesion required at aggregation is one of the first steps in the process of differentiation. Preliminary indications of similar changes in yeast flocculation have also been obtained, possibly suggesting the unmasking or synthesis of an adhesive polymer. In some bacterial consortia, however, it has not been demonstrated that patterns arise by aggregation rather than positioned differentiation.

Again referring to Dictyostelium, a consequence of aggregation is the onset of a process of differentiation leading to formation of spore cells and stalk cells (Garrod, this volume). Evidence that contact per se acts as a trigger for gene activation leading to differentiation is equivocal. There is good evidence, however, for the action of diffusible factors, e.g., cyclic AMP, ammonia, and differentiation-inducing factor (DIF), operating as morphogens (factors controlling morphogenesis) during the aggregated stages of the life cycle. Myxobacteria also undergo developmental processes, but so far the control mechanisms have not been identified.

CELL-CELL COMMUNICATION

Long Distance

The evidence for long-distance communication among cells that have aggregated (or are about to aggregate) is most convincing in Dictyostelium, where cyclic AMP has been amply demonstrated to be a chemotactic signal for cells to aggregate. A pheromone in myxobacteria has been demonstrated, but its chemical nature is not yet known (24). In filamentous fungi, the evidence is far less convincing. Such systems are usually studied as cultures on an agar surface. The demonstration of long-distance cell-cell communication in liquid cultures is more difficult.

The mating factors of Saccharomyces cerevisiae appear to serve several functions, including the arrest of DNA synthesis in the sexually complementary cell as well as the induction of sexual agglutination. Both functions can be brought about even when complementary cells are not allowed to come into contact. The demonstration of induction, of course, requires that the cells of complementary mating types be brought together.

Close Contact

It is much easier to present the case for close-contact communication. Several systems of metabolite exchange are known in which the bacterial components depend to a varying extent upon one another. The obligately syntrophic consortia of the Chlorochromatium type (20) and the alcohol- and fatty acid-degrading methanogenic sludge flocs and rumen content particles (2, 16) are mentioned as instances of obligate mutual dependency. Close contact shortens diffusion paths between partners and thus enhances energy yields to the benefit of both. Metabolic cooperation is facultative in aggregates of cyanobacterial heterocysts and surrounding heterotrophic bacteria (18); similar metabolic relationships are most probably involved in activated-sludge flocs degrading soluble and particulate substrates.

In clearly developmental systems, where an aggregate of cells eventually enters a kind of multicellular state, as in cellular slime molds and myxobacteria, the coordination of millions of cells into an integrated whole is difficult to imagine without conceding at least close-contact communication.

In sexual aggregation systems, as sexual agglutination in yeasts and mating reactions in algae and protozoa, there is no doubt as to the need for communication before and at the time of conjugation and fusion. Indeed, it has been hypothesized that a possible reason for the predominance of pair mating in a floc of 10^5 fission yeast cells is that, once one end of a cell has been spoken for by one of its dozen immediate neighbors, the other end is no longer available for sexual interaction by a third cell. A surface modulation working from one end of the cell to the other must somehow be operative to ensure pair formation and avoid the less productive arrangement of a ménage à trois.

In addition to these systems, the synchrony of cell division in Thiopedia plates may be mentioned as requiring at least close-contact communication. Indeed, any group that tends to act as a social unit rather than as individuals must be presumed to have developed an effective communication system among its members.

ABIOTIC ENVIRONMENTAL INFLUENCE

Maximizing Aggregation

Many factors influence the appearance of aggregation in a culture. These may be intrinsic to the cell: its genetic makeup, its stage in the cell cycle, or its stage in the life cycle. Other factors could be nutritional, the state of health of the cell, or factors that could specifically cause the induction of competence. What was more interesting to our group was the influence of the abiotic environment on the expression of competence to form aggregates; in short, the means to maximize or minimize aggregation.

Artificial aggregation can always be resorted to, if a given system does not tend to be aggregative. For this purpose, artificial flocculants, such as metal ions, polyelectrolytes, lectins, and proteins may be used. Maximization of the aggregation of competent cells may involve adjustment of ionic strength, hydrogen-ion concentration, cation concentration, temperature, mixing characteristics, etc. Maximization could mean an increase in the rate and/or extent of aggregation.

Minimizing Aggregation

If the dispersed state of a system is preferred, minimizing of aggregation is the objective. This can be achieved in various ways. For many systems, raising the temperature increases the rate of aggregation as well as the rate of disaggregation. High temperatures can cause reversible disaggregation in many systems. The use of hydrogen-bond-breaking agents such as guanidinium chloride or urea also causes reversible disaggregation. So do surfactants, such as sodium dodecyl sulfate. Extremes of pH may promote disaggregation. Use of chelators brings about reversible deflocculation of Ca^{2+}-dependent systems, such as flocculation of brewing yeasts. The minimizing of aggregation is particularly useful to both experimenter and process engineer who may prefer to work with the dispersed state of an organism.

THE IMPLICATIONS OF BEING AGGREGATED OR DISPERSED

To the Organism

Dispersed individual organisms have greater access to prey or substrate.

While dispersed, they have reduced exposure via dilution to their own wastes, toxic or otherwise. Dispersion implies distribution in a large volume, and being broadly dispersed enhances opportunities for genetic recombination with individuals from other strains and clones. A dispersed population seems apt to have more survivors from specific onslaughts, such as attack by a predator. It is more apt to have some individuals find their way into new environments.

Aggregation emphasizes the collective over the individual. Aggregated organisms in a substrate-deprived environment have the advantage that the death of any of their number releases material of use to all in close proximity (assured by the aggregation). Nonmotile organisms that depend upon Brownian motion to generate sexual collisions with potential mates require an aggregative mechanism to hold them in close proximity.

In chemostat experiments, it was shown (11) that if a medium containing macromolecular substrates (dextran and gelatin) is inoculated with protozoan-free material (taken from a wastewater treatment plant), the chemostat develops mixed cultures of dispersedly growing bacteria, often dominated by Cytophaga-like bacteria (CLB). If protozoa are included, however, flocs develop consisting either of zoogloeae or of filamentous organisms. The Cytophaga-like bacteria are still there but are now living within the flocs. There they play an essential role by degrading the dextran which the flocs alone could not degrade. Free-living CLB would be eliminated by predating protozoa. Thus, selection pressure goes against fast-growing but predation-sensitive CLB, in favor of slow-growing mutualistic flocs. In activated sludge systems, which depend greatly on the efficiency of separation and recycling of organisms, washout may provide a selection pressure similar in effect to predation.

Myxobacteria have developed several mechanisms to ensure that the cells remain together. Theoretically (and under culture condition practically, too), single cells are able to survive and to grow into a colony. The explanation for this pressure for "social life" may be found in the way myxobacteria obtain their nutrients. They are "micro-predators" lysing other microorganisms or degrading organic polymers, implying that they excrete enzymes. To minimize diffusion losses (loss of enzyme activity, loss of products of enzymic action) and to use optimally the energy invested into enzyme production, the organisms cooperate and act as large populations comprising many thousands of cells. The mechanisms that keep the cells together are not yet fully understood. Chemotaxis is obviously involved (autochemotactical substances not known), in addition to a system of "social gliding" (allowing movement only when cells are in contact with one another, perhaps mediated by fimbriae). Also, fruiting-body formation seems to have mainly the function of ensuring that a new life cycle is started by a cell community rather than an individual cell.

To Research and Industry

To the experimenter, dispersion allows treatment of the organisms as individuals, whether assay is microscopic or microchemical. Dispersion allows the experimenter access to a potentially more homogeneous population. It avoids the physiological heterogeneity of cells within a floc due to difficulties with transport of oxygen, proteins, nutrients, or inhibitory products.

On the other hand, aggregation may be advantageous in facilitating harvest or separation of biomass. In systems that are not under full control of the experimenter, aggregation makes it more likely that part of the population is in the proper state for production of a desired product or reaction (e.g., in antibiotic production where certain limitations may be required for secondary metabolism to work). Aggregation is one way of increasing cell concentration and immobilization (Atkinson, this volume).

Listed in Table 1 are the consequences of microbial aggregation to industry. All these lead to opportunities for process modification and improvement, including possibilities for the development of operations which are robust and have economic advantage. They cover industries such as brewing, production of enzymes and single-cell protein, water and wastewater treatment, production of fine chemicals and pharmaceuticals, and they are relevant in manipulated natural environments.

TABLE 1 - The consequences of aggregation and dispersion to industry.

Dispersion:

1. Allows surface cleanliness due to the absence of film formation.
2. Imposes low sedimentation and flotation rates requiring centrifugation for cell separation.
3. Allows high conversion rates when conditions are such that growth in aggregates would cause diffusion/transport limitation.
4. Allows the manipulation of cells in the liquid phase.

TABLE 1 - Continuation

Aggregation:

1. Facilitates cell separation, e.g., sedimentation, centrifugation, filtration.
2. Allows high cell concentrations in continuous fermenters, e.g., tower fermenters.
3. Leads to internal gradients (in aggregates) of physicochemical conditions, e.g., substrate and product concentrations, oxygen, pH, etc.
4. Leads to possibilities for gaseous product evolution in the center of aggregates.
5. Leads to heterogeneous and possibly syntrophic, though ordered, populations of organisms.
6. Affects overall rates of growth, metabolism, attenuation, etc.
7. Affects overall stoichiometry.
8. Allows continuous fermenters to be operated beyond normal washout flow rate (wastewater treatment).
9. Protects against contamination.
10. Facilitates the capture of suspended solids (sewage treatment).
11. Allows the manipulation of growth rate independent of dilution rate.
12. Allows the manipulation of biomass as a single phase.

1984: LOOKING FORWARD

Among the great variety and number of microbial aggregation systems, only a few have been studied beyond the phenomenological stage. For most systems, work is needed in the following areas:

1. Studies of the structures and activities of aggregates and the mechanisms by which aggregation or disaggregation occurs. Needed is information on forces such as polymer cross-bridging, ion-binding, and hydrodynamic effects; the nature and distribution of organisms and polymers within the aggregate; interactions among cells; and reactions and transport of chemicals within the aggregates.

2. Studies of possible interspecific genetic transfer among members of consortia and activated-sludge flocs; studies of co-evolution and of enhancing mutualistic tendencies in stable consortia.

3. Studies of mechanisms of regional elaboration of extracellular polymers.

4. Identification of the relevant physical and chemical properties of suspensions that will allow the design and/or selection of chemicals to aid their flocculation. Studies of "naturally flocculating" suspensions may lead to the production of effective chemicals.

5. Development of guides and standard protocols for study and assessment of aggregate structure and strength.

6. Development of industrially feasible methods for manipulating non-flocculent suspensions to produce stable and manageable aggregates.

REFERENCES

(1) Atkinson, B., and Daoud, I.S. 1976. Microbial flocs and flocculation in fermentation process engineering. Adv. Biochem. Eng. 4: 41-124.

(2) Bryant, M.P. 1979. Microbial methane production - theoretical aspects. J. Anim. Sci. 48: 193-201.

(3) Calleja, G.B. 1984. Cell aggregation. In The Yeasts, 2nd ed., eds. A.H. Rose and J.H. Harrison, vol. 1. London: Academic Press, in press.

(4) Calleja, G.B. 1984. Microbial Aggregation. Boca Raton, FL: CRC Press.

(5) Calleja, G.B., and Johnson, B.F. 1977. A comparison of quantitative methods for measuring yeast flocculation. Can. J. Microbiol. 23: 68-74.

(6) Carlile, M.J., and Gooday, G.W. 1978. Cell fusion in myxomycetes and fungi. In Membrane Fusion, eds. G. Poste and G.L. Nicolson, pp. 219-265. Amsterdam: North-Holland Publishing.

(7) Chet, I., and Henis, Y. 1975. Sclerotial morphogenesis in fungi. Ann. Rev. Phytopath. 13: 169-192.

(8) Garrod, D.R., and Nicol, A. 1981. Cell behaviour and molecular mechanisms of cell adhesion. Biol. Rev. 56: 199-242.

(9) Gibbons, R.J., and van Houte, J. 1978. Oral bacterial ecology.

In Textbook of Oral Biology, eds. J.H. Shaw, E.A. Sweeney, C.C. Cappuccino, and S.M. Meller, pp. 684-705. Philadelphia: W.B. Saunders.

(10) Goodenough, U.W. 1980. Sexual microbiology: mating reactions of Chlamydomonas reinhardii, Tetrahymena thermophila and Saccharomyces cerevisiae. In The Eukaryotic Microbial Cell, eds. G.W. Gooday, D. Lloyd, and A.P.J. Trinci, pp. 301-328. Cambridge: Cambridge University Press.

(11) Güde, H. 1979. Grazing by protozoa as selection factor for activated sludge bacteria. Microb. Ecol. 5: 225-237.

(12) Harris, R.H., and Mitchell, R. 1973. The role of polymers in microbial aggregation. Ann. Rev. Microbiol. 27: 27-50.

(13) Hirsch, P. 1980. Some thoughts on and examples of microbial interactions in the natural environment. In Aquatic Microbial Ecology, eds. R.R. Colwell and J. Foster, pp. 36-54. College Park, MD: Maryland Sea Grant Publication, University of Maryland Press.

(14) Laubenberger, G., and Hartmann, L. 1971. Physical structure of activated sludge in aerobic stabilization. Water Res. 5: 335-341.

(15) Loomis, W. 1975. Dictyostelium discoideum: A Developmental System. New York: Academic Press.

(16) McInerney, M.J., and Bryant, M.P. 1980. Syntrophic associations of H_2-utilizing methanogenic bacteria and H_2-producing alcohol and fatty acid degrading bacteria in anaerobic degradation of organic matter. In Anaerobes and Anaerobic Infections, eds. G. Gottschalk, N. Pfennig, and H. Werner, pp. 117-126. Stuttgart: G. Fischer Verlag.

(17) O'Day, D.H., and Horgen, P.A., eds. 1981. Sexual Interactions in Eukaryotic Microbes. New York: Academic Press.

(18) Paerl, H.W. 1982. Interactions with bacteria. In The Biology of Cyanobacteria, eds. N.G. Carr and B.A. Whitton, pp. 441-461. Berkeley: University of California Press.

(19) Parker, D.S.; Kaufman, W.J.; and Jenkins, D. 1971. Physical conditioning of activated sludge floc. J. Water Poll. Contr. Fed. 43: 1817-1833.

(20) Pfennig, N. 1980. Syntrophic mixed cultures and symbiotic consortia with phototrophic bacteria: a review. In Anaerobes and Anaerobic Infections, eds. G. Gottschalk, N. Pfennig, and H. Werner, pp. 127-

131. Stuttgart: G. Fischer Verlag.

(21) Reissig, J.L., ed. 1977. Microbial Interactions. London: Chapman and Hall.

(22) Rosenberg, E., ed. 1984. Myxobacteria: Development and Cell Interactions. Berlin: Springer Verlag.

(23) Sezgin, M.; Jenkins, D.; and Parker, D.S. 1978. A unified theory of filamentous activated sludge bulking. J. Water Poll. Contr. Fed. 50: 362-381.

(24) Stephens, K.; Hegeman, G.D.; and White, D. 1982. Pheromone produced by the myxobacterium Stigmatella aurantiaca. J. Bacteriol. 149: 739-747.

(25) Vincent, B. 1974. The effect of adsorbed polymers on dispersion stability. Adv. Coll. Interface Sci. 4: 193-277.

(26) White, D. 1981. Cell interactions and the control of development in myxobacteria populations. Int. Rev. Cyt. 72: 203-227.

Microbial Adhesion and Aggregation, ed. K.C. Marshall, pp. 323-335. Dahlem Konferenzen 1984. Berlin, Heidelberg, New York, Tokyo: Springer-Verlag.

Physiology of Cell Aggregation: Flocculation by Saccharomyces cerevisiae As a Model System

A.H. Rose
Zymology Laboratory
School of Biological Sciences
University of Bath, Bath BA2 7AY, England

Abstract. An understanding of aggregation in a population of living cells is considered to involve six basic concepts or unit processes. These are the molecular architecture of the cell surface, the mechanisms by which dispersed cells are repelled, the forces which cause dispersed cells to aggregate, the changes in cell-surface architecture which cause a switch to aggregation, the structure of aggregates, and the importance of aggregation in the life-style of living cells. These concepts are considered using flocculation by the yeast Saccharomyces cerevisiae as a model.

INTRODUCTION

The study of living organisms has brought to light innumerable examples of cell aggregation. When one recalls the plethora of cell and tissue types, it is hardly surprising that those cell-aggregation phenomena that have been described in any degree of detail, and which have been studied to an extent that makes possible a meaningful physiological interpretation, are quite small in number. This paper, which aims to present a critical overview of the physiological mechanisms by which cells aggregate, would be of little value if it simply rehearsed the multitude of reports which have appeared on aggregation of various microbial, plant, and animal cells. Instead, I have highlighted the various physiological problems, which in many ways can be looked upon as unit concepts that underpin the phenomenon of cell aggregation. To illustrate the extent to which biologists currently have an understanding of the molecular basis of these concepts, I have selected the phenomenon of flocculation in Saccharomyces cerevisiae both as a model and a reasonably well understood cell-aggregation process. Cell aggregation can best be defined (6) as the gathering together of cells to form stable, contiguous, multicellular associations under physiological conditions.

FLOCCULATION IN THE YEAST SACCHAROMYCES CEREVISIAE

Yeasts are fungi that exist under most environmental conditions in the form of single cells (30). The yeast species which has been most extensively studied is S. cerevisiae, the microorganism responsible for converting sugars into ethanol and carbon dioxide in the manufacture of alcoholic beverages and in the leavening of bread. Strains closely related to S. cerevisiae, and previously dubbed S. carlsbergensis or S. uvarum, are now considered to be strains of S. cerevisiae (37). Many strains of S. cerevisiae used in the manufacture of beers, but not of wines or for baking bread, have the ability to form flocs, a term derived from the Latin word floccus for a lock of wool. Floc formation, or flocculation, is defined as the ability of strains of yeast, when in the stationary phase of growth in batch culture, spontaneously to aggregate and form flocs which sediment in the culture (33). The ability to flocculate has traditionally been exploited in the manufacture of beers. By selecting a strain of S. cerevisiae, cells of which remain dispersed in the fermenting malt wort during exponential growth but which form flocs when the culture enters the stationary phase of growth, the brewer is able to effect an efficient fermentation by the dispersed yeast cells and, as a result of flocculation, receive assistance in separating the yeast cells from the newly fermented beer. Some strains of S. cerevisiae have been selected for their ability to remain dispersed throughout all phases of batch growth, and these are referred to as powdery yeasts.

Expression of the ability to flocculate depends to some extent on the chemical composition of the medium in which the strain is grown. Physiological studies are most conveniently carried out using strains of S. cerevisiae that express their flocculating ability when grown in a chemically defined medium, for then the role of medium components in expression of flocculation can more easily be studied. The majority of strains of S. cerevisiae which have been used in flocculation studies derived originally from beer breweries and are diploid or more likely aneuploid. For some time it was believed that haploid strains of S. cerevisiae do not flocculate. This, it is now known, is not true. It would appear that geneticists, the source of most haploid strains of S. cerevisiae, tend to discard flocculent strains since these are somewhat difficult to handle in laboratory experiments; this is yet another example of man-made natural selection, but this time in the laboratory rather than in the environment. Finally, it is important to distinguish flocculation from chain formation. In some strains of S. cerevisiae, a daughter cell fails to separate from the mother cell although a cross septum forms between the cells. Further buds may then be produced on the mother, daughter, and subsequent daughter cells to form a chain. Chain formation differs from flocculation in that flocs are formed from erstwhile separate discrete cells and, as discussed later, can be dispersed to reform a suspension of single cells.

PHYSIOLOGICAL CONCEPTS IN CELL AGGREGATION

To understand the physiological mechanism by which a population of cells aggregates, certain physiological concepts need to be considered. These concepts can be viewed as the unit processes in any explanation of the aggregation process, and they are considered in detail in this section. They apply to any aggregation process, but yeast flocculation will be used to illustrate each concept.

The Molecular Architecture of the Cell Surface

Aggregation of cells, purely on a priori grounds, would appear to involve the surfaces of cells, from which it follows that a knowledge of the molecular architecture of the cell surface is a prerequisite to explaining the physiological basis of the process. The alternative possibility that the aggregation process is directed from within cells must be considered, although it rarely is. From the limited evidence available, it would appear that floc formation by S. cerevisiae does indeed involve the surfaces of cells. Thus, heat-killed cells flocculate, but only if they originated from flocs (27). Moreover, walls isolated from cells form flocs, but only if they are prepared from flocculent cells (8, 9).

A great deal has been published on the chemical composition of the surface layers of cells. However, the majority of these reports simply provide a chemical analysis of the isolated envelope layers and give little or no indication of the nature of the molecules that lie exposed on the envelope surface. There are some notable exceptions. The surface of the mammalian erythrocyte is reasonably well understood (16), as increasingly also is the surface of certain Gram-negative bacteria (28).

Information on the nature of molecules located on the surfaces of cells has come from the use of specific tagging compounds such as lectins that, after being suitably coupled with an electron-dense molecule (e.g., ferritin), can be visualized in the electron microscope. A second technique involves measuring the surface or zeta potential of a population of cells at different pH values, and from these data calculating the average density of charged groups on the cell surface. Knowing the nature of ionogenic molecules in the surface layers, an estimate can then be made of their density on the cell surface. What little is known of the nature of molecules on the surface of walls of S. cerevisiae has come from use of the second of these techniques. Meager though the body of data may be considered when viewed alongside that published on bacteria, information on the chemical composition of the wall of S. cerevisiae is reasonably comprehensive, particularly when compared with that available on fungal walls in general. Chemical analyses of isolated walls of S. cerevisiae have shown them to contain roughly equal proportions of a mixture of β-glucans and α-mannan, accounting altogether for about 80% of the dry weight of the wall. Variable values have been reported for these proportions, and the figures quoted are a rough approximation. The three glucans are neutral molecules, comprising an acetic acid-insoluble (1-3)-glucan, α (1-6)-glucan, and an alkali-soluble (1-3)-glucan (14). The mannan, on the other hand, is an anionic polymer. Whereas the backbone of this polymer is made up of α (1-6)-linked mannose residues, the polymer has side chains of mannose and mannobiose residues which are joined to the backbone by phosphodiester linkages (2).

The remainder of the wall of S. cerevisiae is made up of a mixture of protein, chitin, and lipid. The chitin is thought with some certainty to be located in bud scars (4); it is an electrically neutral component. The wall protein, which has been little investigated (14), is a source of negative charges emanating from C-terminal carboxyl groups and terminal carboxyl groups on aspartate and glutamate residues, as well as of positive charges on N-terminal amino groups and basic groups on residues of arginine, histidine, and lysine. The authenticity of wall lipid in S. cerevisiae is a contentious subject. Many workers claim that, when detected in isolated wall preparations, lipid, which consists mainly of a mixture of phospholipid and triacylglycerols (24, 29), merely represents plasma-membrane lipid that remained associated with the inside of the wall. Nevertheless, it could, through the presence of phosphodiester linkages in phospholipids, be a source of negative charges.

The emphasis on potential sources of negative and positive charges in wall components of S. cerevisiae reflects the generally considered opinion that such charges are at the basis of the ability of populations of yeast cells to remain dispersed and, following changes in the overall density and/or distribution of surface charges, to aggregate. Alternative explanations, not invoking charge distribution on cell surfaces, may well be proposed to explain the mechanism by which other cell types aggregate; these alternatives are considered by Garrod, Hirsch, and Rutter and Vincent (all this volume).

Living cells, including microorganisms, have a net negative charge on the cell surface at physiological pH values. It is generally considered that this net charge is physiologically important in the sequestration of cations, particularly Mg^{2+}, a high concentration of which is required for the activity of some plasma membrane-bound enzymes (31). The presence of a net negative charge on the surface of S. cerevisiae at physiological pH values was established some thirty years ago (17). Since then, the provenance of this net charge has been examined using the refined technique of microelectrophoresis at different pH values (32) and has specifically been examined in relation to the involvement of surface charge in relation to flocculation (3, 11-13, 18). On a priori grounds, only one of the potentially ionogenic wall components in S. cerevisiae can conceivably be excluded as a possible cell-surface generator of charge. This is phospholipid which, in all likelihood, lies at the inner surface of isolated walls, as already suggested. As such, in the intact cell, it is not likely to contribute to cell-surface charge. On the other hand, cell-surface positive charges on S. cerevisiae could be contributed by N-terminal amino groups on proteins and basic groups in arginine, histidine, and lysine residues in these proteins. Negative charges could arise from C-terminal carboxyl groups in wall proteins, from terminal carboxyl groups in aspartate and glutamate residues, as well as from phosphodiester linkages in wall phosphomannan. Experimentally it has been shown that, in a medium at pH 2.0, the net charge on a yeast cell may be zero or slightly negative, depending on the strain and the conditions under which the cells were grown. As the pH value of the medium is raised, the net negative charge increases in density, usually up to pH 7.0, after which it can show a decline (3, 18). The charge attributable to phosphodiester groups is taken to be equivalent to the mobility of cells at pH 4.0, while that attributable to carboxyl groups is equivalent to the difference in mobility of cells at pH 7.0 and pH 4.0 (10).

The Mechanism by Which Cells Are Kept Dispersed

An obvious first question to ask in any physiological study of aggregation is: what are the forces that prevent cells from adhering to one another before the aggregation process begins? Unfortunately, few workers have addressed themselves to this problem, preferring to concentrate on the physiology of the aggregation process. In all likelihood, cells remain dispersed in a population as a result of repulsion of like charges, usually negative, on the cell surface. In S. cerevisiae, one or both of the two types of negative charge could be responsible for keeping cells dispersed. A limited amount of data is available which allows an estimate to be made of the relative contribution to dispersal of these negative charges.

Treatment of yeasts or yeast walls with 1,2-epoxypropane causes esterification of exposed carboxyl groups. When cells of several strains of S. cerevisiae that had the capacity to form flocs were treated with this reagent, the rate of floc formation was decreased (18). However, when such organisms were treated with 60% hydrofluoric acid, a treatment that excises phosphate groups from the wall mannan, their rate of flocculation was greatly increased (18). These findings strongly suggest that populations of S. cerevisiae, in the exponential phase of growth and prior to the expression of flocculating ability, remain dispersed, mainly as a result of the negative charge generated by phosphodiester groups in the wall phosphomannan. Further support might come from experiments involving similar treatments on organisms before they acquire the ability to flocculate.

The Mechanism by Which Cells Aggregate

Many more physiological experiments have been carried out on the mechanism by which cells form aggregates. The nature of the cell-surface groups involved varies considerably, depending on the cell type. What a large number of aggregation processes have in common is that adhesion of one cell to another is frequently mediated by salt bridges, and that Ca^{2+} ions are often involved.

Evidence for the involvement of Ca^{2+} ions in floc formation by S. cerevisiae is fairly substantial, despite the fact that claims have been made for a similar role for other cations. Addition of Ca^{2+} ions to suspensions in distilled water of cells that have been dispersed from flocs results in reflocculation. It has been known for some time that the process of reflocculation is inhibited by ethylene diamine tetra-acetic acid (EDTA) and ethylene glycol-bis-(β-amino ethyl ether)N,N'-tetra-acetic acid (EGTA), the inhibitory effect being reversible by addition of further amounts of Ca^{2+} ions (5). Reports (35, 36) have furnished further evidence for a specific role for Ca^{2+} ions. Thus, the floc-forming ability of this cation is antagonized by the alkaline-earth metal ions Ba^{2+} and Sr^{2+}, whereas sodium salts are weak antagonists. Moreover, the concentration of Ca^{2+} ions required to flocculate a suspension of cells (around 10^{-8}M) is low enough to preclude the possibility that it is acting as a counter-ion.

Until recently, a much more controversial topic than the nature of the cation involved in floc formation in S. cerevisiae was the source of the cell-surface negative charges that participate in formation of flocs. As already indicated, there were originally two candidates for this role, namely, carboxyl groups and phosphodiester groups. A further possibility to consider is that both types of anionic grouping are involved. The balance of evidence currently available is that the cell-surface anionic groups involved in floc formation are carboxyl groups. The main evidence comes from reports of a correlation between the increase in density of cell-surface carboxyl groups, as measured by microelectrophoresis, as cells in growing cultures of several strains of S. cerevisiae acquire the ability to flocculate (3). Second, inhibition of the rate of yeast flocculation following treatment of cells with 1,2-epoxypropane, which leads to esterification of cell-surface carboxyl groups, provides additional evidence for a role for these groups in flocculation. That 1,2-epoxypropane treatment causes only a partial inhibition of flocculation rate is probably attributable to the reported incomplete nature of the esterification process (15). Some years ago, the suggestion was made (22, 23) that the negative charges generated by phosphodiester linkages in the wall phosphomannan were involved in floc formation. The main evidence against this proposal is that acquisition of flocculating ability does not correlate with an increase in the density of cell-surface phosphodiester groups (3), whereas excision of these linkages by treatment with hydrofluoric acid, as already described, leads to an increase in flocculation rate (18).

Having established the most likely participants in floc formation, matters become much less clear when considering the nature of the cell-cell linkage. Some workers believe that there may exist a direct bridge between carboxyl groups on the surfaces of adjacent cells, mediated by a calcium ion. Others believe that the bridge is a longer one, with other molecules, conceivably polypeptides, being involved. There is, however, a total lack of evidence for the existence of these longer bridges, with the possible exception of strains of S. cerevisiae which flocculate only in wort and not in defined media, the wort supposedly providing a peptide which may be involved in bridge formation (34). The structure of the cell-cell bridges in flocs of S. cerevisiae and also the number of bridges required to keep cells in a floc are subjects which are urgently in need of further research.

Two recent contributions to the yeast flocculation literature propose a more involved scenario. Miki and his co-workers (25, 26) implicate cell-wall phosphomannan in bridge formation. Their most persuasive evidence for a role for phosphomannan came from the observation that flocculation can be inhibited by addition to floc suspensions of mannose or α-methyl-D-mannoside (each at 500mM) or of concanavalin A or succinylated concanavalin A, both of which bind specifically to mannan. Although these observations are open to an alternative interpretation, Miki and his co-workers (25) have proposed bridge structures involving carboxyl groups on surface proteins and phosphodiester linkages on surface phosphomannan.

The Mechanism of the Switch to Aggregation

Very few experiments have been reported which illuminate the molecular basis of the switch from dispersal to aggregation in a population of cells. This paucity of both information and, apparently, interest reflects our ignorance of the detailed molecular structure of cell surfaces, and often of techniques for measurement of changes that take place at cell surfaces.

The cell surface and changes in surface structure that are a prerequisite to yeast-cell aggregation are no exception. Embarrassingly few studies have been reported on this aspect of yeast flocculation. Accepting that the onset of aggregation correlates with the appearance on the cell surface of S. cerevisiae of an increased density of carboxyl groups, the need is to explain the origin of these groups. One possibility is that, as the rate at which cell populations divide is retarded, new acidic proteins are inserted into the cell wall, thereby providing an increase in the density of cell-surface carboxyl groups. Acquisition of the ability to flocculate in S. cerevisiae is known to involve protein synthesis, since the process is inhibited by cycloheximide (1, 20). Another possible explanation is that, as the rate of cell growth declines, a rearrangement of the location of acidic cell-wall proteins takes place, as a result of which the density of cell-surface carboxyl groups is increased. Third, one needs to consider the possibility that additional cell-surface carboxyl groups are generated as a result of chemical modification to existing wall proteins. Perhaps because it is easiest to tackle experimentally, available research data are confined exclusively to the last of these three possibilities.

Flocculation in S. cerevisiae is known to be under genetic control, although there has been some confusion over the number and location of flocculation (FLO) genes (19, 21). One assumes, therefore, that the products of these flocculation genes include enzymes that catalyze changes in cell-surface architecture and so bring about the switch to flocculation; they may also regulate the expression of these gene products at the transcription, translation, or latency level. Research on the genetics of yeast flocculation ideally should progress hand in hand with that on the physiology of the process.

In theory, free carboxyl groups can be generated in a wall protein as a result of the action of endopeptidases, leading to formation of one free carboxyl and amino group and of amidases which convert asparagine and glutamine residues in wall proteins to aspartate and glutamate residues. Beavan and his colleagues (3) were the first to address themselves to this problem with S. cerevisiae. They reported (3) that acquisition of flocculating ability in Labatt strains 11 and 30 correlated with an increase in amidase activity, although the amidase activity which they assayed (following release of ammonia from L-asparagine) may not be a true measure of the capacity to deamidate asparagine and glutamine residues in wall proteins. In addition, Beavan and his colleagues reported a correlation between increase in activity of one endopeptidase (proteinase C) but not of two other similar proteinases (proteinases A and B) with endopeptidase activity. An extension of this study by Calvert (7), after confirming some of the data reported by Beavan and his colleagues (3), went on to show that the lytic enzyme activities involved are located in intracellular vesicles in a latent form, thereby exposing the need to explain how they are transformed from the latent to the active form as growth rate is retarded.

Overall, there appears to be a correlation between acquisition of flocculating ability in some strains of S. cerevisiae and the appearance of an increase in the activity of certain enzymes which might explain formation of an increased density of cell-surface carboxyl groups. Justification for further research on this aspect of yeast flocculation is obvious; in fact, the data available only scratch the surface of the problem. Not least of the problems is the need to explain the very sudden appearance of flocs in cultures as they are about to enter the stationary phase of growth, a phenomenon which is readily observed in the laboratory and which has been described graphically (3) in the form of a plot of the flocculating rate of a variety of strains of S. cerevisiae against the density of cell-surface protein carboxyl groups.

The Forces That Keep Cells in Aggregates

Most cell aggregates that are examined in laboratory studies are fairly stable and lose cells at quite slow rates. This is explained quite simply by the fact that most laboratory cultures are stirred, at a slow or fast rate, so that any aggregates that are formed are subject to a liquid-shear stress; only those aggregates that remain stable merit study. Irrespective of the aggregation system, however, it is important to know the nature, and if possible the strength, of the forces that maintain cells in an aggregate. On the one hand, there is the possibility that those forces that caused cells to form an aggregate are solely responsible for the stability of the aggregate. Alternatively, there is a need to consider the possible involvement of additional forces in the maintenance of aggregate stability.

In floc formation by S. cerevisiae, there are reasons to believe that forces in addition to those that caused cells to aggregate are at least partly responsible for the stability of flocs, although as with many other aspects of floc formation by yeast, the available evidence can easily fail fully to convince the more sceptical observer. Strong evidence that hydrogen bonding is involved in floc stability in S. cerevisiae comes from reports that deflocculation can be brought about by urea, a rise in temperature, or an increase in the dielectric constant of the medium or floc suspension (27). Further evidence in support of this contention comes from the ability of mannose, as already described, to break down yeast flocs. Hydrogen bonding is most likely to take place between mannose residues on cell-surface phosphomannan.

The Importance of Aggregation in the Life-style of Living Organisms

Populations of discrete cells do not aggregate in order to provide the cell biologist with convenient experimental systems in the laboratory. More specifically, S. cerevisiae does not flocculate to satisfy the requirements of the beer brewer, even though selected for that characteristic. Why, therefore, should certain strains of S. cerevisiae have acquired and retained the ability to flocculate? Some speculations have been voiced on this matter, mostly with inconclusive results largely because of ignorance of the physiological mechanisms involved in flocculation. On the one hand, there have been proposals to the effect that flocculation represents a vestigial form of sexual agglutination, a view which was most recently readvocated by Miki and his colleagues (25, 26). On the other hand, yeast flocculation is seen as a means whereby cells that find themselves in nutritionally poor environments can stick together, thereby increasing their ability to make maximum use of nutritionally rich microenvironments which they might generate by selective lysis. At this stage of progress in research on yeast flocculation, "yer pays yer money and yer takes yer choice"!!

REFERENCES

(1) Baker, D.A., and Kirsop, B.H. 1972. Flocculation in Saccharomyces cerevisiae as influenced by wort composition and by actidione. J. Inst. Brewing 78: 454-458.

(2) Ballou, C.E. 1976. Structure and biosynthesis of the mannan component of the yeast cell envelope. Adv. Microbial Phys. 14: 93-158.

(3) Beavan, M.J.; Belk, D.M.; Stewart, G.G.; and Rose, A.H. 1979. Changes in electrophoretic mobility and lytic enzyme activity associated with development of flocculating ability in Saccharomyces cerevisiae. Can. J. Microbiol. 25: 888-895.

(4) Cabib, E.; Roberts, R.; and Bowers, B. 1982. Synthesis of the yeast cell wall and its regulation. Ann. Rev. Biochem. 51: 763-793.

(5) Calleja, G.B. 1984. Cell aggregation. In The Yeasts, 2nd ed., eds. A.H. Rose and J.S. Harrison, vol. 1. London: Academic Press, in press.

(6) Calleja, G.B.; Johnson, B.F.; and Yoo, B.Y. 1981. The cell wall as sex organelle in fission yeast. In Sexual Interactions in Eukaryotic Microbes, eds. D.H. O'Day and P.A. Horgen. New York: Academic Press.

(7) Calvert, G.W. 1983. Flocculation in Saccharomyces cerevisiae. Ph.D. Thesis, University of Bath, England.

(8) Eddy, A.A. 1955. Flocculation characteristics of yeasts. III. General role of flocculating agents and special characteristics of a yeast flocculated by alcohol. J. Inst. Brewing 61: 318-320.

(9) Eddy, A.A. 1958. Composite nature of the flocculation process of top and bottom strains of Saccharomyces. J. Inst. Brewing 64: 143-151.

(10) Eddy, A.A., and Rudin, A.D. 1958. Comparison of the respective electrophoretic and flocculation characteristics of different strains of Saccharomyces. J. Inst. Brewing 64: 139-142.

(11) Eddy, A.A., and Rudin, A.D. 1958. Part of the yeast surface apparently involved in flocculation. J. Inst. Brewing 64: 19-21.

(12) Eddy, A.A., and Rudin, A.D. 1958. The structure of the yeast cell wall. I. Identification of charged groups at the surface. Proc. Roy.

Soc. Lond. B148: 419-432.

(13) Fisher, D.J. 1975. Flocculation - some observations on the surface charges of yeast cells. J. Inst. Brewing 81: 107-110.

(14) Fleet, G.H. 1984. Yeast cell walls. In The Yeasts, 2nd ed., eds. A.H. Rose and J.S. Harrison, vol. 2. London: Academic Press, in press.

(15) Gittens, G.J., and James, A.M. 1963. Some physical investigations of the behaviour of bacterial surfaces. VI. Chemical modifications of surface components. Biochim. Biophys. Acta 66: 237-249.

(16) Harrison, R., and Lunt, G.G. 1980. Biological Membranes, 2nd ed. Glasgow: Blackie.

(17) Jansen, H.E., and Mendlik, F. 1951. A study on yeast flocculation. Proceedings European Brewing Convention Congress, pp. 59-81, Brighton.

(18) Jayatissa, P.M., and Rose, A.H. 1976. Role of wall phosphomannan in flocculation of Saccharomyces cerevisiae. J. Gen. Microbiol. 96: 165-174.

(19) Johnstone, J.R., and Reader, H.P. 1983. Genetic control of flocculation. In Yeast Genetics: Fundamental and Applied Aspects, eds. J.F.T. Spencer, D.M. Spencer, and A.R.W. Smith, pp. 205-224. New York: Springer-Verlag.

(20) Lands, W.E.M., and Graff, G. 1981. Effects of fatty acid structure on yeast cell physiology. In Current Developments in Yeast Research, eds. G.G. Stewart and I. Russell, pp. 331-343. Toronto: Pergamon Press.

(21) Lewis, C.W.; Johnston, J.R.; and Martin, P.A. 1976. The genetics of yeast flocculation. J. Inst. Brewing 82: 158-160.

(22) Lyons, T.P., and Hough, J.S. 1970. Flocculation of brewer's yeast. J. Inst. Brewing 76: 564-571.

(23) Lyons, T.P., and Hough, J.S. 1971. Further evidence for the cross-bridging hypothesis for flocculation of brewer's yeast. J. Inst. Brewing 77: 189-203.

(24) McMurrough, I., and Rose, A.H. 1967. Effect of growth rate and substrate limitation on the composition and structure of the cell wall of Saccharomyces cerevisiae. Biochem. J. 105: 189-203.

(25) Miki, B.L.A.; Poon, N.H.; James, A.P.; and Seligy, V.L. 1982. Possible mechanisms for flocculation interactions governed by gene FLO1 in Saccharomyces cerevisiae. J. Bacteriol. 150: 878-889.

(26) Miki, B.L.A.; Poon, N.H.; and Seligy, V.L. 1982. Repression and induction of flocculation interactions in Saccharomyces cerevisiae. J. Bacteriol. 150: 890-899.

(27) Mill, P.J. 1964. The nature of the interaction between flocculent cells in the flocculation of Saccharomyces cerevisiae. J. Gen. Microbiol. 35: 61-68.

(28) Nikaido, H., and Nakae, T. 1979. The outer membrane of Gram-negative bacteria. Adv. Microbial Phys. 20: 163-250.

(29) Nurminen, T., and Suomalainen, H. 1973. On the enzymes and lipid of cell envelope fractions from Saccharomyces cerevisiae. Proceedings of the 3rd International Specialized Symposium on Yeasts, eds. H. Suomalainen and C. Waller. Helsinki: Print Oy.

(30) Phaff, H.J.; Miller, M.W.; and Mrak, E.M. 1978. The Life of Yeasts, 2nd ed. Cambridge, MA: Harvard University Press.

(31) Rose, A.H. 1977. Chemical Microbiology, 3rd ed. London: Butterworths.

(32) Sherbet, G.V. 1978. The Biophysical Characterization of the Cell Surface. London: Academic Press.

(33) Stewart, G.G., and Russell, I. 1981. Yeast flocculation. In Brewing Science, ed. J.R.A. Pollock, vol. 2. London: Academic Press.

(34) Stewart, G.G.; Russell, I.; and Garrison, I.F. 1975. Some considerations of the flocculation characteristics of ale and lager yeast strains. J. Inst. Brewing 81: 248-257.

(35) Taylor, N.W., and Orton, W.L. 1973. Effect of alkaline-earth metal salts on flocculation in Saccharomyces cerevisiae. J. Inst. Brewing 79: 294-297.

(36) Taylor, N.W., and Orton, W.L. 1975. Calcium in flocculence of Saccharomyces cerevisiae. J. Inst. Brewing 81: 53-57.

(37) Yarrow, D. 1984. Genus Saccharomyces Meyen ex Rees. In The Yeasts, a Taxonomic Study. Amsterdam: North-Holland, in press.

Microbial Adhesion and Aggregation, ed. K.C. Marshall, pp. 337-349. Dahlem Konferenzen 1984. Berlin, Heidelberg, New York, Tokyo: Springer-Verlag.

Aggregation, Cohesion, Adhesion, Phagocytosis, and Morphogenesis in Dictyostelium – Mechanisms and Implications

D.R. Garrod
Cancer Research Campaign Medical Oncology Unit
Southampton General Hospital, Southampton SO9 4XY, England

Abstract. gpl26, the contact site of vegetative Dictyolstelium cells, may be a receptor for bacterial phagocytosis. It is also involved in cell-cell adhesion and plays a role in size regulation during morphogenesis. It seems to work by a nonspecific mechanism because univalent anti-gpl26 antibody inhibits phagocytosis of latex beads and adhesion to polystyrene surfaces. It is suggested that protein-carbohydrate interaction may be involved in bacterial phagocytosis but not in cell-cell adhesion.

INTRODUCTION

The life cycle of the cellular slime mold, Dictyostelium discoideum, has been fully described many times (see (3) for details). Here we need note only the following aspects.

1. In the vegetative or growth phase the cells are free-living individuals undergoing division and feeding on other microorganisms such as yeasts and bacteria. In the laboratory the cells usually are supplied with bacteria such as Escherichia coli B/r or Klebsiella aerogenes.

2. Strains such as Ax2 and Ax3, though still able to feed on bacteria, have been adapted for growth under axenic conditions.

3. Exhaustion of food leads to onset of the multicellular or developmental stage of the life cycle. The first morphologically recognizable step is the chemotactic aggregation stage. The cells move into an aggregation

center, guided by chemotactic signalling involving cyclic AMP (23) and by cell cohesion (26). From this stage onwards until the formation of the fruiting body, Dictyostelium is a multicellular organism.

Here I shall be concerned with the adhesion mechanism possessed by vegetative cells and retained until well into the developmental phase. We have recently identified a molecule, a cell surface glycoprotein of M_r 126,000 (gp126), which is wholly or partly responsible for adhesion of vegetative cells (6). I shall discuss the biological role of this molecule and possible mechanisms of cell adhesion. First, however, I will review the other D. discoideum contact sites.

In 1968 Gerisch (13) reviewed his evidence for the existence of two adhesion mechanisms being present on the surface of D. discoideum cells at the aggregation stage. This was confirmed by Beug et al. (1) and Beug, Katz, and Gerisch (2) who raised antisera against vegetative and aggregating cells and showed that the two adhesion mechanisms could be specifically inhibited by univalent antibody fragments. A summary of their conclusions is that one mechanism (contact sites A or CSA) resides in an antigen which appears at the onset of the aggregation stage. CSA-mediated adhesion is resistant to ethylene diamine tetra acetic acid (EDTA) treatment and responsible for end-to-end cohesion of cells in aggregation streams (Fig. 1). It is this mechanism which is involved in the guidance of cells in aggregation streams, a process called "contact following" by Shaffer (26). The other mechanism (contact sites B or CSB) is present on vegetative cells and persists at least until the aggregation stage where it appears to be responsible for side-to-side adhesion between aggregating cells. CSB-mediated cohesion is sensitive to EDTA.

Subsequently, CSA has been identified as a glycoprotein of M_r 80,000 (21). Its amino acid and carbohydrate composition have been analyzed (22), the protein has been partially sequenced, and monoclonal antibodies have been raised against it (24).

Two other adhesion molecules concerned with cell-cell adhesion at the aggregation stage or beyond have been subsequently identified. These are gp95 (27) and gp150 (12).

Identification of each of these molecules has depended upon the use of univalent antibody fragments to inhibit cell-cell adhesion. This powerful technique which was introduced by Gerisch and his colleagues (2) has now been successfully applied to vertebrate cells as well as slime molds

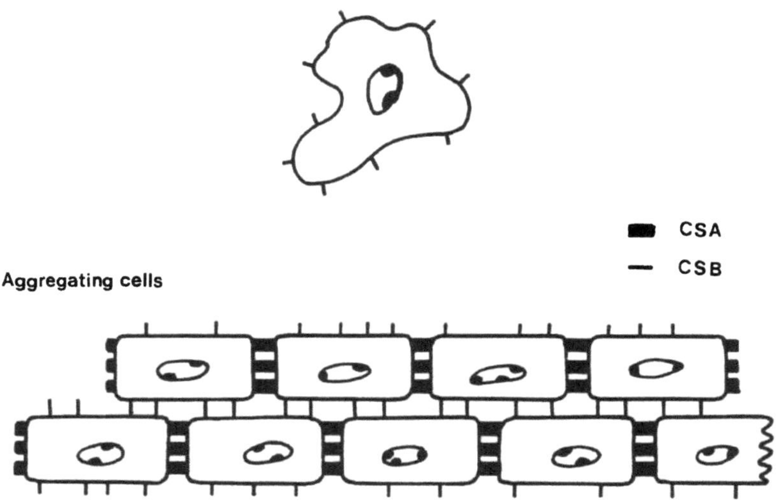

FIG. 1 - Diagram illustrating distribution contact sites on vegetative and aggregating D. discoideum cells suggested by EDTA-sensitivity.

(7, 11).

We have used the same technique to identify the vegetative cohesion molecule, gp126, in D. discoideum Ax2 cells. A summary of our method is given in Table 1. We have been able to obtain an antigen-specific polyclonal antibody against the vegetative contact site, gp126, and have used this to begin to investigate the biological role of the molecule and the mechanism involved in cell adhesion. Although we have no formal proof, we assume that gp126 and CSB are equivalent (6).

SOME CHARACTERISTICS OF D. DISCOIDEUM ADHESION PRIOR TO THE AGGREGATION STAGE

I shall now list certain observations on the cohesion of D. discoideum cells, which I believe to be characteristic of the vegetative mechanism. In most cases these observations have been made on cells which have been allowed to cohere or aggregate (not chemotactically) in agitated suspension in some type of simple salt or buffer solution after being removed from their vegetative growth conditions by washing away bacteria or axenic medium.

TABLE 1 - Identification of GP126 as contact site.

1) Raise antibodies against Ax2 cells.
2) Prepare Fab' fragments and test for cohesion-inhibiting activity.
3) Label cell surface with ^{125}I and immunoprecipitate antigens with inhibitory antibody. Obtain four or five labeled bands.
4) Absorb inhibitory Fab' with gel slices containing individual bands.
5) One slice, corresponding to M_r 126,000, absorbs activity.
6) Raise antibody against gp126 cut from gel. Fab' fragment of this inhibits cohesion.

1. The rate of cohesion increases with increasing ionic strength being zero in distilled water (4). I believe that this tells us that electrostatic repulsion can be a limiting factor in cohesion, but I do not think that it tells us anything positive about the mechanism of cohesion. It is doubtful whether this knowledge has any physiological significance. Having said that, I should point out that the cells become more sensitive to increasing ionic strength as they approach the aggregation stage (4), and this is correlated with a decrease in surface charge density (10, 19). It is possible that this change contributes the increased adhesiveness of cells at aggregation.

2. The vegetative cohesion mechanism is inhibited by EDTA (13), and this effect is not mediated by an increase in electrostatic repulsion (14). Divalent cations have many roles in cellular physiology, making it difficult to determine whether their role in cell adhesion is a direct one such as the maintenance of the functional configuration of an adhesion molecule or contact site. The latter must be considered a possibility in relation to the vegetative cohesion mechanism of Dictyostelium.

3. Cohesion is inhibited at 2°C (9). Our hypothesis which still seems reasonable was that this is because of inhibition of cell motility, preventing cells from flattening against each other and thereby increasing the area of mutual cohesion. This suggests that the adhesion mechanism is of low strength or affinity so that cell cohesion cannot be maintained by a point contact but requires interaction between cells over a large area.

4. Cohesion is inhibited by a low molecular weight factor present in the culture medium of axenic cells which have attained the stationary phase (18, 28-30). This factor gives complete, readily reversible inhibition

of cohesion and appears to be specific for vegetative cohesion since it, like EDTA, only partially inhibits the cohesion of aggregation stage cells. It is not a chelating agent but, unfortunately, it has not been chemically characterized. We suggested that it might be the terminal portion of one component of a ligand-receptor type cohesion mechanism (18) but have no good evidence to support this view.

5. Cohesion is inhibited by plasma membranes derived from vegetative cells (17). Again the effect appears specific because membranes from vegetative Ax2 cells completely inhibit cohesion of vegetative cells but only partially inhibit cohesion of aggregation-competent cells. These results seemed consistent with our ligand-receptor hypothesis proposed in relation to the low molecular weight factor. However, attempts to isolate the molecules involved from lithium di-iodo salicylate (LIS) extracts of plasma membrane proved unsuccessful.

6. Unlike other workers (20), we obtain no inhibition of vegetative cohesion with monosaccharides, including glucose, except at very high concentrations (> 100mM) (28). Furthermore, we have found that concanavalin A-binding extracts of plasma membranes are without effect on cell cohesion (Ellison and Garrod, unpublished). These results may suggest that cell surface carbohydrate is not directly involved in some type of ligand-receptor interaction binding mechanism between cells. On the other hand, such a mechanism is involved in phagocytosis of bacteria (see below).

7. Cohesion is completely and specifically inhibited by monospecific Fab' against gp126 at a concentration of 0.7 mg/ml (5, 6). It is not inhibited by control Fab' which binds to the cell surface.

BIOLOGICAL ROLE OF GP126

Why should vegetative D. discoideum cells possess a cohesion mechanism when they lead a solitary existence and have no requirement to stick to each other? I now believe that the answer to this question has been staring me in the face for several years. Let me first mention some simple observations. Actively growing log phase axenic cells stick to each other in axenic medium and cohere very rapidly when harvested and shaken in phosphate buffer (29). By contrast wild-type (NC-4) cells, when harvested from their agar plates whilst they are still in log phase, are poorly cohesive (8). The important difference between the two is the presence of bacteria on agar plates and absence of bacteria in axenic medium.

Following the method of Glynn (15), we have shown that anti-gp126 Fab' greatly inhibits the binding of radioactively labeled E. coli B/r to vegetative axenic amoebae (5). This result can be demonstrated by using fluorescently labeled bacteria (Fig. 2). Furthermore, immuno-precipitation with anti-gpl26 antibody shows that there is much gp126 on the surface of axenically-grown Ax2 cells, but very little on vegetative NC-4 cells and bacterially-grown Ax2 cells.

Our hypothesis concerning these facts is as follows. gpl26 is a receptor for bacterial phagocytosis (there are others). During phagocytosis the receptor is internalized with the bacteria, resulting in a diminution of the amount of gp126 on the cell. Thus NC-4 cells and bacterially-grown Ax2 cells have little gp126 on their surfaces while axenically-grown Ax2 cells have much. gp126 also mediates cell cohesion so that the low cohesiveness of vegetative NC-4 cells and the high cohesiveness of log phase axenic cells are directly related to the amount of gp126 on their surfaces (5) (Fig. 3).

gp126 appears to be a molecule with a dual function, in phagocytosis and in cell-cell adhesion. Is the latter role, the one we have been studying intensively, an incidental one, or does it have a role in development of D. discoideum? Bacterially-grown cells recover their cohesiveness very quickly (less than 2 h) after being removed from bacteria (8). gp126 is present on the cell surface well beyond the aggregation stage of the life cycle as demonstrated by immunoprecipitation and fluorescent antibody staining (unpublished observations). Anti-gp126 Fab' disrupts aggregation streams, frequently leaving end-to-end contacts (presumably mediated by CSA) (6). Cells can complete development in the presence of cohesion-inhibitory concentrations of gp126 Fab' but form tiny fruiting bodies in comparison with controls. The aggregates formed at the aggregation stage are correspondingly very small (5). We conclude that gp126 makes an important contribution to the cohesiveness of cells at the aggregation stage. Blocking it abolishes one of the cohesion mechanisms available to the cells at this stage, thereby reducing their cohesiveness. Since the aggregation streams are less cohesive, smaller aggregates are formed.

(It has been suggested (20) that CSB is a developmental trigger in the sense that CSB-mediated cohesion triggers the activation of developmentally important genes. That cells develop in the presence of inhibitory gp126 Fab' strongly contradicts this view.)

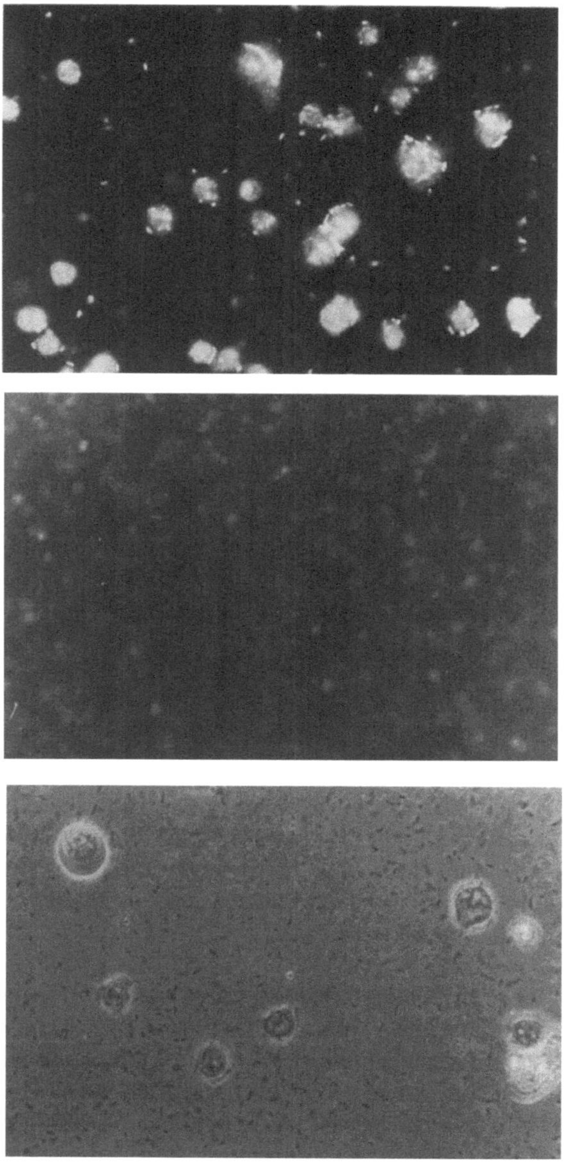

FIG. 2 - (a) Fluorescent E. coli B/r adhering to the surface of log phase Ax2 cells. (b) In the presence of anti-gp 126 Fab' bacteria do not bind to cells. (c) Same field as (b) by phase contrast showing the cells.

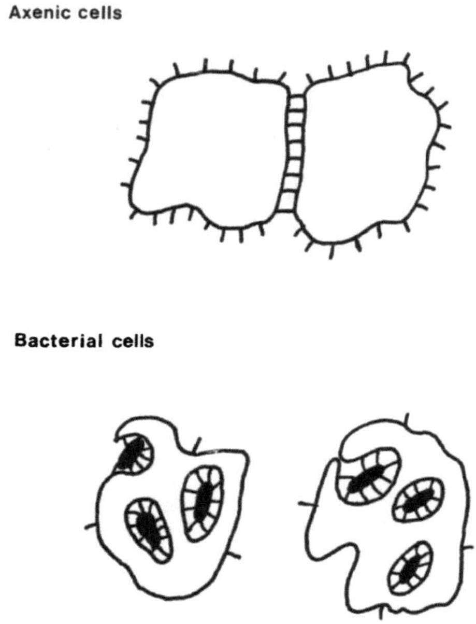

FIG. 3 - Suggested distribution of gp126 in vegetative axenic and bacterially-grown cells.

gp126 thus has a dual function. Its primary role is probably in phagocytosis, but it has been secondarily involved in the morphogenetic process. It may be speculated that the evolution of multicellularity may have involved the acquisition by unicellular organisms of a feeding receptor which could also mediate cell cohesion.

In proposing that cell-cell adhesion involves the same receptors as phagocytosis, the question arises as to whether slime mold cells eat each other. Indeed, cannibalism amongst slime mold amoebae has been recorded (16). It is probable that large amoebae eat small ones. When two amoebae of similar size adhere, the pressure to eat and be eaten must counteract each other, resulting simply in adhesion. This presents the social phase of the slime mold amoebae existence as a precarious one with the ever-present danger of being gobbled up by a larger colleague. A cannibalistic slime mold species, Dictyostelium caveatum, has recently been described (32). It will be interesting to know how this fits into the picture.

TABLE 2 - Inhibition of cell adhesion to glass and polystyrene by anti-gp126 Fab'.

	% cells attached after 30 min.	
	Glass	Polystyrene
Control	82.5	95
Fab' (1mg/ml)	1.7	25

MECHANISMS

Although four different contact sites are known for Dictyostelium, gp80 (CSA), gp95, gp126, and gp150, nothing is yet known about the mechanism which binds cells together. When purified gp80 is added to cells, it has no effect on cohesion (21). This may argue against a ligand-receptor interaction dependent upon, say, protein carbohydrate interaction (see (7)). The observation (see above) that cell surface glycoprotein fractions also have no effect on cohesion may suggest a similar conclusion. I feel that our investigations with gp126 may be starting to suggest some possibilities about how this molecule operates, though I stress that this may mean nothing in relation to other contact sites.

Following the work of Vogel et al. (31) (whose results will be considered in more detail below), we tested the effect of anti-gp126 Fab' on the phagocytosis of latex beads (1.1μm polystyrene latex, Sigma). Qualitative experiments show that the Fab' inhibits phagocytosis of latex (Fig. 4). Furthermore, the Fab' greatly diminishes the adhesion of Ax2 cells to the surface of polystyrene bacteriological dishes (Table 2).

Thus gp126 is involved in the binding of a) bacteria, b) latex particles, and c) other cells to D. discoideum amoebae. If we assume that the same mechanism is involved in all three types of adhesion, it seems most unlikely that a "ligand-receptor interaction" in the normally accepted sense of that phrase can be involved. However, the surface of latex particles is hydrophobic. Can it therefore be that gp126 is a receptor for hydrophobicity? Presumably it must be a receptor for some property which is common to the surfaces of bacteria, other cells, and latex particles. Are there hydrophobic regions on the surfaces of cells and bacteria which could form the basis for gp126-mediated binding?

Vogel et al. (31) selected D. discoideum mutants which were defective in phagocytosis. Such mutants also showed minimal cell-cell adhesion. Mutant amoebae were defective in phagocytosis of latex particles

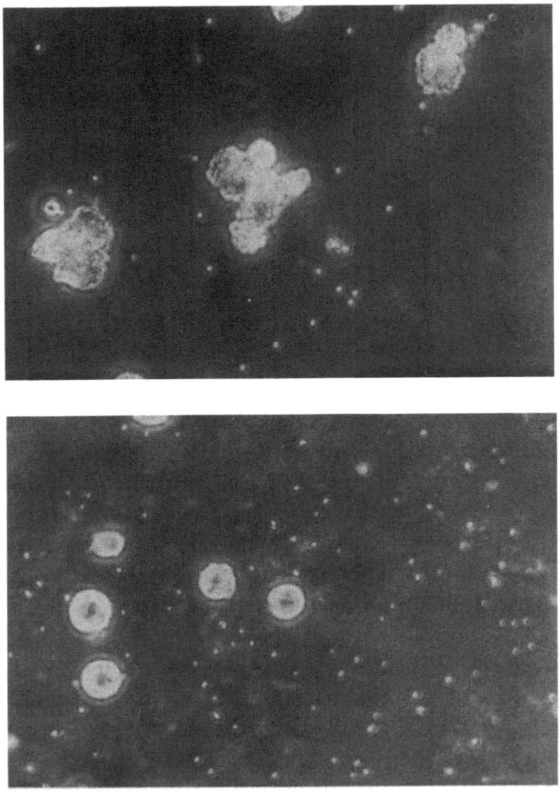

FIG. 4 - (a) 1.1μm latex spheres adhere to vegetative axenic cells in phosphate buffer. (b) In anti-gp126 Fab' binding of latex to cells is greatly reduced.

(inhibited by yeast extract) and of bacteria not having terminal glucose residues in their lipopolysaccharide. They therefore suggested the hypothesis that there are two classes of phagocytosis receptors on the surface of the amoebae: a) a carbohydrate receptor with glucose-binding activity and b) a "nonspecific" hydrophobic receptor. In the mutants the latter was thought to be defective, suggesting that it is the nonspecific receptor which is involved in cell-cell adhesion. I suggest that gp126 is the molecule in which this receptor activity resides. Unless there are terminal glucose residues on the surface of slime mold amoebae, it seems unlikely that the glucose receptor would also be involved in cell-cell adhesion. I suggest that the glucose receptor has a role only in phagocytosis. I further suggest that there may be other carbohydrate receptors on the cell surface which have a role in phagocytosis. (This possibility does not seem to be ruled out by the work on slime mold lectins, e.g., discoidins I and II which are possible candidates for such receptors (25). They were originally believed to play a role in cell-cell adhesion, but the evidence for this is not strong.)

In summary, I believe that gp126 is a specific receptor for a nonspecific, generalized, or widely distributed surface property. It seems to be crucial now to determine the structure of the molecule, particularly of its active site, and to determine how it binds the surface of other cells and bacteria.

Acknowledgements. Original work supported by the Science Research Council. Experimental work was done by J. Ellison and C. Chadwick.

REFERENCES

(1) Beug, H.; Gerisch, G.; Kempff, S.; Riedel, V.; and Cremer, G. 1970. Specific inhibition of cell contact formations in Dictyostelium by univalent antibodies. Exp. Cell. Res. 63: 147-158.

(2) Beug, H.; Katz, F.E.; and Gerisch, G. 1973. Dynamics of antigenic membrane sites relating to cell aggregation in Dictyostelium discoideum. J. Cell Biol. 56: 647-658.

(3) Bonner, J.T. 1967. The Cellular Slime Molds. Princeton, NJ: Princeton University Press.

(4) Born, G.V.R., and Garrod, D. 1968. Photometric demonstration of aggregation of slime mould cells showing effects of temperature and ionic strength. Nature 220: 616-618.

(5) Chadwick, C.M.; Ellison, J.E.; and Garrod, D.R. 1983. A dual role for Dictyostelium contact site B in phagocytosis and developmental size regulation. Nature 307: 646-647.

(6) Chadwick, C.M., and Garrod, D.R. 1983. Identification of the cohesion molecule, contact sites B, of Dictyostelium discoideum. J. Cell Sci. 60: 251-266.

(7) Edwards, J.G. 1983. The biochemistry of cell adhesion. Prog. Surf. Sci. 13: 125-196.

(8) Garrod, D.R. 1972. Acquisition of cohesiveness by slime mould cells prior to morphogenesis. Exp. Cell. Res. 72: 588-591.

(9) Garrod, D.R., and Born, G.V.R. 1971. Effect of temperature on the mutual adhesion of preaggregation cells of the slime mould,Dictyostelium discoideum. J. Cell Sci. 8: 751-765.

(10) Garrod, D.R., and Gingell, D. 1970. A progressive change in the electrophoretic mobility of preaggregation cells of the slime mould, Dictyostelium discoideum. J. Cell Sci. 6: 277-284.

(11) Garrod, D.R., and Nicol, A. 1981. Cell behavior and molecular mechanisms of cell adhesion. Biol. Rev. 56: 199-242.

(12) Geltosky, J.E.; Weseman, J.; Bakke, A.; and Lerner, R.A. 1979. Identification of a cell surface glycoprotein involved in cell aggregation in D. discoideum. Cell 18: 391-398.

(13) Gerisch, G. 1968. Cell aggregation and differentiation in Dictyostelium discoideum. Curr. Top. Dev. Biol. 3: 157-197.

(14) Gingell, D., and Garrod, D.R. 1969. Effect of EDTA on electrophoretic mobility of slime mould cells and its relationship to current theories of cell adhesion. Nature 221: 192-193.

(15) Glynn, P.J. 1981. A quantitative study of the phagocytosis of Escherichia coli by myxamoebae of the slime mould Dictyostelium discoideum. Cytobios 30: 153-166.

(16) Huffman, D.M., and Olive, L.S. 1964. Engulfment and anastamosis in the cellular slime molds (Acrasiales). Am. J. Botany 51: 465-471.

(17) Jaffé, A.R., and Garrod, D.R. 1979. Effect of isolated plasma membranes on cell-cell cohesion in the cellular slime mould. J. Cell Sci. 40: 245-256.

(18) Jaffé, A.R.; Swan, A.P.; and Garrod, D.R. 1979. A ligand-receptor model for the cohesive behaviour of Dictyostelium discoideum axenic cells. J. Cell Sci. 37: 157-167.

(19) Lee, K.-C. 1972. Cell electrophoresis of the cellular slime mould Dictyostelium discoideum. II. Relevance of the changes in cell surface charge density to cell aggregation and morphogenesis. J. Cell Sci. 10: 249-265.

(20) Marin, F.T.; Goyette-Boulay, M.; and Rothman, F.G. 1980. Regulation of development in Dictyostelium discoideum III. Carbohydrate-specific intercellular interactions in early development. Dev. Biol. 80: 301-312.

(21) Müller, K., and Gerisch, G. 1978. A specific glycoprotein as the target site of adhesion blocking Fab in aggregating Dictyostelium cells. Nature 274: 445-449.

(22) Müller, K.; Gerisch, G.; Fromme, I.; Mayer, H.; and Tsugita, A. 1979. A membrane glycoprotein of aggregating Dictyostelium cells with the properties of contact sites. Eur. J. Biochem. 99: 419-426.

(23) Newell, P.C. 1977. Aggregation and cell surface receptors in cellular slime molds. In Microbial Interactions: Receptors and Recognition, ed. J.L. Reissig, Series B, vol. 3, pp. 3-57. London: Chapman and Hall Ldt.

(24) Ochai, H.; Schwarz, H.; Merkl, R.; Wagle, G.; and Gerisch, G. 1982. Stage-specific antigens reacting with monoclonal antibodies against contact site A. A cell-surface glycoprotein of Dictyostelium discoideum. Cell Differ. 11: 1-13.

(25) Rosen, S.D., and Barondes, S.H. 1978. Cell adhesion in the cellular slime molds. In Specificity of Embryological Interactions. Receptors and Recognition, ed. D.R. Garrod, Series B, vol. 4, pp. 233-264. London: Chapman and Hall Ltd.

(26) Shaffer, B.M. 1962. The Acrasina. Adv. Morphogen. 2: 109-182.

(27) Steinemann, C., and Parish, R.W. 1980. Evidence that a developmentally regulated glycoprotein is target of adhesion-blocking Fab in reaggregating Dictyostelium. Nature 286: 621-623.

(28) Swan, A.P. 1978. An inhibitor of cell cohesion from Dictyostelium discoideum. Ph. D. Thesis, University of Southampton.

(29) Swan, A.P., and Garrod, D.R. 1975. Cohesive properties of axenically grown cells of the slime mould Dictyostelium discoideum. Exp. Cell Res. 93: 479-484.

(30) Swan, A.P.; Garrod, D.R.; and Morris, D. 1977. An inhibitor of cell cohesion from axenically grown cells of the slime mould, Dictyostelium discoideum. J. Cell Sci. 28: 107-116.

(31) Vogel, G.; Thilo, L.; Schwarz, H.; and Steinhart, R. 1980. Mechanism of phagocytosis in Dictyostelium discoideum: Phagocytosis is mediated by different recognition sites as disclosed by mutants with altered phagocytotic properties. J. Cell Biol. 86: 456-465.

(32) Waddell, D.R. 1982. A predatory slime mould. Nature 298: 464-466.

Microbial Adhesion and Aggregation, ed. K.C. Marshall, pp. 351-371. Dahlem Konferenzen 1984. Berlin, Heidelberg, New York, Tokyo: Springer-Verlag.

Consequences of Aggregation

B. Atkinson
The Brewing Research Foundation
Nutfield, Surrey RH1 4HX, England

Abstract. Microbial aggregation, i.e., the formation of flocs and films, is considered in the context of biological process engineering. Special attention is given to the microbial artefacts that can be achieved by adhesion or flocculation. The significance of these artefacts is explored in relation to the development of biological reactors and solid-liquid separation systems.

INTRODUCTION

For present purposes, aggregation will be interpreted as the accumulation of organisms either on surfaces or as flocs. In addressing the consequences of such aggregation, it is necessary to identify the areas of science, technology, and their applications around which consideration will be centered. Furthermore, it is necessary to recognize that the consequences of aggregation can involve technical opportunities as well as technical difficulties requiring technical solutions. Here the context is biological process engineering - the extent to which the consequences of aggregation in this area are also relevant to other areas is a matter for exploration elsewhere.

Given the above, there is the matter of emphasis to be considered, characterized here by the following proposition: From an awareness of the occurrence of adhesion or flocculation and the ubiquity of the phenomena, together with some knowledge of the scientific principles involved, what possibilities exist for the invention of new process equipment and for process innovation which will progress biological process engineering for the benefit of the community?

In developing the views demanded by these considerations, a degree of speculation is clearly necessary, if only as a guide to the very considerable research and development efforts being undertaken under the heading "biotechnology." The fact that much of the speculation will turn out to be in error is inevitable and has to be accommodated. It has to be recognized that turning speculation into process equipment and processes takes considerable will, dedication, and time, and it is not to be undertaken lightly.

The successful exploitation of microbial aggregation within a process places a heavy demand on the local conditions throughout the process, particularly with regard to the physical environment and the fluid mechanics (shear). Both process and equipment development in these circumstances require a "green-fingered" understanding of the principles of microbial growth, adhesion and flocculation, and of the interaction of particles and fluids. It will be suggested that the beneficial consequences and potential of microbial aggregation are large, but that present research effort needs to be adjusted to "open the door." Unfortunately, for a variety of reasons, research-based exploratory development studies are very difficult to justify presently in all technological sectors, whether university, research organization, or plant supplier, and, perhaps most surprisingly in the places where invention and innovation usually occur, that is, within the process operating companies.

PHYSICAL PROPERTIES OF MICROBIAL AGGREGATES (see (1-4))

Types

Flocs

Figures 1 - 3 illustrate the variety of macromorphologies encountered when organisms agglomerate "naturally." The particle shapes range from spherical to ragged. Both size and shape are affected by shear.

Films on solid supports

Figure 4 provides an idealized sketch of a microbial film "attached" to a support surface and exposed to fluid shear forces. In Fig. 4 the film thickness (L) is steady, reflecting an equilibrium between film growth and the removal of organisms by the action of the fluid. Organisms in

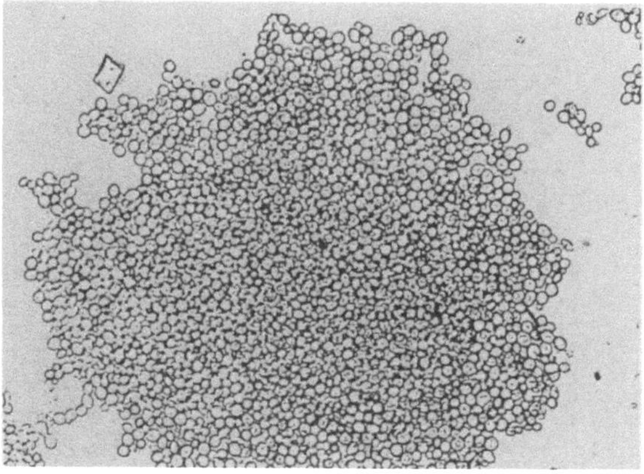

FIG. 1 - Yeast flocs taken from a tower fermenter.

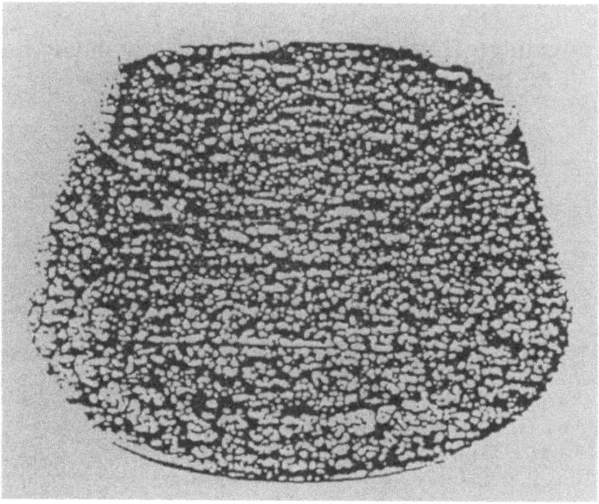

FIG. 2 - Flocs of Aspergillus niger prepared by the batch culture (4 days) of a spore inoculum.

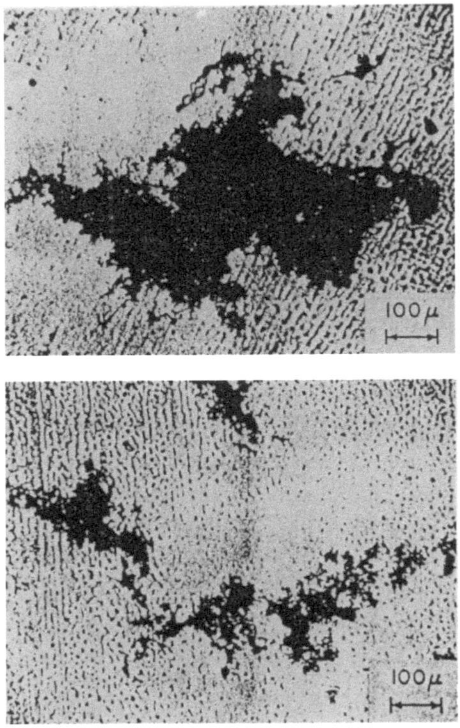

FIG. 3 - Activated sludge flocs (low shear, top; high shear, bottom).

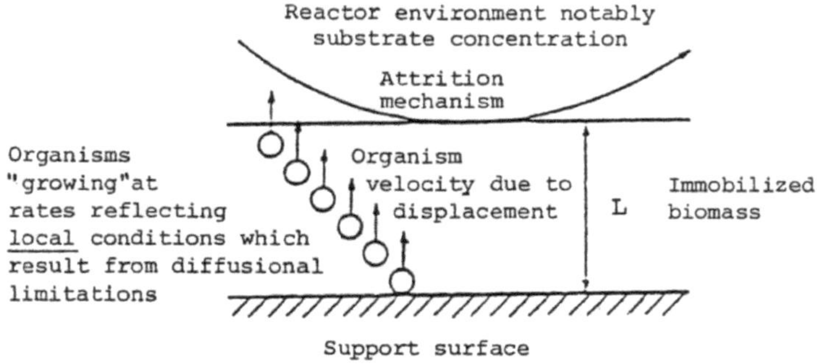

FIG. 4 - Idealized view of a microbial film.

TABLE 1 - Occurrence and "size" of microbial films.

Description	Thickness (mm)
Uncontrolled zoogloeal film typical of mixed cultures	0.2 - 4
Zoogloeal film subject to mechanical or hydrodynamic control	0.07 - 0.2
Pure cultures	0.001- 0.01
Casual deposition	0.001

the depths of the film grow at a rate reflecting the local conditions and, in so doing, displace adjacent organisms towards the free surface of the film. The occurrence of such films is summarized in Table 1 and it is suggested that the films form most readily and develop most prolifically under mixed culture conditions.

Biomass support particles (see (5))
Figure 5 shows a number of reticulated structures with porosities in excess of 80% and filled with microbial growth. The particles are prefabricated, fill under fermentation conditions, and have a preselected shape, size, and density. Microbial overgrowth is controlled by combinations of particle-particle contact and fluid shear.

Gel entrapment
Concentrated microbial suspensions can be entrapped within the matrix of organic polymers such as alginates, polyacrylamide, etc., and formed into spheres, cylinders, and sheets of a required size.

Size and shape
The size and shape characteristics of various types of microbial aggregate are summarized in Tables 2 and 3.

Density
Freely suspended microbial flocs and gel particles have densities similar to those of aqueous solutions. In contrast, biological artefacts based

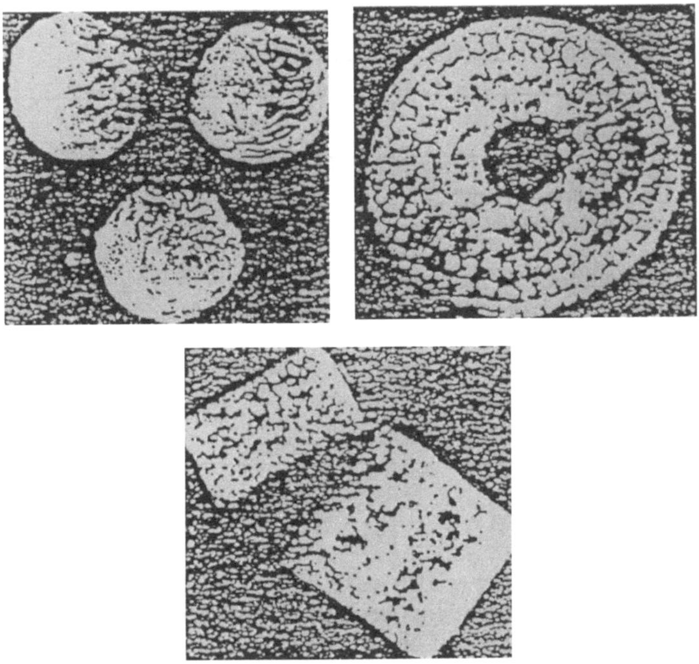

FIG. 5 - Biomass support particles. (a) Stainless steel wire spheres (6mm diameter). (b) Polypropylene toroids (53mm overall diameter with 20mm diameter torus). (c) Surface and side view of reticulated polyester foam particles (25 x 25 x 10mm).

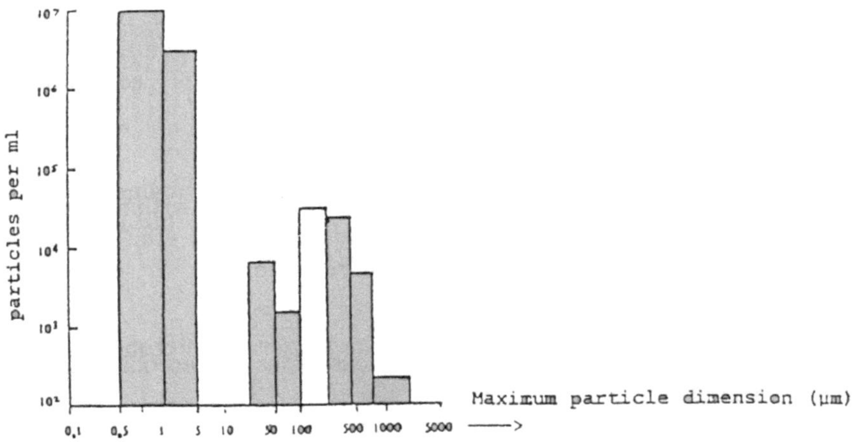

FIG. 6 - Size distribution of activated sludge flocs in a stirred tank (high shear).

TABLE 2 - Geometric characteristics of biological particles and artefacts.

	Freely suspended microbial flocs	Biomass support particles and gels	Microbial films
Size	Effective size unknown (to be determined experimentally; see Table 3 for typical values)	Predetermined by support	Thickness unknown (to be determined experimentally)
Size distribution	See Fig. 6	Supports of uniform size	Depends on fermenter configuration and operation, e.g., tubular fermenters with low fluid shear contain a range of thickness
Shape	Irregular (see Figs. 1-3)	Regular, e.g., spherical, cuboid, or sheets	Follows shape of support, e.g., planar or near-spherical

TABLE 3 - Characteristic size of flocs.

Microorganism	Floc size and range (diam. in mm)
Agaricus blazei (mushroom)	2.8 - 25
Aspergillus niger	0.2 - 2.0
Brewers' yeast	
Physically limited	< 2
Fermentation-limited	< 13
Mixed bacterial cultures	0.025 - 5

on solid particulate supports or biomass support particles have a range of densities depending on the support density, support porosity, and microbial film thickness (see Figs. 7 and 8). It appears that in practice the L/r_s values for solid supports and pure cultures are very small (<< 1), whereas the values for mixed cultures can range up to 10. The porosities of the biomass support particles currently in use range from 0.80 to 0.98.

REACTION PROPERTIES OF MICROBIAL AGGREGATES

It is suggested that the overall rate of "reaction" of a biological particle depends upon size, but that diffusion limitations can be either advantageous or disadvantageous (Fig. 9).

Diffusional limitations are usually discussed in terms of substrate ingress. There are, of course, diffusional limitations on the egress of the products, which may affect the rates of reaction within the aggregates. Examples include product inhibition by ethanol and carbon dioxide, and both

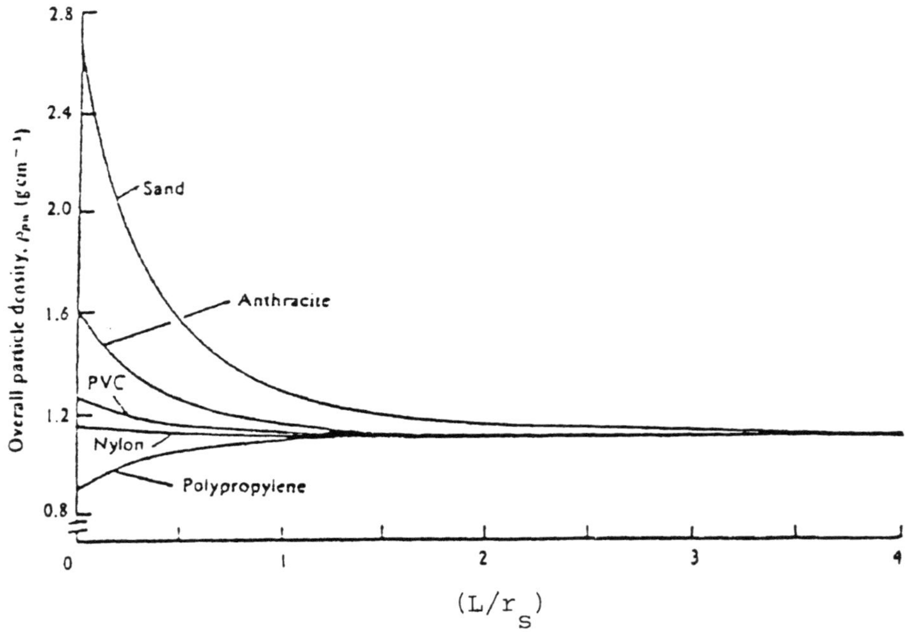

FIG. 7 - Overall densities of solid support particles with attached biomass of wet density 1.1g cm^{-3}. L: microbial film thickness, r_s: support particle radius.

advantageous and disadvantageous effects due to the local pH values within the aggregates. In addition, the production of sparingly soluble gases such as carbon dioxide and nitrogen can result in gas evolution within aggregates, causing their breakup.

Equations 1 - 4 are a series of empirical relationships often used to describe microbial kinetics. These equations do not contain particle size explicitly, but since they all can be applied to the same microbe-substrate system, although in different aggregate forms, there is the implication that the size and shape of the particle has a significant effect.

$$R = \frac{a\,S^*}{b + S^*} \tag{1}$$

$$R = f \tag{2}$$

$$R = g\,S^* \tag{3}$$

$$R = h\,(S^*)^{1/2} \quad , \tag{4}$$

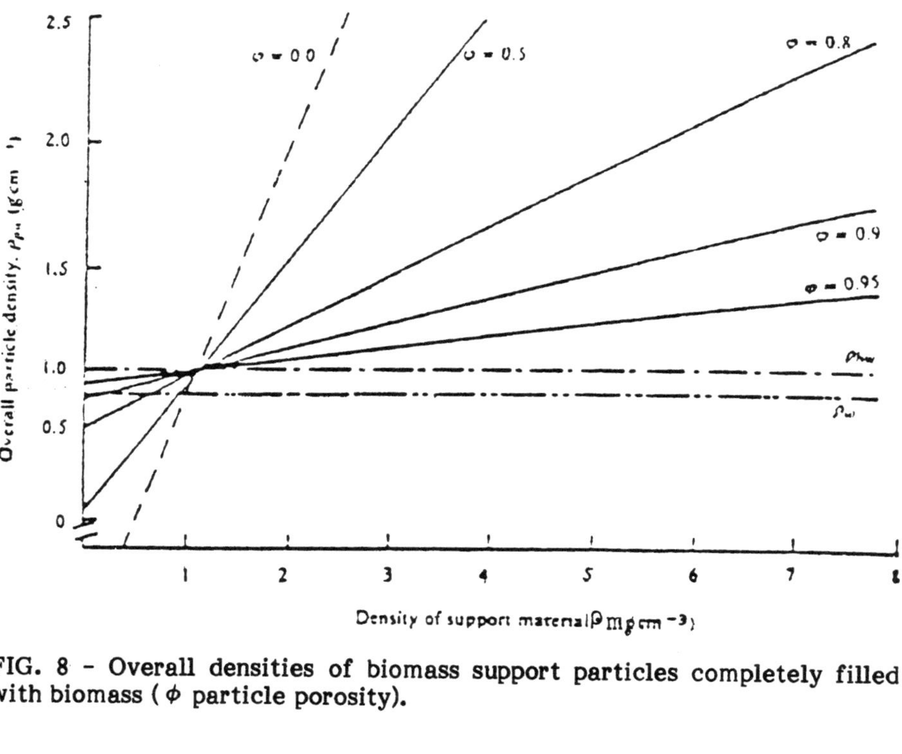

FIG. 8 - Overall densities of biomass support particles completely filled with biomass (ϕ particle porosity).

where R is the specific rate of reaction; S* is the substrate concentration in the liquor; and a, b, f, g, and h are empirical coefficients.

It appears that, for the various types of microbial aggregate under consideration, i.e., flocs, films on solid supports, biomass support particles, and gel particles, the application of a given conceptual model of diffusion with biological reaction will lead to the same algebraic form of equation when applied to each aggregate in turn. This allows graphical representations of generalized algebraic equations to be used to represent the diffusion-limited microbial kinetics of immobilized microorganisms independent of the method of immobilization (3).

Disadvantageous Diffusional Limitations

With Monod kinetics (Eq. 1), diffusional limitations are likely to be disadvantageous because the lower substrate concentrations within the microbial aggregate lead to reduced rates of reaction when compared with the dispersed state, where all the organisms are exposed to the higher fermenter substrate concentrations. Figure 10 illustrates the typical size effects to be anticipated when biological artefacts are used

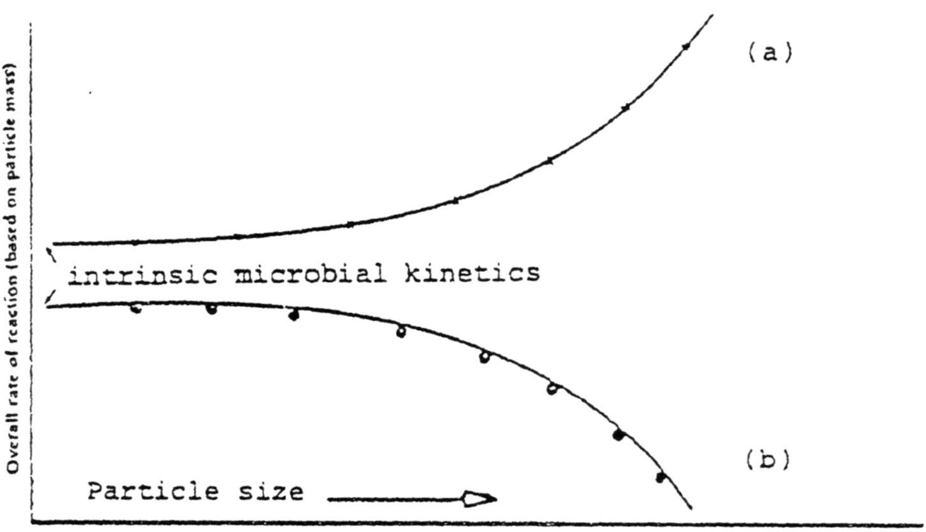

FIG. 9 - Effect of reduced particle size on the overall rate of reaction of a biological particle (a) when diffusional limitations are advantageous (b) when diffusional limitations are disdvantageous.

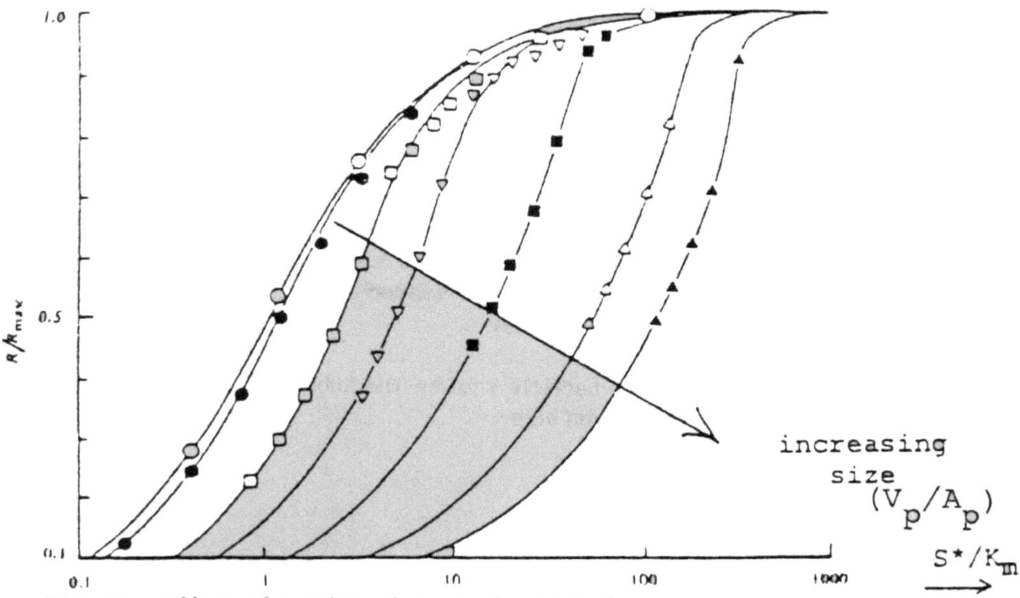

FIG. 10 - Effect of particle size on microbial kinetics.

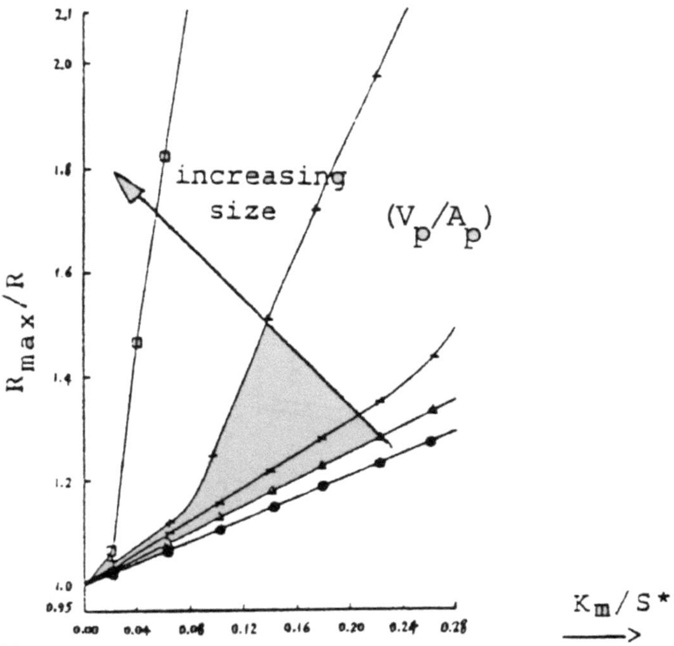

FIG. 11 - Lineweaver-Burk plot of data in Fig. 10.

with a given microbe–substrate system. The data contained in Fig. 10 were obtained from the biological rate equation which describes diffusion-limited Monod kinetics (3). This equation expresses the relative specific rate of substrate uptake (R/R_{max}) versus fermenter substrate concentration (S*) with the size of the microbial aggregate (V_p/A_p) as a parameter. The results of plotting data such as that in Fig. 10 in the Lineweaver-Burk format are illustrated in Fig. 11, and the dangers of using this format for biological particles are clear. Figure 12 further illustrates the effect of size on the Lineweaver-Burk format, this time for a range of particle size distributions.

Figure 13 shows the effects of particle size on the substrate concentration in the center of a particle of given size.

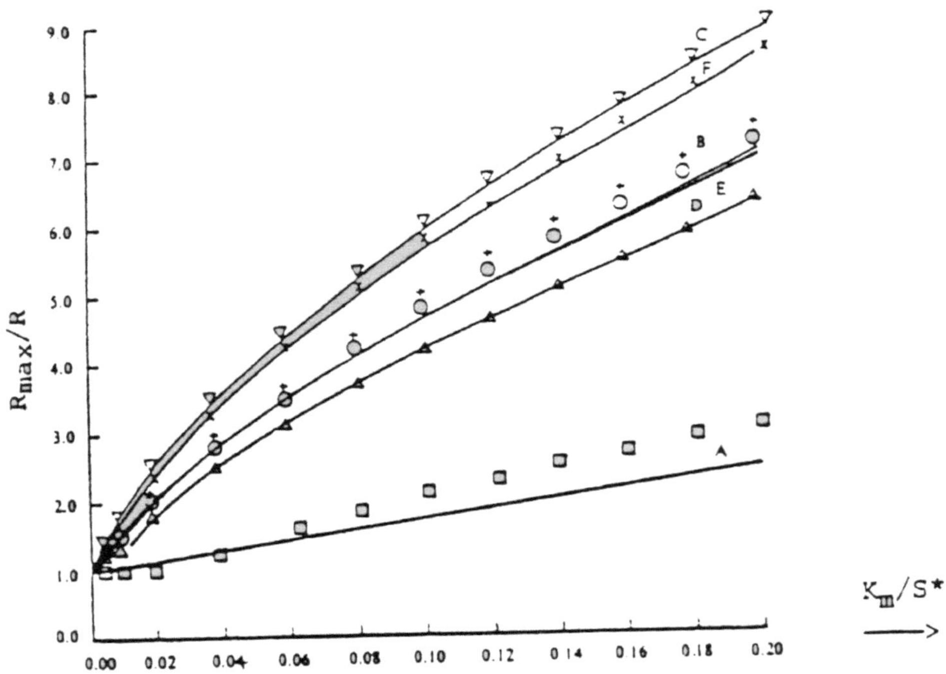

FIG. 12 - Lineweaver-Burk plot for a variety of assumed particle size distributions (A-F).

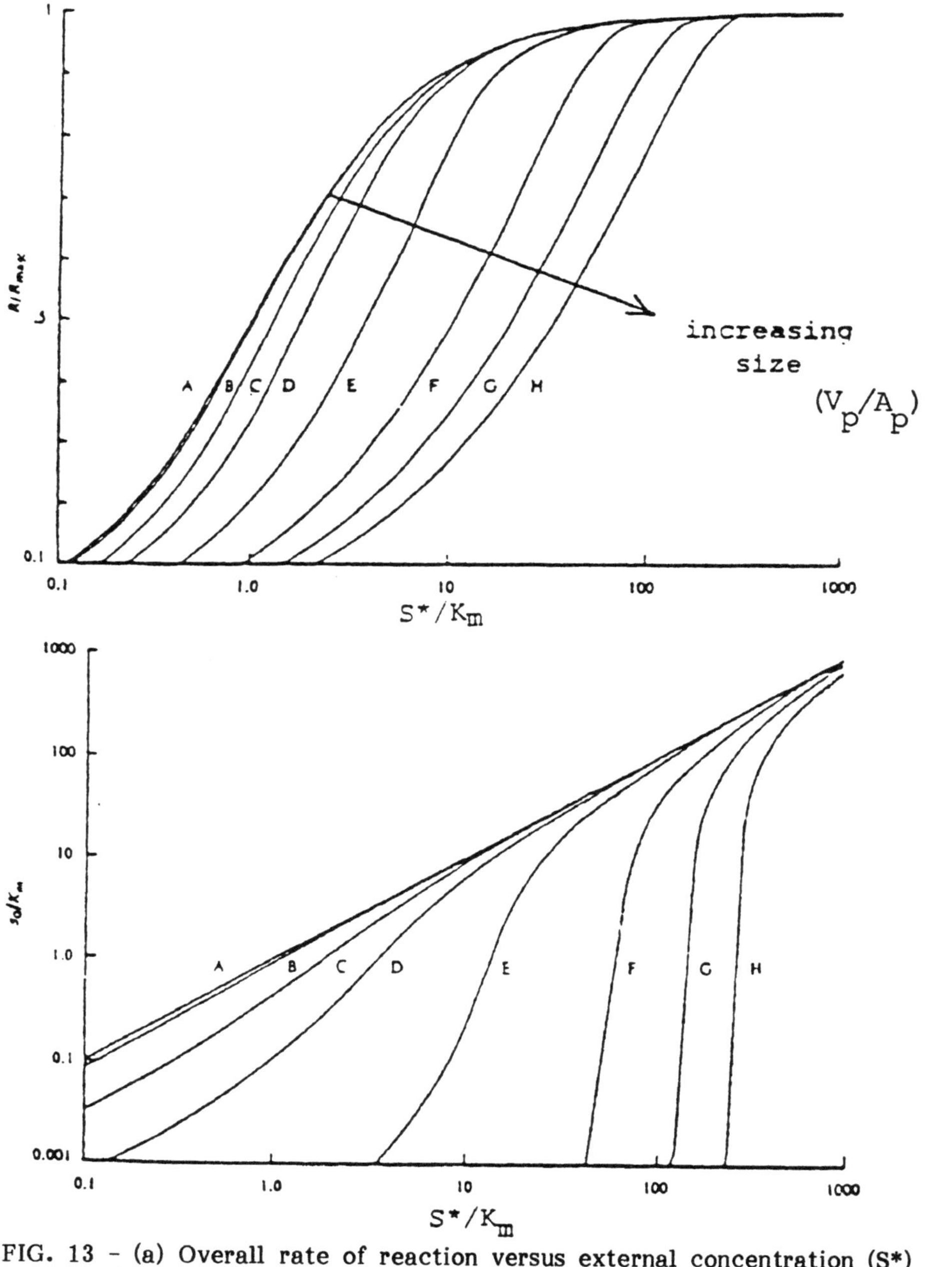

FIG. 13 - (a) Overall rate of reaction versus external concentration (S*) with particle size (V_p/A_p) as a parameter. (b) Center-line concentration (S_O) versus external concentration (S*) with particle size as a parameter.

Advantageous Diffusional Limitations

Monod kinetics modified to account for substrate inhibition can be described by:

$$\frac{R}{R_{max}} = \frac{S^*/K_m}{1 + S^*/K_m + \alpha^2 (S^*/K_m)^{1/2}} , \quad (5)$$

where $\alpha = (K_m/K_i)^{1/2}$.

Figure 14 shows the effect of diffusional limitations when the intrinsic kinetics are described by Eq. 5. The features to note in Fig. 14 are: a) at "low" substrate concentrations the diffusional limitation leads to a reduced rate of reaction, b) at "high" substrate concentrations large particles are superior to small particles, and c) for a given particle size there is an optimum substrate concentration corresponding to a "maximum" rate of reaction.

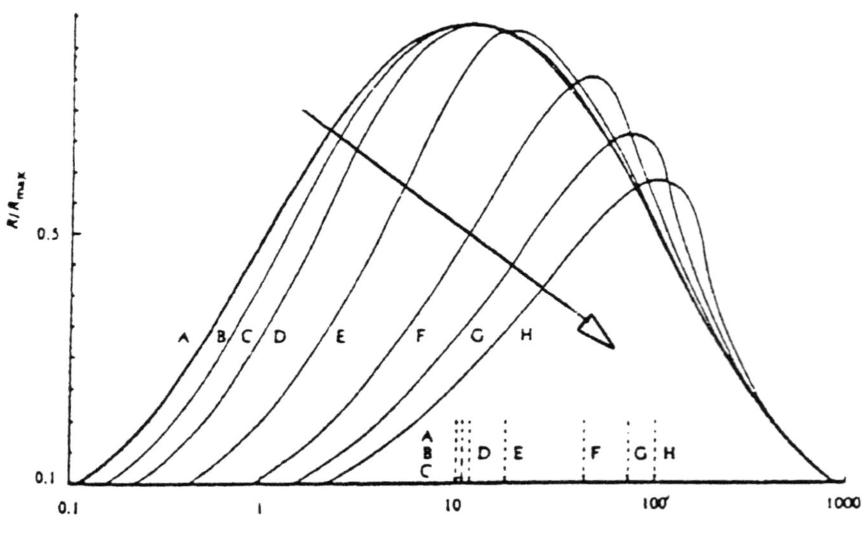

FIG. 14 - Substrate-inhibited kinetics - effects of particle size on the overall rate of substrate uptake for diffusion-limited spherical particles (α = 0.1 in Eq. 5; —, optimum S^*/K_m; ⟶ , increasing size (V_p/A_p)).

The reasons for this situation lie in the fact that as the concentration falls with distance within the particle, the effects of substrate inhibition become locally less significant.

IMPLICATIONS

Reactors (see (3))

Some of the advantages, in addition to increased throughout, of the use of yeast flocs in a continuous tower fermenter are summarized in Table 4. It can be seen from Table 4 that there are clear advantages in the use of microbial particles and artefacts. Unfortunately, freely suspended flocs fail to meet the general needs of large-scale biological process engineering because organisms, even when they flocculate, do not usually do so in a convenient form (size, shape, and density).

Microbial films provide a way of overcoming density limitations (see Fig. 7) but suffer from the fact that adhesion occurs only to a limited extent (see Table 1) and that the film thickness cannot be preselected.

It follows from the above that biomass support particles and gel entrapment provide the most flexible basis for the development of biological artefacts. It appears that any organism, independent of its adhesion or flocculation characteristics, can be made to occupy a biomass support particle (5) or can be gel-entrapped.

The volumetric rate of reaction ($ML^{-3}T^{-1}$) in a fermenter is given by:

$$R_v = R\, n_p\, m_p, \tag{6}$$

where R is the specific overall rate of reaction by a biological artefact, n_p is the number of particles per unit fermenter volume, and m_p is the biomass holdup in a particle.

It is clear from Eq. 6 that the productivity of a reactor depends upon diffusion-limited microbial kinetics (since this determines R), the physical arrangement of the reactor (since this determines n_p), and the quantity of biomass (m_p) which can be accumulated by a single particle under given conditions. Since m_p depends upon a balance between microbial growth and attrition, it can either be steady, i.e., when these factors are in equilibrium, or growth can accumulate when the rate of microbial attrition is insufficient to compensate for the rate of microbial growth. Thus, it can be anticipated that the magnitude of m_p is affected by those factors influencing growth, for example, fermenter substrate concentration and particle size, as well as those factors affecting attrition, for example, particle-particle contacts and fluid shear.

TABLE 4 - Characteristics of tower fermenters.

Technical improvement	Achievement	Comment
Increased biomass concentration (i.e., holdup)	Yes	Use of flocculated micro-organism in a fluidized bed
Predetermined biomass	Partially over restricted range	Use of knowledge of liquid-phase fluidization
Biomass concentration (i.e., holdup) independent of process throughput	No	Bed expansion and elutriation depend on process throughput
Predetermined biomass size and shape	Partially	By selection of strains which give a restricted range of size and shape, although there will be a maximum size for a given strain
Improved biomass recovery	Yes	Because flocculant strains are used
Improved yield coefficients		No information
Biomass/sludge "age" independent of process throughput	Partially	Due to retention of biomass by flocculation
Multiple reactions in a single reactor due to local variations in environment and/or species	Yes	Because liquid phase is in plug flow and carbon compounds are attacked sequentially
Retention by fluid phase of dilute, suspension-free, aqueous properties, especially viscosity	Yes	Because liquid and "solid" phases are segregated

The magnitude of n_p depends upon the reactor arrangement, particle-fluid properties, and the interaction between particles and fluid. This is an area for independent study. The magnitude of m_p depends largely upon the particle type. In use, gel-type particles may rupture or overgrow, and biomass support particles may overgrow.

The consequences of these various factors is that appropriate reactor configurations have to be developed for each particle type. A basic requirement for a successful configuration lies in "robustness," i.e., the ability to operate for a long period (months) without, other than routine, cause for concern as regards blockage and particle deterioration (physically or biologically), while maintaining a steady conversion efficiency.

As an example of the above considerations, Fig. 15 shows a fermenter configuration developed for the use of stainless steel biomass support particles containing a fungal organism. The spouted-bed arrangement prevents particle overgrowth by providing a high shear zone through which all the particles have to pass at frequent intervals. In practice, the particle movement is achieved by operation of the reactor on a recycle basis, which allows any required level of microbial attrition to be used independently of the feed-flow rate to the fermenter.

Solid-liquid Separations (see (3, 4))
Biological artefacts of the type under consideration provide opportunities for improved biomass recovery procedures as listed in Table 5.

TABLE 5 - Biomass recovery.

Type	Procedure
Flocs	Sedimentation
Films on solid supports	Decantation and stripping
Biomass support particles	Decantation and squeezing (see Fig. 16)

Processes

Biological artefacts with a density significantly different from that of aqueous solutions can be moved throughout a process by well established procedures based upon 'particle-fluid mechanics. This allows for the possibility that such artefacts could be moved between "reaction" zones as required, e.g., from a "growth" zone to a "conversion" zone. The process configuration shown in Fig. 17 represents a development in this direction. The two reactors contain sand particles with different microbial species "attached"; these species are in "equilibrium" with the local chemical environment and this prevents cross-contamination.

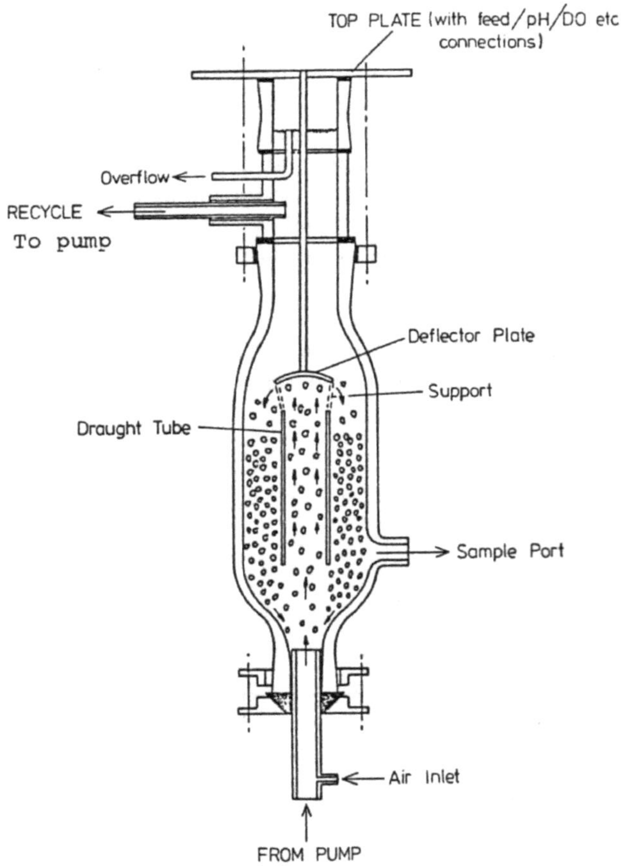

FIG. 15 - Spouted bed fermenter.

FIG. 16 - Biomass being removed from a support particle.

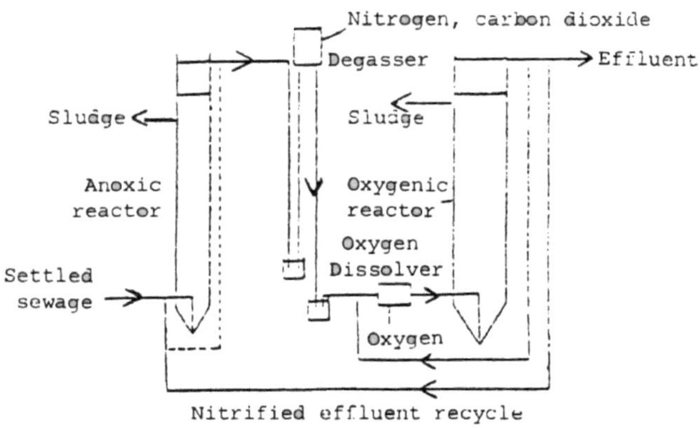

FIG. 17 - Two-reactor fluidized-bed system for complete sewage treatment, including denitrification, using the carbon present in settled sewage (4).

GENERAL REFERENCES

(1) Atkinson, B., and Daoud, I.S. 1976. Microbial flocs and flocculation. In Advances in Biochemical Engineering, eds. T.K. Ghose, A. Fiechter, and N. Blakeborough, vol. 4. Berlin: Springer-Verlag.

(2) Atkinson, B., and Fowler, H.W. 1974. The significance of microbial film in fermenters. In Advances in Biochemical Engineering, eds. T.K. Ghose, A. Fiechter, and N. Blakeborough, vol. 3. Berlin: Springer-Verlag.

(3) Atkinson, B., and Mavituna, F. 1983. Biochemical Engineering and Biotechnology Handbook. London: Macmillan.

(4) Cooper, P.F., and Atkinson, B. 1981. Biological Fluidised Bed Treatment of Water and Wastewater. Chichester: Ellis Horwood.

SPECIFIC REFERENCE

(5) Atkinson, B.; Black, G.M.; Lewis, P.J.S.; and Pinches, A. 1979. Biological particles of a given size, shape and density for use in biological reactors. Biotechn. Bioeng. 21: 193.

NOMENCLATURE

Symbol	Description	Dimensions
a, b, f, g, h	empirical coefficients (defined in Eqs. 1 - 4)	various
A_p	external area of a wet microbial aggregate	L^2
K_i	substrate inhibition coefficient (defined in Eq. 5)	ML^{-3}
K_m	Monod coefficient	ML^{-3}
L	thickness of a microbial film	L
m_p	biomass associated with a particle (defined in Eq. 6)	M
n_p	number of particles per unit bed volume (defined in Eq. 6)	L^{-3}

Symbol	Description	Dimensions
r_s	diameter of a solid support particle	L
R_{max}	maximum specific rate of carbon conversion	T^{-1}
R_V	overall rate of carbon conversion per unit reactor volume	$ML^{-3}T^{-1}$
S_o	substrate concentration in the center of a biomass support particle	ML^{-3}
S^*	substrate concentration adjacent to a microbial aggregate	ML^{-3}
V_p	wet volume of a microbial aggregate	L^3
α	parameter defined in Eq. 5	-
ρ_{bw}	density of drained wet biomass within a biomass support particle	ML^{-3}
ρ_m	density of matrix material of a biomass support particle	ML^{-3}
ρ_{pw}	overall density of a support particle with associated wet biomass	ML^{-3}
ρ_w	density of water	ML^{-3}
ϕ	porosity of a biomass support particle	-

Microbial Adhesion and Aggregation, ed. K.C. Marshall, pp. 373-393. Dahlem Konferenzen 1984. Berlin, Heidelberg, New York, Tokyo: Springer-Verlag.

Microcolony Formation and Consortia

P. Hirsch
Institut für Allgemeine Mikrobiologie
Universität Kiel (Biozentrum)
2300 Kiel, F.R. Germany

Abstract. Pure and mixed microcolonies (MC), as well as permanent consortia of microorganisms, are discussed. The contact of cells within MC's varies from strand connections remaining after division to postdivisional adhesion and common polymer encasements of fairly distant cells. The amount and location of polymer produced as well as morphogenetic events determine the final colony shape. MC formation may also be influenced by chemotactic attraction of the organisms in question by a common polymer. Properties of MC's and their ecology are discussed briefly. Metabolic activity is homogeneously distributed only among members of very young colonies. Syntrophic relationships in mixed cultures may eventually lead to the formation of consortia, where co-evolution is assumed to occur. Microcolonies and consortia offer their component members fairly stable conditions and protection against environmental stress. The protective role of the common polymer is discussed.

INTRODUCTION

After division of microorganisms, the daughter cells often stay and multiply together and thus form a "colony." Contrary to an aggregate, the cells in a pure culture colony represent the offspring of initially one cell; they can be considered as a clone. The present paper deals with "microcolonies," which are here defined as the microscopically small initial stages of a clone colony where all or most of the cells can still be recognized, i.e., grow one-layered. Motility of daughter cells, especially in the aquatic habitat, may prevent the formation of microcolonies (MC's), and it has been assumed that many aquatic microorganisms normally live dispersed and that their colonial aggregates may be an artifact caused by the observation techniques. Microcolony formation on surfaces may thus be considered as a passive process, a failure of the daughter cells to disperse. Swarms of myxobacteria, in contrast, stay together due to active processes.

Microcolony formation (MCF) is best observed in the laboratory and with solid media, although MC's do occur in nature on most surfaces. Whereas laboratory MC's of pure cultures consist of only one species (a clone), natural MC's may also be formed by mixed populations with varying degrees of interactions. Such syntrophic, mixed cultures may consist of two or more microorganisms with similar ecological niches but different physiologies that complement each other. If the syntrophic partners stay together more or less permanently, with normal cell-to-cell contact, we talk about "consortia." The present paper describes various types of microcolonies and consortia and their ecological consequences. Attempts will be made to understand the processes that lead to their formation.

METHODS FOR THE OBSERVATION OF MICROCOLONIES AND CONSORTIA

Microcolony formation in the laboratory is usually observed with slide cultures which, of course, need to be adapted to the needs of the organisms under study (Figs. 1 to 3). Development of colonies may be followed with cinemicrographic techniques (41). Another technique, especially useful for actinomycetes, consists of inoculating the edges of a ditch in an agar plate and placing a coverslip over the inoculated edges so that it bridges both sides. The organisms growing out of the agar attach to the lower side of the coverslip and may then be viewed by direct microscopy. Microcolony surfaces can be viewed with a scanning electron microscope with or without prior in situ fixation of the colony with glutaraldehyde/OsO_4, dehydrating, and critical point drying (8). The natural arrangement of cells in an MC can be studied by the following technique. The MC is cut from the agar plate as a cube and placed upside down onto a clean glass slide. The slide is then heated from below to fix and attach the MC's edge cells to the glass surface. Finally, the agar block is carefully removed without lateral dislocation, the slide is dried and remoistened for microscopic observation (Fig. 4).

Natural microcolony formation in soil or water is studied best by exposing

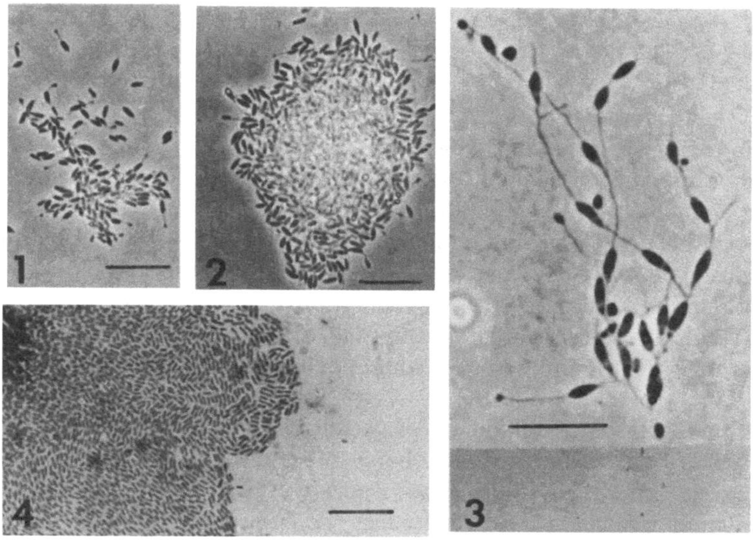

FIGS. 1, 2 - Slide culture microcolonies of Hyphomicrobium B-522. FIG. 3 - MC of Rhodomicrobium sp. FIG. 4 - Edge of an agar MC of Bactoderma sp. "printed" onto a glass slide. Magnification bars = 10 µm.

glass slides in situ for 1-7 days (15). Sand, in the case of soil exposure, and algae, in the case of water exposure, have to be removed before the slides can be studied with a phase microscope. The slides may be precoated with a specific medium or with 2% water agar (17).

A SURVEY OF VARIOUS TYPES OF MICROCOLONIES

Coenobia, Nets, and Mycelia - Where Cells Retain Direct Contact After Cell Division

Cell plates

Thiopedia rosea, an anaerobic purple sulfur bacterium, normally has coenobia of sixteen cells which divide synchronously and which are arranged in a regular 4 x 4 plate (Fig. 5). After alternating division in two planes, the cells remain connected by a large number of strands of unknown nature (18). Cell elongation before division spreads these strands apart so that the following division step reduces their number approximately by one half. After further reductions during each second division, the moment comes when there are no further connecting strands remaining, and thus the whole cell plate is divided into two daughter plates (20). A certain degree of clumping keeps these daughter plates together to form an MC which may be aided by the profusely produced surface polymers. Similar processes have been observed in the case of Lampropedia hyalina (25) and Deinococcus (Micrococcus) radiodurans (7).

Bacteria forming permanently bundled coenobia

Pelosigma spp. are S-shaped, slender bacteria that form permanent flat, sigmoid coenobia of four or multiples of four (19). The attachment usually occurs on the cell pole and is thought to be mediated by a polymer. The coenobium is motile by means of an organelle interpreted as a bundle of polar flagella. All cells divide synchronously, giving rise to two daughter bundles that are often held together for a long time. Similar conditions may be found with Peloploca spp. which, however, remain nonmotile.

Microcolonies arising from groups of propagative cells

In the genus Streptosporangium, chains of several spores remain together after being liberated from the sporangium. These may result in microcolonies which - strictly speaking - do not represent a clone, although this spore chain originated from a clone colony (coenobium). Hyphal fragmentation in Nocardia spp. leads to the same result. Even gliding hormogonia of cyanobacteria (Oscillatoria spp.) could be mentioned here.

Net-forming microorganisms

Microcolonies in the form of networks can be found in many microbial groups. The green bacterium, Pelodictyon clathratiforme, produces three-dimensional nets by simultaneous polar branching of two adjacent cells of a chain. This leads to the formation of two polarly touching, Y-shaped cells resulting in a small ring structure between them which subsequently enlarges by cell elongation. The Y-shaped cells finally undergo ternary fission to yield three daughter cells each, which are now members of a ring, the beginning of a net. After fission, the cells of a net are held together by extracellular polymer of unknown nature.

Nets of Thiodictyon elegans may be formed in a similar fashion, but Y-shaped stages have not been found as yet. It appears more probable that the cells are held together by a biterminally produced polymer. Net formation may be a chance effect where those cells join that happen to touch each other at the proper site. Masses of cells without net formation have been observed.

The predatory rod-shaped bacteria described by Perfil'ev and Gabe (30) seemed to have "sticky poles," arising from extracellular polymer production. "Dictyobacter rampax" cells formed nets that could subsequently constrict around a bacterial prey cell; multiplication could occur by division of the whole net (Fig. 6). Teratobacter sp. formed irregular, ropey nets with the cells usually arranged in parallel fashion. Cyclobacter constrictor cell chains could form a "lasso net" by curling up terminally, resulting in the formation of one or two catching loops. All of these predatory bacteria attached to each other in an unknown way, although polymer production (and subsequent modification) can be assumed to be responsible for this MC formation.

The case of Hydrodictyon reticulatum (a green alga) is somewhat different. The nucleus divides mitotically to form up to 2000 planospores which lose their flagella while still remaining within the mother cell. A net is then formed by aggregation of the planospores in such a way that three

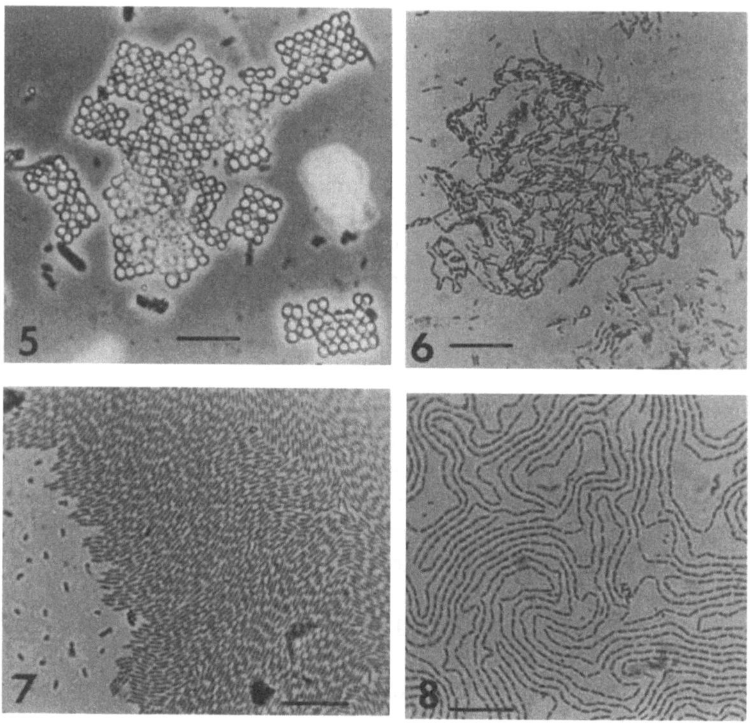

FIG. 5 - Thiopedia rosea: MC of coenobia attached to a slide. FIG. 6 - Net forming aquatic bacteria (Dictyobacter?). FIGS. 7,8 - MC's of Bactoderma (FIG. 7) and related bacteria (FIG. 8): Magn. bars = 10 µm.

spores always join by means of an unknown holdfast substance (11). The hollow spheres of Volvox globator arise by ingrowth of a part of the surface area to form a daughter sphere, a motile microcolony which becomes liberated upon disintegration of the mother sphere.

Microcolonies of mycelial actinomycetes

The Mc of actinomycetes, such as Nocardia or Streptomyces spp., arise by (often multiple) germination of propagative cells (segmentation or fragmentation spores, conidia). The primary hypha branches profusely, and the branches grow outward in a radial fashion. After cross wall formation, fragmentation may occur (Nocardia), and the MC center (the site of the oldest cells) may undergo lysis. Cell attachment or separation are controlled, in part, by environmental factors.

Microcolonies with Mostly Parallel, Rod-shaped Cells

The parallel arrangement of mxyobacteria in a swarm is well-known. The rod-shaped cells remain closely apposed to each other as they glide, continually surrounded by slime (10) and continually growing. They thus form a sheet; their microcolonies are extremely flat. The same holds true for Bactoderma spp. (Figs. 7 and 8). Many cyanobacteria, especially those forming trichomes, glide on surfaces. Microcolonies of such forms are again flat and consist of closely packed parallel bundles of trichomes. Beggiatoa cell filaments and those of many flexibacteria likewise form flat colonies of parallel cells. Even bacteria such as Spaerotilus or Leptothrix spp., which normally grow as cell chains, form similar flat MC's with parallel cell chains. Peripheral cell chains are often curved but still closely apposed.

An interesting phenomenon is the migration of Bacillus circulans colonies and that of related species. Movement occurs in a circular fashion with the main colony leading in front and smaller colonies and MC's trailing behind. Although one can assume that close contact between the cells is mediated by a polymer, information on the nature of this polymer and the reasons for colony gliding are still lacking.

Some rod-shaped bacteria form cell chains with the individual cells slightly out of the main axis of the chain and just polarly touching the adjacent cells (Fig. 9). The cell chains radiate outward, and the microcolony is flat. Such bacteria have been fairly common in anaerobic soil and groundwater enrichments (23). Giving an even stranger appearance are rod-shaped bacteria that arrange themselves in a parallel fashion to form "palisade microcolonies" (Figs. 10 and 11). The palisades are often bent, and thus one may get the impression of seeing Chinese letters. This kind of microcolony can be observed in the anaerobic microbiota of the hypolimnion of many lakes; here the cells have the shape of cigars (4) and carry gas vesicles. They can also be found in soil enrichments where approximately 1 g of soil is suspended in 100 ml of distilled water, and they occur among coryneform bacteria. Spherical microcolonies of laterally gas-vacuolated anaerobic hypolimnetic bacteria have their cells arranged in such a way that the direction of the gas vesicles points to the colony center (Fig. 12).

Microcolonies with Some Distance Between Individual Cells

Bacteria of the genera Siderocapsa, Arthrobacter, and Azotobacter form flat microcolonies which are embedded in or surrounded by a thick capsule. There is space for only a certain number of cells within one joint capsule, and this structure appears to be formed by the first cell of such an MC. The individual cells within a capsule do not seem to touch each other.

Anaerobic, hypolimnetic bacteria of the genus Brachyarcus are semicircular rods embedded in a thick, flat layer of slime. Normally they occur in groups of two, four, eight, or sixteen cells, with the groups arranged symmetrically opposite each other. There do not seem to be connections between these cells other than the common slime sheet (Fig. 13). A similar arrangement, but three-dimensional, is found in the cyanobacterium, Eucapsis. Microcolonies of Merismopedia resemble coenobia of Thiopedia, except that the cells are individually embedded in a flat sheet of slime, without interconnecting strands. The slightly curved cells of Nevskia secrete a polymer laterally which can be considered as a slime stalk. Upon division, this slime stalk also branches, and eventually a flat microcolony arises with the cells on the periphery and the common, branched polymer in the center.

In some cases, bacteria forming cell chains remain connected by interlacing pili which are formed on both cell poles. Such organisms have been found in groundwater (23).

Bacterial Rosettes As Microcolonies

A bacterial three-dimensional Rosette is a microcolony or aggregate where all cells adhere polarly to one common spot, usually a central polymer. Rosette formation, therefore, is a consequence of polymer production at localized points on the cell surface. Such polymers may be flagella, glycolipids, or carbohydrates; in the latter case, they can be localized by staining procedures (ruthenium red, alcian blue,

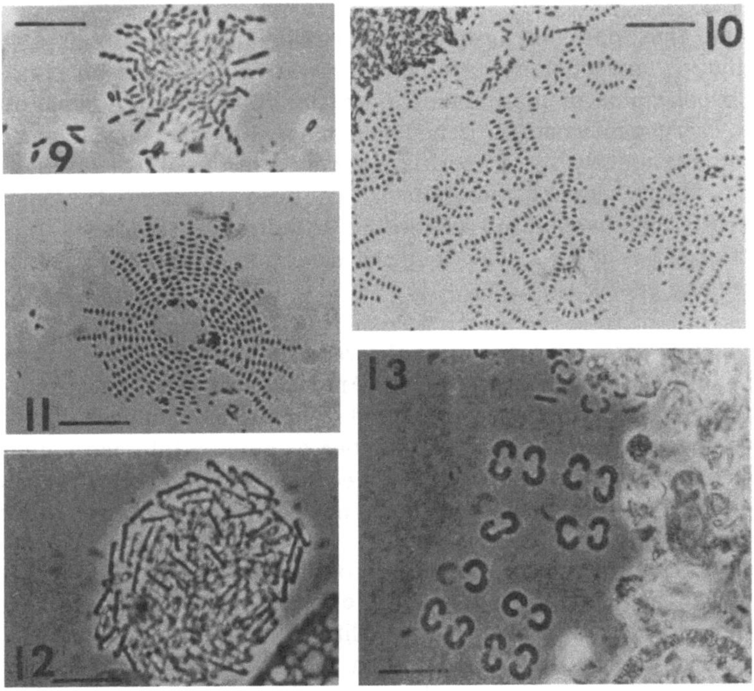

FIG. 9 - MC of anaerobic groundwater bacteria. FIGS. 10, 11 - Palisade MC's of soil bacteria, partly in the "Chinese letter" arrangement. FIG. 12 - Anaerobic, gas-vacuolated hypolimnetic bacteria forming an MC where the gas vesicles are all directed toward the center. FIG. 13 - Pretzel-shaped Brachyarcus sp. from a bog lake with symmetric, attached MC. Magn. bars = 10 µm.

phosphotungstic acid, etc.). Localized, restricted adhesive substance production has the advantage of leaving a maximum of cell surface still exposed. The formation of three-dimensional rosettes is quite common among a variety of attaching bacteria; they can be observed in the genera Planctomyces, Pasteuria (sensu Staley), Hyphomicrobium, Caulobacter, Blastobacter, Methylosinus, Rhodopseudomonas, Agrobacterium, Seliberia, etc. The cells of a rosette may all be equal in size and age (Fig. 14, and see (39)), or they may be partly older, partly younger (Hyphomicrobium ZV-580, Fig. 15; Caulobacter spp., Fig. 16). Some bacteria are obligatory rosette formers (Planctomyces bekefii, Blastobacter spp.), and others may also live in the dispersed state, depending on growth conditions. Rosette formation by Hyphomicrobium ZV-580 was inhibited by galactose, mannose, concanavalin A, or chloramphenicol (29). In most cases the number of cells per rosette is limited, due to spatial restrictions. Rosette formation is quite common among algae and even protozoa. Many planktonic algae remain together after cell division, either with direct contact (Ankistrodesmus, Fig. 17) or connected by slime stalks (Dictyosphaerium, Fig. 18).

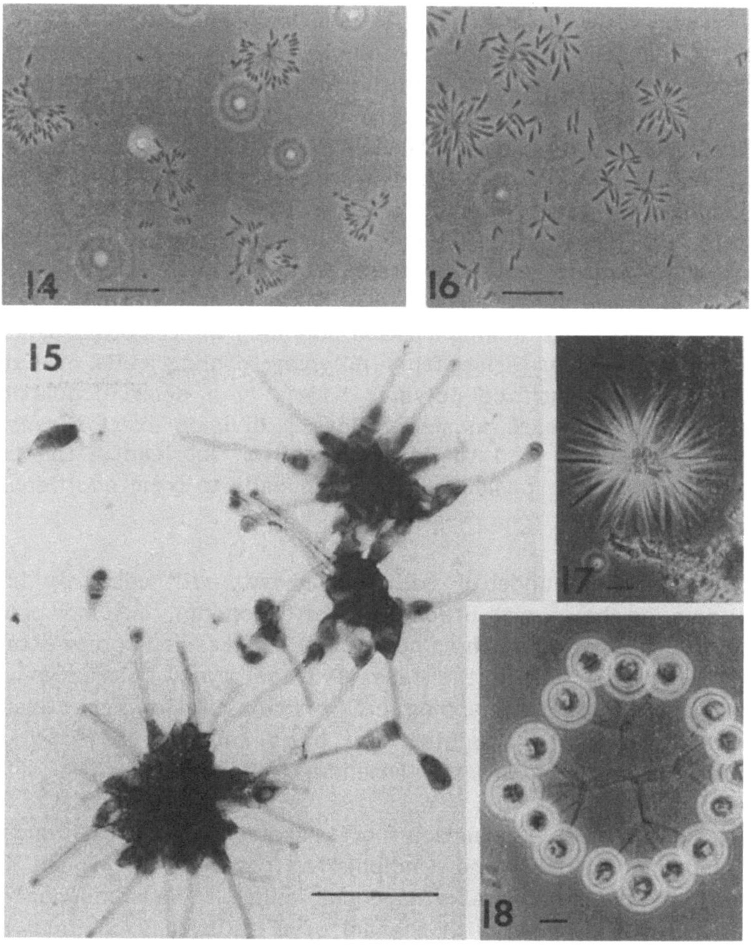

FIGS. 14, 16 - Caulobacter rosettes with mainly cells of equal age (Fig. 14) or cells of different size and age (Fig. 16). Both strains were isolated from Lake Plußsee, F.R. Germany. FIG. 15 - Rosettes of Hyphomicrobium ZV-580. FIG. 17 - Ankistrodesmus rosette. FIG. 18 - Rosette of Dictyosphaerium sp. from a bog lake. Magn. bars = 10 µm (Figs. 14, 16-18), or = 5 µm (Fig. 15).

Naturally Occurring Attached Microcolonies

We can find dispersed or more concentrated bacterial MC's on glass slides that were exposed to water or soil. Caulobacter often spreads over large areas, probably due to an extended motility of the swarmers. Bacteria of the Planctomyces group were densely attached, despite some swarmer motility (Fig. 19). Hyphomicrobium NQ-521Gr normally formed irregular clumps on the slide surface, except when exposed to visible light of high intensity (Caldwell, personal communication). Certain wave lengths of this light probably interfere with the formation of adhesive substances.

SOME PROCESSES OF MICROCOLONY FORMATION

It now becomes obvious that the many types and shapes of microcolonies have different origins. Flat MC's arise from sheet-forming organisms (Nevskia), or from gliders and filamentous forms (Myxobacteria, Sphaerotilus). Raised colonies stem from division and/or subsequent movement in three planes. But flat, folded, and "rough" colonies are formed by variant cells lacking surface polymer, whereas raised, "smooth," and glistening colonies arise from polymer-forming cells. What would be the role of the surface polymer here? In a dense population, cell elongation or an increase in size following division exert a pressure on the neighboring cells, and the presence of a "lubricant," the polymer, may well result in the pushing up of some cells to form additional layers and a raised colony.

The cell shape and mode of cell wall growth will certainly influence the MC form, as shown with Brachyarcus or Thiopedia. Polymer production at specific surface sites (such as cell poles) may result in polar attachment or in net formation, as in the case of Thiodictyon, Pelodictyon, or the predatory bacteria. The snapping cell division of Corynebacterium spp. gives rise to V-shaped daughter cell pairs and may lead to palisade formation or to arrangements that resemble Chinese letters.

In most cases the formation of rosettes is little understood. Stove-Poindexter (39) studied Caulobacter rosettes and found that all of the rosette cells had stalks of equal length. The component cell of a mature rosette adhered with the tip of a stalk to a common holdfast material. She concluded "that the initiation of rosette formation normally occurred through adhesion of swarmer cells which were all at a common stage of development." Rosette formation by swarmer cells was temperature-dependent: at 5°C no rosettes were found; when heated to 30°C, the swarmer rosettes formed within a few minutes. She also discussed random collision and chemotactic attraction as two possible mechanisms for aggregation; in both cases these were followed by adhesion to the holdfast polymer. She ruled out chemotactic attraction since nonmotile mutants also formed rosettes - even in liquid media. Also, adhesion in rosettes was not strain-specific, since mixed rosettes were formed in high density, mixed strain populations.

Our own observations contradict some of these findings. Caulobacter cultures from Lake Plußsee formed rosettes with cells of different stalk lengths and hence of different age (Fig. 16). Many Caulobacter cultures show slightly enlarged stalk tips at the site where the holdfast polymer appears. it seems possible, therefore, that metabolic activity at the stalk tip results in the production of a chemotactic attractant which is produced to lure the Caulobacter swarmers to the center of rosette formation. A requirement for cellular associations in Caulobacter development has been claimed by Croffy et al. (6).

Rosette formation by Hyphomicrobium ZV-580 is due to the monopolar formation of a stainable polymer (5). It is the distal, blunt cell pole where the polymer appears, and consequently, the mother cells are located in the center of the rosette with their hyphae radiating outwards (Figs. 15, 20). The nature of this polymer has been investigated by Moore and Marshall (29) who believe it to be a glycoprotein or a peptidopolysaccharide. They also found that only young cells were attached to floating coverslips and could thus form rosettes, and that only swarmers could undergo irreversible adhesion.

My own unpublished observations with several Hyphomicrobium strains revealed a remarkable fact. Strain ZV-580 showed a cell polarity that was different from the other strains. The proximal, pointed cell pole - formerly attached to the mother hypha - was the site of new hyphal outgrowth. Therefore, the distal, blunt pole could produce polymer and attach. In the other strains (B-522, NQ-521Gr), new hyphae emerged from the distal pole, and none of these produced a polar adhesive as did ZV-580. Likewise, bipolar hyphal outgrowth was rare in rosette-forming ZV-580, whereas it was common in the other strains. Kjelleberg and Marshall (personal communication) have strong evidence to indicate that rosettes are formed via polar hydrophobic interactions in Hyphomicrobium ZV-580.

PROPERTIES OF MICROCOLONIES

Microcolony Shape

The arrangement of cells in a microcolony must determine the MC shape.

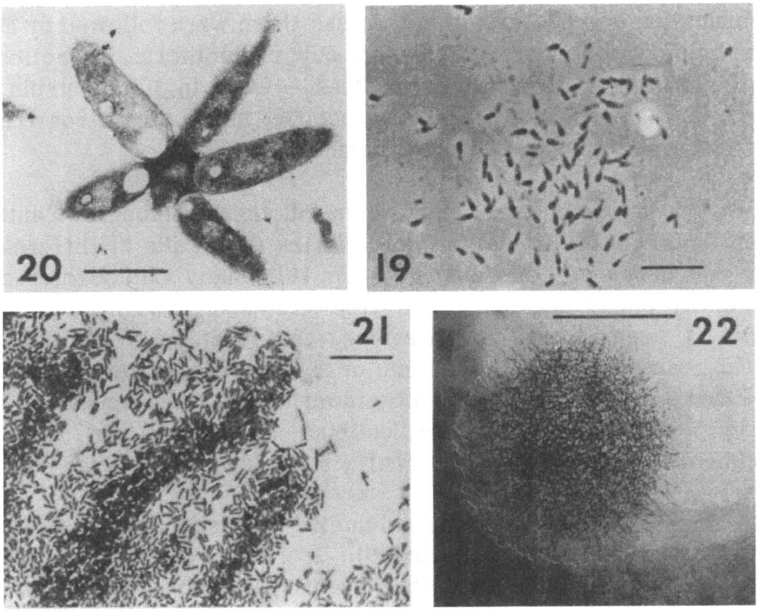

FIG. 19 - Budding bacteria of the Planctomyces group: MC formed within three days on a slide exposed in a forest pond. Magn. bar = 10 µm. FIG. 20 - Rosette of Hyphomicrobium ZV-580: thin section with central polymer (courtesy of S.F. Conti, Amherst); magn. bar = 3 µm. FIG. 21 - Edge of the colony of Fig. 22: print on glass slide. Note ridges of denser growth corresponding to colony folds. Magn. bar = 10 µm. FIG. 22 - Magn. bar = 1 mm.

The marginal folds of the colony in Fig. 22 could be well reproduced by "printing" the colony upside down on a slide (Fig. 21); they appear as ridges. This colony is thought to have originated in a manner similar to the one shown in Fig. 9.

Microcolony Fine Structure

Various studies have demonstrated the inhomogeneity of bacterial colonies. Peripheral cells resemble those from liquid cultures, but cells from the center may be irregular in shape or even lysing (8), as is especially prevalent in Nocardiae.

Growth zones in bacterial colonies may be studied by sectioning ^{3}H-labeled colonies and subjecting these to autoradiography (35). In aerobic pseudomonads, the growth zones corresponded to the region of oxygen penetration, the periphery, and the upper central portion. In Staphylococcus aureus, growth inhibition by local accumulation of toxic substances is also important (35). Wimpenny and Parr (42) sectioned colonies of Enterobacter cloacae and determined enzyme activities in the sections. The top 120 µm, where oxygen supply was sufficient, had the highest enzyme activities. We can expect physiologically older cells in the center and bottom to become leaky for growth factors, vitamins, etc. Microorganisms with immature swarmer cells (Hyphomicrobium) therefore show cross-feeding effects. A single cell of these, or an MC, do not grow well unless older cells or larger colonies are present. In this case the swarmers obtain vitamin B_{12} from older cells.

Genetic Exchange in Bacterial Rosettes

Several pseudomonads were shown to undergo genetic recombination while in the rosette stage (16). The cells were interconnected by fimbriae and polar holdfast polymer. But recombination did not occur in Pseudomonas rosettes studied by other authors (26). More work is needed to understand reasons for and consequences of rosette formation. In which way is the genetic material transferred; is competence needed? Does this also occur in nature, or is it only a laboratory phenomenon?

ECOLOGICAL CONSEQUENCES OF MICROCOLONY FORMATION

The first cell of a microcolony encounters great problems on a new surface. Nutrient supply may be limiting; extracellular enzymes, ionophores, and growth factors may be missing. Subsequent cells find protection against such environmental stress as drought, toxic compounds, predators, or abrupt changes in environmental parameters. MCF may protect against being washed away in lotic systems (rivers, groundwater, the oral cavity, or intestines). A profound disadvantage is the competition by sister cells; waste products may reach inhibitory concentrations. A suspended MC that grows too large may become sedimented.

An area that deserves study is the type of surface that supports MC formation. In the aquatic habitat, polymer-coated surfaces are preferred sites for MC formation. The basal holdfast polymer of a choanoflagellate (Fig. 23) or one side of a diatom frustrule (Fig. 24) are given as examples. There is evidence that initial attachment even to rock surfaces in the desert is mediated by polymer bridging (38).

MIXED MICROCOLONIES WITH POSSIBLE SYNTROPHY OF THE COMPONENTS

In nature, "pure colonies" are quite rare, but organisms occurring side

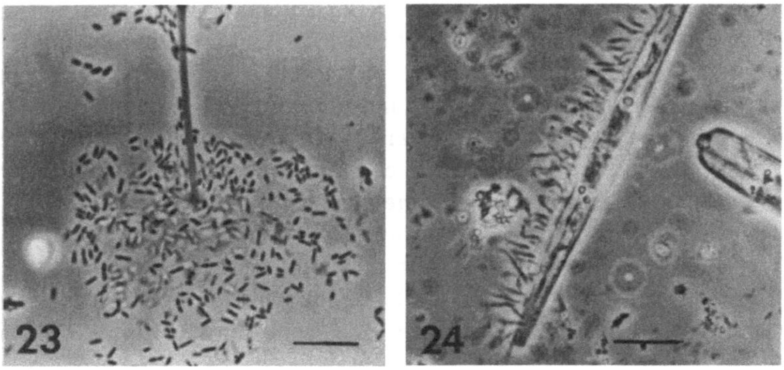

FIG. 23 - A bacterial microcolony attached to the holdfast of a choanoflagellate (not visible here). Magn. bar = 10 µm. FIG. 24 - A bacterial microcolony attached to a diatom frustrule found in the Solar Lake (Sinai). Note attachment only to one side. Magn. bar = 10 µm.

by side in the same location may support each other biochemically. As long as there is no direct and lasting contact between partners which have a nutritionally beneficial relationship, we talk about "syntrophy." Recent advances in the study of anaerobic fatty acid degradation and syntrophy with methanogens have shown that in the anaerobic environment such relationships are of great importance (27, 43). Kaspar (unpublished Ph.D. Thesis) reported a main residence time for H_2 in a mature anaerobic digester to be less than 0.4 sec. The mean free path of a molecule ($\bar{x}$) can be calculated as $\bar{x} = \sqrt{2\text{-}D \cdot t}$, where D = diffusivity (L^2/t) and t = time. For H_2 at 33°C and a 0.4 sec time period, $\bar{x}$ = 75 µm. Therefore, the site of H_2 consumption and H_2 production in a digester are less than 75 µm apart. This suggests very close associations between these two groups of organisms in the digester (Bryers, personal communication). For transient mixed MC's to be formed in aquatic habitats, the syntrophic partners must adhere to each other, probably by formation of one or several adhesive polymers. Such conditions have been studied with Streptococcus sanguis and Actinomyces viscosus (34). Sacculi of the former bacterium would still adhere to A. viscosus after extensive purification and even lysozyme treatment. Enrichments of groundwater bacteria contained mixed aggregates (microcolonies) of Caulobacter spp. and Seliberia spp. where cell-to-cell contact was initiated by contact between the holdfasts of both bacterial components (Hirsch and Rades-Rohkohl, unpublished results). The freshwater red alga, Kyliniella latvica, occasionally carries in the thick sheath Hyphomicrobium cells lined up parallel and perpendicular to the long filament axis (13). Since bacteria-free algal filaments occur, this cannot be called a consortium. Loose symbioses between chrysomonads and budding bacteria are apparently common in bogwater (12). The bacteria lie attached to the alga. In most of these syntrophic associations, joint polymers have been observed to mediate transient adhesion of the partners.

MICROBIAL CONSORTIA

Types, Properties, and Formation of Consortia

Consortia have been defined as close associations of two different organisms with possibly syntrophic relationships (40). A variety of such consortia is listed in Table 1. Since the members of consortia stay together for a long time during their life cycle, beneficial interactions between them can be assumed. But the degree of interdependence varies. The Chlorobium component of "Chlorochromatium aggregatum" can be cultured alone (28), while the central organism has never been found without its green partner. In a Cryptomonas-Caulobacter consortium, the association appears to be transient but does influence the morphology of the central flagellate (24). Since the central organism is considerably larger in most cases, multiplication of consortia occurs by division of the central organisms and redistribution of the already attached, smaller symbionts. In such a "cyclic" consortium, division of both partners needs to be well synchronized to assure always the proper ratio of peripheral to central cells.

For an "acyclic" consortium to become newly established, certain conditions have to be met, and a sequence of steps has to be followed: a) The partners have to occur in the same environmental location, which already guarantees a high degree of physiological similarity. b) Both partners should have a need for each other, as in a stress situation. Extreme conditions often favor the production of a polymer on the cell surface. Hirsch (21) observed clumping of anaerobic hypolimnetic bacteria as a result of sample shaking (oxygenation). c) The partners will have to find each other. A motile central cell may be chemotactically attracted by excretions of "peripheral" cells. Or, the peripheral organisms may accidentally touch a surface polymer of a central cell and remain there permanently. Adhesiveness of at least one partner must play an importnat role in the initial formation of consortia. The presence of adhesive polymers has been demonstrated by thin sectioning of "C. aggregatum," where the central cell is covered by a cottony and/or cup-shaped polymer, and the chlorobia lie embedded in another polymer (4). A polymer has also been found in the case of a CH_4-consuming, floc-forming consortium in flush-tanks (15). It would be interesting to find out if polymer

TABLE 1 - Types of consortia and their components

Common names[1)]	Outside organisms	Central organism	Reference
"Chlorochromatium aggregatum" Lauterborn 1906	green Chlorobium sp. 2-4 layers of cells	a large, motile, colorless bacterium with capsule and s.t. stalked	(40)
"Chlorochromatium glebulum" Skuja 1956	green, gas-vacuolated bacteria (Pelodictyon luteolum?)	a colorless, bent Microspira-like bacterium with capsule	(36)
"Pelochromatium roseum" Lauterborn 1913	several pinkish-brown Chlorobium cells	a colorless bacterium with capsule ("Endosoma palleum")	(40)
"Pelochromatium roseo-viridis" Gorlenko and Kusnezov 1972	several pinkish-brown Chlorobium cells covered by layers of green Pelodictyon luteolum	a large, motile, colorless bacterium with capsule	(14)
"Cylindrogloea bacterifera" Perfil'ev 1914	one layer of green bacteria, possibly with gas vesicles	a chain of short, colorless rods within a thick capsule	(36)
"Chloroplana vacuolata" Dubinina and Kusnezov	green rods (Pelodictyon?) alternating with colorless, bright rods to form flat bands; both partners with gas vesicles		(9)
"Peloploca undulata" Lauterborn 1913	colorless chains of rod-shaped bacteria, alternating with chains of colorless, gas-vacuolated, rod-shaped bacteria		(22)
"Thermophilic, CH_4-forming granular consortium"	Methanosarcina sp. as a thick layer	colorless, ovoid and rod-shaped bacteria producing CH_4	(1)
"Metapolytoma bacteriferum" Skuja 1958	one layer of colorless, rod-shaped bacteria	a colorless flagellate cell (Polytoma-like)	(37)
"Cryptomonas-Caulobacter consortium" Klaveness 1982	Caulobacter sp., attached to median and indented region of flagellate	a large, flagellated Cryptomonas cell	(24)

1) Without standing in nomenclature

production could be initiated by contact with the other partner. d) Chemical/biochemical interactions have to begin. Synchronous division may depend on syntrophic interactions, on cooperative metabolism of one substrate, and on a constant environment. e) A temporary consortium may eventually become permanent (cyclic) when co-evolution leads to an increasing interdependency, especially of the central organism. Mixed, three-membered consortia may arise when the central organism's adhesive polymer becomes less specific.

Ecological Consequences of Consortium Formation

Presently, none of the consortia can be grown in pure culture, and therefore information on the types of interactions (syntrophy) is still scarce. It has been suggested that the symbiosis of consortium-containing green bacteria may be based on mutual provision of products of the sulfur cycle (32). The central bacterium could be an SO_4^{2-} or S^o-reducer. The chlorobia, on the other hand, are known to excrete large amounts of organic substances (up to 30% of their primary production (31)). However, the beneficial effect could also consist of only polymer formation, and the polymer attract nutrients from the environment. Polymer formation and excretion of soluble nutrients by cyanobacteria (Fischerella) have been shown to stimulate Legionella pneumophila in a naturally occurring consortium (2). The removal of a toxin (a trihydroxamate chelate?) from Microcystis aeruginosa mucilage by Zoogloea sp. has been suggested (3). Finally, active phototactic transport of the "Chlorochromatium" to the light through the motility of the central bacterium was observed by Pfennig (33). Here the chlorobia acquired motility and phototaxis by joining the central organisms.

Some Questions Left to Be Answered

As in microcolony formation, what is the role of the polymer? Is adhesive substance production a prerequisite for or a consequence of consortium formation? Specific surface polymer receptor sites (or charges) may play an important role in these processes. These sites could not be modified by the environment, or else the partners would lose contact. What is the chemical nature of these polymers; are they produced by one or both partners? Is consortium formation an irreversible process, or would one or both partners be able to redissolve the connections? Many "Chlorochromatium aggregatum" consortia observed in natural samples show the central bacterium to be extremely pale, possibly dead, while the peripheral chlorobia remain attached. A damaging effect of these chlorobia on the central organisms seems to be possible. Certainly, many more observations on well-growing consortia are needed. As a first step, "Chlorochromatium aggregatum" enrichment cultures have been successfully kept in the laboratory for six months by employing a Perfil'ev convection chamber gradient technique (Hirsch, unpublished). The possibility of producing artificial consortia has been investigated by this author. Nutritionally stressed cells of a Hyphomicrobium culture were mixed with cells of a coccal cyanobacterium. In a mineral salts medium lacking organic carbon, the Hyphomicrobium produced a polar holdfast polymer and attached perpendicularly to the cyanobacterium to form a rosette. This consortium persisted for over a year without any addition of carbon or energy compounds; both partners multiplied, and the cyanobacterium cocci were all covered with hyphomicrobia.

REFERENCES

(1) Bochem, H.P.; Schobert, S.M.; Sprey, B.; and Wengler, P. 1982. Thermophilic biomethanation of acetic acid: morphology and ultrastructure of a granular consortium. Can. J. Microbiol. 28: 500-510.

(2) Bohach, G.A., and Snyder, I.S. 1983. Adherence of Legionella pneumophila to cyanobacteria as a possible means for acquiring nutrients. Abstract. Am. Soc. Microbiol. Q-106: 278.

(3) Caldwell, D.E. 1980. Association between cyanobacteria and Zoogloea spp. during plankton blooms. Abstract. Amer. Soc. Microbiol. N-47: 171.

(4) Caldwell, D.E., and Tiedje, J.M. 1975. A morphological study of anaerobic bacteria from the hypolimnia of two Michigan lakes. Can. J. Microbiol. 21: 362-376.

(5) Conti, S.F., and Hirsch, P. 1965. Biology of budding bacteria. III. Fine structure of Rhodomicrobium and Hyphomicrobium spp. J. Bacteriol. 108: 515-525.

(6) Croffy, B.R.; Mansour, I.D.; and Stachow, C.S. 1977. Requirement for cellular associations in the development of Caulobacter crescentus. Abstracts. Am. Soc. Microbiol. I-79: 167.

(7) Driedger, A.A. 1970. The ordered growth pattern of microcolonies of Micrococcus radiodurans: first generation sectoring of induced lethal mutations. Can. J. Microbiol. 16: 1133-1135.

(8) Drucker, D.B., and Whittaker, D.K. 1971. Microstructure of colonies of rod-shaped bacteria. J. Bacteriol. 108: 515-525.

(9) Dubinina, G.A., and Kusnezov, S.I. 1976. The ecological and morphological characteristics of microorganisms in Lesnaya lamba (Karelia). Int. Rev. ges. Hydrobiol. 61: 1-19.

(10) Dworkin, M. 1972. The Myxobacteria: new directions in studies of prokaryotic development. Crit. Revs. Microbiol. 2: 435-452.

(11) Esser, K. 1976. Kryptogamen. Blaualgen, Algen, Pilze, Flechten. Praktikum und Lehrbuch. Berlin: Springer-Verlag.

(12) Geitler, L. 1948. Symbiosen zwischen Chrysomonaden und knospenden Bakterien-artigen Organismen sowie Beobachtungen über Organisationseigentümlichkeiten der Chrysomonaden. Oester. Botan. Z. 95: 300-324.

(13) Geitler, L. 1965. Ein Hyphomicrobium als Bewohner der Gallertmembran der Süßwasser-Rhodophycee Kyliniella. Arch. Mikrobiol. 51: 309-400.

(14) Gorlenko, W.M., and Kusnezov, S.J. 1972. Über die photosynthesierenden Bakterien des Kononjer-Sees. Arch. Hydrobiol. 70: 1-13.

(15) Henrici, A.T., and Johnson, D.E. 1935. Studies of freshwater bacteria. II. Stalked bacteria, a new order of Schizomycetes. J. Bacteriol. 30: 61-93.

(16) Heumann, W. 1956. Der Sexualzyklus sternbildender Bakterien. Arch. Mikrobiol. 24: 362-395.

(17) Hirsch, P. 1972. Neue Methoden zur Beobachtung und Isolierung ungewöhnlicher oder wenig bekannter Wasserbakterien. Z. Allg. Mikrobiol. 12: 203-218.

(18) Hirsch, P. 1973. The fine structure of Thiopedia spp. Abstracts, pp. 184-185. Symposium on Prokaryotic Photosynthetic Microorganisms, Freiburg, F.R. Germany.

(19) Hirsch, P. 1974. Genus Pelosigma Lauterborn 1913,100. In Bergey's Manual of Determinative Bacteriology, 8th ed., eds. R.E. Buchanan and N.E. Gibbons, pp. 215-216. Baltimore: The Williams & Wilkins Co.

(20) Hirsch, P. 1977. Ecology and morphogenesis of Thiopedia spp. in ponds, lakes, and laboratory cultures. In Proceedings of the Second International Symposium on Photosynthetic Prokaryotes, 1976, eds. G.A. Codd and W.D.P. Stewart, pp. 13-15. Dundee, Scotland.

(21) Hirsch, P. 1980. Some thoughts on and examples of microbial interactions in the natural environment. In Aquatic Microbial Ecology. Proceedings of the Conference of the American Society of Microbiology, eds. R.R. Colwell and J. Foster, pp. 36-54. Maryland Sea Grant Publication, University of Maryland, College Park, MD.

(22) Hirsch, P. 1981. The family Pelonemataceae. In The Prokaryotes. A Handbook on Habitats, Isolation and Identification of Bacteria, eds. M.P. Starr et al., pp. 412-421. Berlin: Springer-Verlag.

(23) Hirsch, P., and Rades-Rohkohl, E. 1983. Microbial diversity in a groundwater aquifer in Northern Germany. In Developments in Industrial Microbiology, ch. 14, pp. 183-200. Arlington, VA: Society of Industrial Microbiology.

(24) Klaveness, D. 1982. The Cryptomonas-Caulobacter consortium: facultative ectocommensalisms with possible taxonomic consequences? Nord. J. Bot. 2: 183-188.

(25) Kuhn, D.A., and Starr, M.P. 1965. Clonal morphogenesis of Lampropedia hyalina. Arch. Mikrobiol. 52: 360-375.

(26) Mayer, F. 1970. Elektronenmikroskopische Untersuchung der Zellverbindung bei einem sternbildenden Bakterium. Z. Allg. Mikrobiol. 10: 329-334.

(27) McInerney, M.J.; Bryant, M.P.; and Pfennig, N. 1979. Anaerobic bacterium that degrades fatty acids in syntrophic association with methanogens. Arch. Microbiol. 122: 129-135.

(28) Mechsner, K. 1957. Physiologische und morphologische Untersuchungen an Chlorobakterien. Arch. Mikrobiol. 26: 32-51.

(29) Moore, R.L., and Marshall, K.C. 1981. Attachment and rosette formation by hyphomicrobia. Appl. Envir. Microbiol. 42: 751-757.

(30) Perfil'ev, B.V., and Gabe, D.R. 1969. Capillary Methods of Investigating Micro-organisms (Translated by J.M. Shewan). University of Toronto Press.

(31) Pfennig, N. 1978. General physiology and ecology of photosynthetic bacteria. In The Photosynthetic Bacteria, eds. R.K. Clayton and W.P. Sistrom, ch. 1, pp. 3-18. New York: Plenum.

(32) Pfennig, N. 1978. Syntrophic associations and consortia with phototrophic bacteria. Abstracts. 12th International Congress on Microbiology, p. 16. Munich, F.R. Germany.

(33) Pfennig, N. 1980. Syntrophic mixed cultures and symbiotic consortia with phototrophic bacteria: a review. In Anaerobes and Anaerobic Infections, eds. G. Gottschalk et al., pp. 127-131. Stuttgart: G. Fischer Verlag.

(34) Reusch, V.M.; Foster, J.L.M.; and Haberkorn, D.S. 1983. Specific coaggregation and the cell wall of Streptococcus sanguis. J. Bacteriol. 155: 869-899.

(35) Reyrolle, J., and Letellier, F. 1979. Autoradiographic study of the localization and evolution of growth zones in bacterial colonies. J. Gen. Microbiol. 111: 399-406.

(36) Skuja, H. 1956. Taxonomische und biologische Studien über Phytoplankton schwedischer Binnengewässer. Nova Acta Regiae. Soc. Sci. Upsal. Ser. IV,16: 1-104.

(37) Skuja, H. 1958. Eine neue vorwiegend sessil oder rhizopodial auftretende synbakteriotische Polytomee aus einem Schwefelgewässer. Svensk. Botan. Tidskr. 52: 379-390.

(38) Staley, J.T.; Palmer, F.; and Adams, J.B. 1982. Microcolonial fungi: common inhabitants of desert rocks? Science 215: 1093-1095.

(39) Stove-Poindexter, J. 1964. Biological properties and classification of the Caulobacter group. Bacteriol. Rev. 28: 231-295.

(40) Trüper, H., and Pfennig, N. 1971. Family of phototrophic green sulfur bacteria: Chlorobiaceae Copeland, the correct family name; rejection of Chlorobacterium Lauterborn; and the taxonomic situation of the consortium-forming species. Int. J. Syst. Bacteriol. 21: 8-10.

(41) Wilkins, J.R.; Darnell, W.L.; and Boykin, E.H. 1972. Cinemicrographic study of the development of subsurface colonies of Staphylococcus aureus in soft agar. Appl. Microbiol. 24: 786-797.

(42) Wimpenny, J.W.T., and Parr, J.A. 1979. Biochemical differentiation in large colonies of Enterobacter cloacae. J. Gen. Microbiol. 114: 487-489.

(43) Wolfe, R.S., and Pfennig, N. 1977. Reduction of sulfur by Spirillum 5175 and syntrophism with Chlorobium. Appl. Envir. Microbiol. 33: 427-433.

Epilogue

The reasonable microbiologist adapts himself to the microbial world. The unreasonable one persists in trying to adapt the microbial world to himself. Therefore, all progress depends on the unreasonable microbe.

– D. Mirelman

Glossary

ADHERENCE - This term is not a legitimate alternative to adhesion. The attachment of microorganisms to surfaces involves physicochemical interactions between the organism and the substratum and should only be described in terminology acceptable to the physical chemist.

ADHESIN - A surface macromolecule of a cell known to be responsible for specific adhesion.

ADHESION - This term can be defined unambiguously only in terms of the energy involved in the formation of the adhesive junction. Thus, two surfaces may be said to have adhered when work is required to separate them to their original condition.

ADSORPTION - The accumulation of molecules at a fluid interface at a concentration exceeding that in the bulk fluid, brought about as a result of random Brownian motion.

AGGREGATE - A collection of microbial cells in intimate contact. This definition, avoiding any connotation of mechanism as to origin and genesis of the collection, is treated here as a structure - the product of a process.

AGGREGATION - The process leading to aggregate formation.

ASSOCIATION - This is a term which describes the interaction between a microorganism and a surface, without specifying a mechanism for this interaction. The initial indication of a reaction between microorganisms and a complex biological surface is usually the observation of an increased retention time of the bacterium at the surface (compared to that of a random nonsticky collision). Such increased retention time may be due to adhesion, as defined above, or to nonadhesive mechanisms (e.g., trapping in mucus gel). Until the precise mechanism of the interaction has been elucidated, it may be properly described as microbial association with a surface.

BIOFILM - Microorganisms and extracellular products associated with a substratum.

DEPOSITION - This term is normally used to describe the accumulation of particles at a fluid interface brought about by the application of an external force, e.g., gravity.

DISPERSION - To the organism, dispersion implies distribution in a large volume, and being broadly dispersed enhances opportunities for genetic recombination with individuals from other strains and clones. It also allows for more survivors from specific onslaughts, such as attack by predators, and tends to result in more individuals finding their way into new environments. To the experimenter, dispersion allows treatment of the organisms as individuals, whether assay is microscopic or microchemical, and allows the experimenter access to a potentially more homogeneous population, avoiding the physiological heterogeneity of cells within a floc due to difficulties with transport of oxygen, proteins, nutrients, or inhibitory products.

FIBRILLAE - This term describes the more amorphous surface adhesive appendages which appear to lack the regular filamentous forms of fimbriae.

FIMBRIAE - Discrete nonflagella, filamentous structures of uniform diameter which probably function as adhesive organelles.

HYDROPHOBICITY AND HYDROPHILICITY - The structure of water in the region near any surface is perturbed over distances of maybe up to several tens of molecular layers. Near a so-called hydrophobic surface

HYDROPHOBICITY AND HYDROPHILICITY (continued) - the water is less structured in terms of intermolecular hydrogen bonding between the water molecules, whereas near a hydrophilic surface water is more structured. If two similar surfaces come together such that the two perturbed water layers overlap, then there must be a displacement of perturbed water molecules into bulk water. This will lead to a decrease in free energy (i.e., an attraction) in the case of a hydrophobic surface, but to an increase in free energy (i.e., a repulsion) in the case of a hydrophilic surface. These structural interactions, which clearly vary with separation between the surfaces, are an addition to the Van der Waals interactions between the particles concerned. If one brings a hydrophobic surface up to a hydrophilic surface, one cannot predict, a priori, whether there will be a net repulsion or attraction.

NONSPECIFIC ADHESION - Microbial adhesion involves macromolecules on the surface of the microorganism concerned. These may interact with the surface of the substratum or the macromolecules present on that surface. A number of classes of interactions may be involved (ionic, dipolar, H-bonds, hydrophobic interactions). The number

NONSPECIFIC ADHESION (continued) - and strength of these interactions vary considerably from case to case, so that the level of adhesion of a given microbe to a variety of surfaces (or a range of microbes to a given surface) will also vary considerably. This, however, does not imply specificity.

PILI - Those organelles involved in conjugation and DNA transfer between cells.

RECEPTOR - A component on the surface which is bound by the active site of an adhesin during the process of specific microbial attachment.

SELECTIVITY - This implies the ability of microorganisms to achieve different degrees of adhesion to a variety of surfaces. For example, this could be achieved by means of a series of substrata with different Van der Waals attractive properties. Alternatively, selectivity may result from quantitative differences in the numbers of specific adhesive molecules on the surfaces. Consequently, it is not possible to decide whether adhesion has a specific molecular basis merely by observing relative numbers of cells which attach to surfaces or by measuring their strength of attachment. Specificity requires some demonstration of molecular stereochemical complementarity.

SPECIFIC ADHESION - An additional requirement for specific adhesion is some form of stereochemical constraint which brings more than one (normally several) pair of neighboring, interacting groups on the microorganism and the substratum into contact. An example would be sugar residue-protein interactions on two complementary polymers, i.e., a lock-and-key mechanism is required. Monomers and oligomers which sufficiently mimic the combining sites of either interacting surface polymer can occupy the combining sites and block specific attachment, reducing it to a nonspecific level of adhesion characteristic for that system.

SUBSTRATE - A material utilized by microorganisms, generally as a source of energy (plural: substrates).

SUBSTRATUM - A solid surface to which a microorganism may attach (plural: substrata). In some instances, the solid substratum may serve as a substrate for certain microorganisms (e.g., cellulose, chitin).

List of Participants with Fields of Research

ATKINSON, B.
Brewing Research Foundation
Lyttel Hall
Nutfield, Surrey RH1 4HX
England

Biological process engineering, brewing

BAZIN, M.J.
Dept.of Microbiology
Queen Elizabeth College
Campden Hill Road
London W11
England

Microbial population dynamics, computer-control of bioreactors

BREZNAK, J.A.
Dept. of Microbiology and
Public Health
Michigan State University
East Lansing, MI 48824
USA

General: animal-microbe symbioses; Specific: biochemistry of symbiosis between termites and their intestinal microbiota

BRYERS, J.D.
Applied Biology
Swiss Federal Institute for Water
Resources and Water Pollution Control
EAWAG
Swiss Federal Institutes of Technology
8600 Dübendorf
Switzerland

Kinetics and stoichiometry of applied and environmental processes; effects of biofilms in natural, industrial fermentation, and water/wastewater treatment systems; transport processes in biofilms

CALDWELL, D.E.
Dept. of Applied Microbiology
and Food Sciences
University of Saskatchewan
Saskatoon, Saskatchewan S7N OW0
Canada

Colonization and degradation of particulate substrates by procaryotic microorganisms

CALLEJA, G.B.
Molecular Genetics Group
National Research Council of
Canada
Room 3141, 100 Sussex Drive
Ottawa K1A OR6
Canada

Microbial physiology and biochemistry, cellular differentiation and development

CHARACKLIS, W.G.
Institute for Biological and
Chemical Process Analysis
Montana State University
Bozeman, MT 59717
USA

Microbial and chemical process engineering

COOKSEY, K.E.
CE & EM
Cobleigh Hall
Montana State University
Bozeman, MT 59717
USA

Adhesion of diatoms to clean surfaces, biosynthesis of adhesive polymers, role of cytoskeleton in adhesion and motility

DAZZO, F.B.
Dept. of Microbiology and
Public Health
Michigan State University
East Lansing, MI 48824
USA

Bacterial attachment to plant root surfaces using the Rhizobium-legume symbiosis as a model system, also mechanism of root hair infection

ECKHARDT, F.E.W.
Institut für Allgemeine Mikrobiologie
der Universität Kiel
Olshausenstrasse 40-60
2300 Kiel 1
Federal Republic of Germany

Microbial weathering of minerals, stones, and building materials (plaster, fresco, etc.) in the soil and in buildings and monuments, respectively; isolation and description of "Tetramicrobium" spp., a new genus of hyphal budding water bacteria

FILIP, Z.
Institut für Wasser-, Boden-, und Lufthygiene des Bundesgesundheitsamtes
Aussenstelle Langen
Heinrich-Hertz-Strasse 29
6070 Langen
Federal Republic of Germany

Environmental microbiology

FLETCHER, M.
Dept. of Environmental Sciences
University of Warwick
Coventry CV4 7AL
England

Mechanisms of bacterial attachment to surfaces and physiology of attached bacteria

FRETER, R.
Dept. of Microbiology
University of Michigan
6734 Medical Science Bldg. 2
Ann Arbor, MI 48109
USA

Role of bacterial adhesion to epithelial cells and to other bacteria and of bacterial association with mucus gel, in the colonization of the mammalian body by indigenous and pathogenic bacteria

GARROD, D.R.
Cancer Research Campaign
Medical Oncology Unit
Southampton General Hospital
Southampton SO9 4XY
England

Cell adhesion - dictyostelium, epithelial cells, desmosomes, focal contacts, matrix interactions

GIBBONS, R.J.
Forsyth Dental Center
140 Fenway
Boston, MA 02115
USA

Bacterial attachment to oral tissues, mechanisms and role in the colonization process

GINGELL, D.
Dept. of Anatomy and Biology
Applied to Medicine
Middlesex Hospital Medical School
Cleveland Street
London W1P 6DP
England

Cell adhesion

GÜDE, H.
Institut für Seenforschung und
Fischereiwesen
Untere Seestrasse 81
7994 Langenargen
Federal Republic of Germany

Microbiology of freshwater and activated sludge; bacteria-protozoa interactions and their consequences on the ecosystem structure

HAMILTON, W.A.
Dept. of Microbiology
Marischal College
University of Aberdeen
Aberdeen AB9 1AS
Scotland

Sulphate-reducting bacteria and the offshore oil industry, studies on basic physiology (hydrogenase), and role in microbial communities affecting hydrocarbon breakdown, H_2S production, and anaerobic microbial corrosion

HATTORI, T.
Institute for Agricultural Research
Tohoku University
Sendai
Japan

Effect of solid surface on microbial activities and growth; microbial life in the microstructure of soil

HIRSCH, P.
Institut für Allgemeine Mikrobiologie
der Universität Kiel, Biozentrum
Olshausenstrasse 40-60
2300 Kiel 1
Federal Republic of Germany

Biology and taxonomy of budding bacteria; groundwater microbiology; microbial endolithis ecosystems in Antarctic rocks; neuston microbiology

HOPPE, H.-G.
Institut für Meereskunde
der Universität Kiel
Düsternbrooker Weg 20
2300 Kiel
Federal Republic of Germany

Bacterial activity in the marine environment; relation between exoenzymatic activity and bacterial abundance

JENKINS, D.
Sanitary Engineering and
Environmental Health Research
Laboratory
University of California
47th and Hoffman Boulevard
Richmond, CA 94804
USA

Solids separation in the activated sludge process, especially factors influencing the growth of filamentous microorganisms

JOHNSON, B.F.
Division of Biological Sciences
National Research Council of Canada
Ottawa, Ontario K1A OR6
Canada

1) Cellular morphogenesis and differentiation; 2) growth and cell division; 3) cell cycle regulation; 4) conjugation and sporulation. All four with yeast; numbers 1 and 4 involve flocculation

JONES, G.W.
Dept. of Microbiology and Immunology
6643 Medical Science II
University of Michigan
Medical School
Ann Arbor, MI 48109
USA

Functions and genetics of bacterial adhesions which interact with the surfaces of animal cells or the glycoprotein pellicle of the tooth surface

KJELLEBERG, S.
Dept. of Marine Microbiology
University of Göteborg
Carl Skottsbergs Gata 22
413 19 Göteborg
Sweden

Survival mechanisms of copiotrophic and oligotrophic bacteria, bacterial adhesion and activity at interfaces

LUND, D.B.
Dept. of Food Science
University of Wisconsin
1605 Linden Drive
Madison, WI 53706
USA

Deposition of materials on heat transfer surfaces

MARSHALL, K.C.
School of Microbiology
University of New South Wales
P.O. Box 1
Kensington, N.S.W. 2033
Australia

Mechanistic and ecological aspects of microbial adhesion

MATTHYSSE, A.G.
Dept. of Biology
Coker Hall 010A
University of North Carolina
Chapel Hill, NC 27514
USA

Attachment of Agrobacterium tumefaciens to plant cells; surface interactions of phytopathogenic bacteria with plant cells

McFETERS, G.A.
Dept. of Microbiology
Montana State University
Bozeman, MT 59717
USA

Aquatic microbiology and the microbial ecology of oligotrophic systems; the physiology of bacteria in water and aquatic ecosystems; the metabolism of bacteria attached to carbon particles and their susceptibility to biocides

MIRELMAN, D.
Dept. of Biophysics
Weizmann Institute of Science
Rehovot 76100
Israel

Adherence of enteropathogenic microorganisms (bacteria and parasitic amoeba) to the intestinal tract; mechanisms of virulence and host intestinal mucosal defenses

MITCHELL, R.
Laboratory of Microbial Ecology
Division of Applied Sciences
Harvard University
Cambridge, MA 02138
USA

Microbial ecology of surfaces

MROZEK, H.
Henkel KGaA
Postfach 1100
4000 Düsseldorf
Federal Republic of Germany

Food microbiology, especially plant hygiene and disinfection

RADES-ROHKOHL, E.
Institut für Allgemeine Mikrobiologie
der Universität Kiel
Olshausenstrasse 40-60
2300 Kiel 1
Federal Republic of Germany

Groundwater microflora, interactions with fecal bacteria and environment

REICHENBACH, H.
Gesellschaft für Biotechnologische Forschung
Abt. Mikrobiologie
Mascheroder Weg 1
3300 Braunschweig
Federal Republic of Germany

Myxobacteria: general biology, morphogenesis, taxonomy, pigments, antibiotics produced by them; gliding bacteria, molecular and classical taxonomy

ROBB, I.D.
Unilever Research Laboratory
Port Sunlight
Wirral, Merseyside L63 3JW
England

Water soluble polymers and polymer absorption, colloid stability

ROSE, A.H.
Zymology Laboratory
School of Biological Sciences
University of Bath
Bath BA2 7AY
England

Composition-function relationships in the yeast envelope, growth of the yeast envelope, yeast flocculation

RUTTER, P.R.
BP Research Center
Sunbury-on-Thames,
Middlesex TW16 7LN
England

Colloid and surface science applied to problems arising in the coal and oil industries

SAVAGE, D.C.
Dept. of Microbiology
University of Illinois
131 Burrill Hall
407 South Goodwin Street
Urbana, IL 61801
USA

Mechanisms by which microorganisms associate with (adhere to) gastrointestinal epithelial surfaces

SCHINK, B.
Fakultät für Biologie
der Universität Konstanz
Postfach 5560
7750 Konstanz
Federal Republic of Germany

Limits and principles of anaerobic degradation of organic matter; especially, syntrophic associations of anaerobic microorganisms and degradation systems in fixed-film bioreactors or sediments

SCHUBERT, R.H.W.
Zentrum für Hygiene
der Universität Frankfurt
Paul Ehrlich Strasse 40
6000 Frankfurt am Main
Federal Republic of Germany

Adherence and growth of bacteria on solid surfaces in relation to aquatic environments

SHILO, M.
Division of Microbial and Molecular Ecology
Institute of Life Sciences
The Hebrew University
Jerusalem
Israel

Physiology of benthic cyanobacteria

SILVERMAN, M.
Agouron Institute
505 South Coast Boulevard
La Jolla, CA 92037
USA

Molecular biology

TANAKA, T.
Room 13-2153
Dept. of Physics
Massachusetts Institute of Technology
Cambridge, MA 02139
USA

Phase transitions in polymer gels, coil-globule transition in single polymer chains, laser light scattering spectroscopy of biological macromolecules and gels, microscopic laser light scattering spectroscopy of biological cells (red blood cells and lens cells)

TYLEWSKA, S.
Dept. of Bacteriology
National Institute of Hygiene
00-791 Warsaw
Poland

Adherence of group A Streptococci to epithelial cells

VINCENT, B.
School of Chemistry
University of Bristol
Cantock's Close
Bristol BS8 1TS
England

Polymers at interfaces; aggregation of colloidal dispersions in the presence of polymers; emulsions; preparation and characterization of novel dispersions

WHITE, D.C.
310 Nuclear Research
Florida State University
Tallahassee, FL 32306
USA

Sedimentary microbial ecology

WILDERER, P.A.
Technische Universität
Hamburg-Harburg
Eissendorferstrasse 38
2100 Hamburg 90
Federal Republic of Germany

Selection pressure provided by periodic stress affecting the capacity and settle-ability of microbial biocommunities such as activated sludge

Subject Index

Author Index

Dahlem Workshop Reports

Life Sciences Research Reports (LS)

LS 32 Exploitation of Marine Communities
Editor: R. M. May (1984, in press)

LS 31 Microbial Adhesion and Aggregation.
Editor: K. G. Marshall (1984)

LS 30 Leukemia.
Editor: I. L. Weissman (1984, in press)

LS 29 The Biology of Learning.
Editors: P. Marler, H. S. Terrace (1984)

LS 28 Changing Metal Cycles and Human Health.
Editor: J. O. Nriagu (1984)

LS 27 Minorities: Community and Identity.
Editor: C. Fried (1983)

LS 26 The Origins of Depression: Current Concepts and Approaches.
Editor: J. Angst (1983)

LS 25 Population Biology of Infectious Diseases.
Editors: R. M. Anderson, R. M. May (1982)

LS 24 Repair and Regeneration of the Nervous System.
Editor: J. G. Nicholls (1982)

LS 23 Biological Mineralization and Demineralization.
Editor: G. H. Nancollas (1982)

LS 22 Evolution and Development.
Editor: J. T. Bonner (1982)

LS 21 Animal Mind - Human Mind.
Editor: D. R. Griffin (1982)

LS 20 Neuronal-Glial Cell Interrelationships.
Editor: T. A. Sears (1982)

Physical, Chemical and Earth Sciences Research Reports (PC)

PC 5 Patterns of Change in Earth Evolution.
Editors: H. D. Holland, A. F. Trendall (1984, in press)

PC 4 Atmospheric Chemistry.
Editor: E. D. Goldberg (1982)

PC 3 Mineral Deposits and the Evolution of the Biosphere.
Editors: H. D. Holland, M. Schidlowski (1982)

Springer-Verlag Berlin Heidelberg New York Tokyo

MIX
Papier aus verantwortungsvollen Quellen
Paper from responsible sources
FSC® C105338

If you have any concerns about our products,
you can contact us on
ProductSafety@springernature.com

In case Publisher is established outside the EU,
the EU authorized representative is:
Springer Nature Customer Service Center GmbH
Europaplatz 3, 69115 Heidelberg, Germany

Printed by Libri Plureos GmbH
in Hamburg, Germany